TROPICAL RAIN FOREST IN SOUTHEAST ASIA

At one time, throughout the world, tropical forests cloaked regions that were, collectively, twice the size of Europe. For ten thousand years or more, the forests endured in rich complexity, as the homes of an estimated fifty to ninety percent of the world's land-dwelling species.

In less than four decades, we destroyed more than half of these ancient forests, and most of their spectacular arrays of species may be lost forever. With each passing year, we log over an additional 38 million acres for fuel, lumber, and other products. That is the equivalent of leveling thirty-four city blocks every minute. At this writing, the human population has exceeded 5.6 billion. It is still growing, exponentially, and so is the demand for forest products.

Rampant deforestation extends beyond the tropics. Today, highly mechanized logging operations are proceeding in temperate forests throughout the United States, Canada, Europe, Siberia, and elsewhere.

Biologists rightly decry the mass extinction, the assault on species diversity, the depletion of a major portion of the world's genetic reservoir. But something else is going on here. Too many of us experience a vague sense of unease when we hike or drive through a logged-over part of our own country, when we see before-and-after photographs of wholesale destruction. Is it because we are losing the comfort of our evolutionary heritage, a connection with our own past?

Many millions of years ago, our primate ancestors moved into the trees of tropical forests. Through countless generations, novel connections formed among increasing numbers of cells in their brains; and vision, hearing, and other senses became highly responsive to information-rich, arboreal worlds.

Does our neural wiring still resonate with rustling leaves, with shafts of light and mosaic shadows? Are we innately attuned to the forests of Eden—or have time and change buried recognition of home?

DIVERSITY OF LIFE

CECIE STARR

BELMONT, CALIFORNIA

RALPH TAGGART

MICHIGAN STATE UNIVERSITY

BIOLOGY:

THE UNITY AND DIVERSITY OF LIFE

SEVENTH EDITION

WADSWORTH PUBLISHING COMPANY

I(T)P™ AN INTERNATIONAL THOMSON PUBLISHING COMPANY

Belmont • Albany • Bonn • Boston
Cincinnati • Detroit • London • Madrid
Melbourne • Mexico City • New York
Paris • San Francisco • Singapore
Tokyo • Toronto • Washington

BIOLOGY PUBLISHER: Jack C. Carey

ASSISTANT EDITOR: Kristin Milotich

EDITORIAL ASSISTANT: Kerri Abdinoor

PRINT BUYER: Randy Hurst

PRODUCTION SERVICES COORDINATOR: Sandra Craig

PRODUCTION: Mary Douglas, Rogue Valley Publications

TEXT AND COVER DESIGN, ART DIRECTION: Gary Head,
Gary Head Design

EDITORIAL PRODUCTION: Myrna Engler-Forkner, Melissa Andrews,
Rosaleen Bertolino, Marilyn Evenson, Susan Gall, Ed Serdziak,
Karen Stough

ARTISTS: Raychel Ciemma, Robert Demarest, Hans & Cassady, Inc.
(Hans Neuhart), Darwen Hennings, Vally Hennings, Betsy
Palay, Precision Graphics (Jan Flessner), Nadine Sokol, Kevin
Somerville, Lloyd Townsend

PHOTO RESEARCH AND PERMISSIONS: Marion Hansen

COVER PHOTOGRAPH: © Frans Lanting/Minden Pictures

COMPOSITION: American Composition & Graphics, Inc.
(Jim Jeschke, Jody Ward, and Valerie Norris)

COLOR PROCESSING: H & S Graphics, Inc. (Tom Anderson,
Nancy Dean, and John Deady)

PRINTING AND BINDING: R. R. Donnelley & Sons Company/Willard

BOOKS IN THE WADSWORTH BIOLOGY SERIES

For more information, contact Wadsworth Publishing Company:

Wadsworth Publishing Company
10 Davis Drive, Belmont, California 94002, USA

International Thomson Publishing Europe
Berkshire House 168-173, High Holborn
London, WC1V 7AA, England

Thomas Nelson Australia
102 Dodds Street
South Melbourne 3205, Victoria, Australia

Nelson Canada
1120 Birchmount Road
Scarborough, Ontario, Canada M1K 5G4

International Thomson Editores
Campos Eliseos 385, Piso 7
Col. Polanco, 11560 México D.F. México

International Thomson Publishing GmbH
Königswinterer Strasse 418
53227 Bonn, Germany

International Thomson Publishing Asia
221 Henderson Road, #05-10 Henderson Building
Singapore 0315

International Thomson Publishing Japan
Hirakawacho Kyowa Building, 3F
2-2-1 Hirakawacho, Chiyoda-ku, Tokyo 102, Japan

CONTENTS IN BRIEF

Highlighted chapters are included in DIVERSITY OF LIFE.

21 THE ORIGIN AND EVOLUTION OF LIFE

In the Beginning . . .

Some evening, watch the moon as it rises from the horizon and think about the 380,000 kilometers between it and you. *Five billion trillion* times the distance between the moon and you are galaxies—systems of stars—at the boundary of the known universe. Wavelengths of light travel faster than anything else, 300 million meters per second, yet long wavelengths from those faraway galaxies are only now reaching the earth.

Near and distant galaxies are suspended in the vast space of the universe. By every known measure, they are all moving away from one another—which means the universe is expanding. And the prevailing theory of that expansion accounts for every bit of matter in the universe, in every living thing.

Think about how you rewind a video tape on a VCR, then imagine "rewinding" the universe. As you do, the galaxies start moving back together. After 10 to 20 billion years of rewinding, all galaxies, all matter, and all of space have become compressed into a hot, dense volume about the size of the sun. You have arrived at time zero.

That enormously hot, dense state lasted only for an instant. What happened next is known as the "big bang"—a truly inadequate name for the stupendous, nearly instantaneous distribution of all matter and energy everywhere, through all of the known universe. About a minute later, temperatures dropped to a billion degrees and fusion reactions produced most of the light elements, including helium, which are still the most abundant elements throughout the universe. Radio telescopes have detected the relics of the big bang— cooled and diluted background radiation, left over from the beginning of time.

Over the next billion years, stars started forming by the gravitational contraction of gaseous material. When the stars became massive enough, nuclear reactions were ignited in their central region, and they gave off tremendous light and heat. As contraction continued, many stars may have become dense enough to promote the formation of heavier elements.

All stars have a life history, from birth to a sometimes spectacularly explosive death. In what might be

a

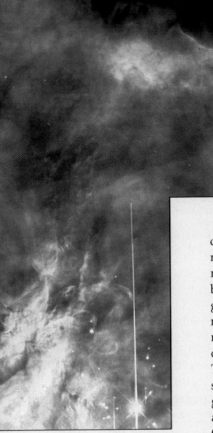

b

called the original stardust memories, the heavier elements released during the explosions became swept up during the gravitational contraction of new stars—and they became raw materials for the formation of even heavier elements. Today, the Hubble space telescope is providing astounding glimpses of such star-forming activity in the dust clouds of Orion and other constellations (Figure 21.1).

Long ago, explosions of dying stars ripped through our own galaxy and left behind a dense cloud of dust and gas that extended trillions of kilometers in space. As the cloud cooled, countless bits of matter gravitated toward one another. By 4.6 billion years ago, the cloud had flattened out into a slowly rotating disk. At the dense, hot center of that disk, the star of our solar system—the sun—was born.

The remainder of this chapter is a sweeping slice through time, one that cuts back to the formation of the earth and the chemical origins of life. It is the starting point for the next four chapters, which will take us along major lines of descent that lead to the present range of species diversity. The story is not complete. Even so, all of the available evidence points to a principle that can be used to organize separate bits of information about the past: *Life is a magnificent continuation of the physical and chemical evolution of the universe, of galaxies and stars, and of the planet earth.*

Figure 21.1 (**a**) In winter in the Northern Hemisphere, brilliant stars of the constellation Orion dominate the southern sky. (**b**) Shown here, the great nebula in Orion, a cloud of dust, silicates, ice, and water. (The Latin *nebula* means mist.) The Hubble space telescope provided this remarkable detailed view of part of the nebula—a hotbed of star formation.

KEY CONCEPTS

1. The available evidence suggests that life originated more than 3.8 billion years ago. Its origin and subsequent evolution have been linked to the physical and chemical evolution of the universe, the stars, and the planet earth.

2. All of the inorganic and organic compounds necessary for self-replication, membrane assembly, and metabolism —that is, for the structure and functioning of living cells— could have formed spontaneously under conditions that existed on the early earth.

3. Not long after the time of life's origin, divergences led to three great prokaryotic lineages—the archaebacteria, the eubacteria, and the ancestors of eukaryotes. A theory of endosymbiosis helps explain the profusion of specialized organelles that characterize eukaryotes.

4. Earth history is divided into five great eras, the boundaries of which reflect times of the largest mass extinctions. Prokaryotes dominated the first two eras, the Archean and Proterozoic. Eukaryotes originated during the late Proterozoic and became spectacularly diverse. All five kingdoms of organisms are characterized by the persistences, extinctions, and radiations of different lineages.

5. Throughout the history of life, asteroid impacts, drifting and colliding continents, and other environmental insults have had profound impact on the direction of evolution.

21.1 CONDITIONS ON THE EARLY EARTH

Origin of the Earth

Cloudlike remnants of stars, including those shown in Figure 21.1, are mostly hydrogen gas. But they also contain water, iron, silicates, hydrogen cyanide, methane, ammonia, formaldehyde, and many other simple inorganic and organic substances. Most likely, the contracting cloud from which our solar system evolved was similar in composition. Between 4.6 and 4.5 billion years ago, the outer regions of the cloud cooled. Swarms of mineral grains and ice orbited in parallel around the sun. By electrostatic attraction, by gravitational pull, they started clumping together (Figure 21.2). In time, the larger, faster clumps started colliding and shattering. Some of the larger ones became more massive by sweeping up asteroids, meteorites, and other rocky remnants of collisions—and gradually they evolved into planets.

While the early earth was forming, much of its inner rocky material melted. Most likely, stupendous asteroid impacts generated the required heat, although heat released during the radioactive decay of minerals contributed to it. As rocks melted, nickel, iron, and other heavy materials moved toward the interior, and lighter minerals floated toward the surface. This process of differentiation resulted in the formation of a crust of basalt, granite, and other types of low-density rock, a rocky region of intermediate density (the mantle), and a high-density, partially molten core of nickel and iron.

Four billion years ago, the earth was a thin-crusted inferno (Figure 21.3a). Yet within 200 million years, life had originated on its surface! We have no record of the event. As far as we know, movements in the mantle and crust, volcanic activity, and erosion obliterated all traces of it. Still, we can put together a plausible explanation of how life originated by considering three questions:

1. What were the prevailing physical and chemical conditions on earth at the time of life's origin?

2. Based on physical, chemical, and evolutionary principles, could large organic molecules have formed spontaneously and then evolved into molecular systems displaying the characteristics of life?

3. Can we devise experiments to test whether living systems could have emerged by chemical evolution?

The First Atmosphere

When patches of crust were forming, heat and gases blanketed the earth. This first atmosphere probably consisted of gaseous hydrogen (H_2), nitrogen (N_2), carbon monoxide (CO), and carbon dioxide (CO_2). Were gaseous oxygen (O_2) and water also present? Probably not. Rocks release very little oxygen during volcanic eruptions. Even if oxygen were released, those small amounts would have reacted at once with other elements, and any water would have evaporated because of the intense heat.

When the crust finally cooled and solidified, water condensed into clouds and the rains began. For millions of years, runoff from rains stripped mineral salts and other compounds from the earth's parched rocks. Salt-laden waters collected in depressions in the crust and formed the early seas.

Without an oxygen-free atmosphere, the organic compounds that started the story of life never would have formed on their own. (As described on page 92, free oxygen would have attacked them.) Without liquid water, cell membranes never would have formed. Cells, recall, are the basic units of life. Each has a capacity for independent existence.

The Synthesis of Organic Compounds

When we reduce cells to their lowest common denominator, we are left with proteins, complex carbohydrates, lipids, and nucleic acids. Today, cells assemble these molecules from small organic compounds—simple sugars, fatty acids, amino acids, and nucleotides. Energy from the environment drives the synthesis reactions. Were small organic compounds also present on the early earth? Were there sources of energy that drove their spontaneous assembly into the large molecules of life?

Mars, meteorites, the earth's moon, and the earth formed at the same time, from the same cosmic cloud. Rocks collected from Mars, meteorites, and the moon contain precursors of biological molecules—so the same precursors must have been present on the early earth.

Figure 21.2 Representation of the cloud of dust, gases, and clumps of rock and ice around the early sun.

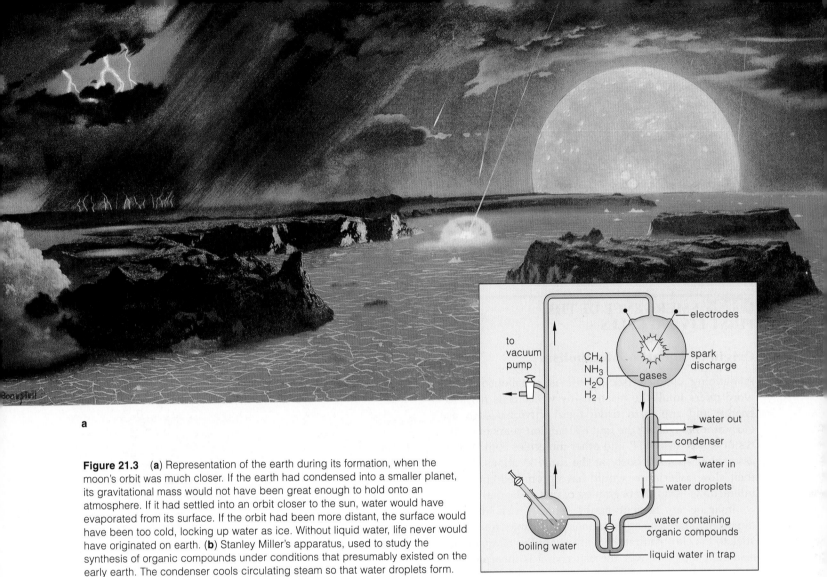

Figure 21.3 (**a**) Representation of the earth during its formation, when the moon's orbit was much closer. If the earth had condensed into a smaller planet, its gravitational mass would not have been great enough to hold onto an atmosphere. If it had settled into an orbit closer to the sun, water would have evaporated from its surface. If the orbit had been more distant, the surface would have been too cold, locking up water as ice. Without liquid water, life never would have originated on earth. (**b**) Stanley Miller's apparatus, used to study the synthesis of organic compounds under conditions that presumably existed on the early earth. The condenser cools circulating steam so that water droplets form.

Sunlight, lightning, or heat escaping from the earth's crust could have supplied the energy to drive their condensation into complex organic molecules.

In the first of many tests of that hypothesis, Stanley Miller mixed hydrogen, methane, ammonia, and water in a reaction chamber (Figure 21.3b). He recirculated the mixture and bombarded it with a spark discharge to simulate lightning. Within a week, amino acids and other small organic compounds had formed. In other experiments that simulated conditions on the early earth, glucose, ribose, deoxyribose, and other sugars were produced from formaldehyde. Adenine was produced from hydrogen cyanide. Adenine plus ribose are present in ATP, NAD, and other nucleotides.

Even if amino acids did form in the early seas, they wouldn't have lasted long. In water, the favored direction of most spontaneous reactions is toward hydrolysis, not condensation (page 36).

Maybe more lasting bonds formed at the margins of seas. By one scenario, clay in the rhythmically drained muck of tidal flats and estuaries served as templates (structural patterns) for the spontaneous assembly of proteins and other organic compounds. Clay consists of thin, stacked layers of aluminosilicates with metal ions at its surface. Clay and metal ions attract amino acids. When clay is first warmed by sunlight, then alternately dried out and moistened, it actually promotes condensation reactions that yield complex organic compounds. We know this from experiments.

Suppose proteins that formed on some clay templates had the shape and chemical behavior necessary to function as weak enzymes in hastening bonds between amino acids. If certain templates promoted such bonds, they would have selective advantage over other templates in the chemical competition for available amino acids. Perhaps selection was at work before the origin of cells, favoring the chemical evolution of enzymes.

Many experiments provide indirect evidence that the complex organic molecules characteristic of life could have formed under conditions that existed on the early earth.

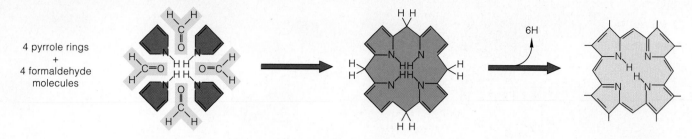

Figure 21.4 How formaldehyde, a substance present on the early earth, might have undergone chemical evolution into porphyrin. Today, porphyrin is the light-trapping and electron-donating component of chlorophyll molecules. It also is a component of cytochrome, which has roles in electron transport systems of a variety of metabolic pathways.

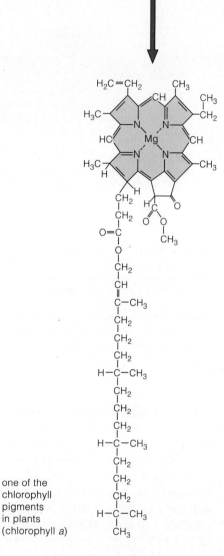

porphyrin ring system

one of the chlorophyll pigments in plants (chlorophyll *a*)

21.2 EMERGENCE OF THE FIRST LIVING CELLS

Origin of Agents of Metabolism

A defining characteristic of life is metabolism. The word refers to all the reactions by which cells harness energy and use it to drive their activities, such as biosynthesis. During the first 600 million years of earth history, enzymes, ATP, and other molecules could have assembled spontaneously at the same locations. If so, their close association would have promoted chemical interactions—and the beginning of metabolic pathways.

Imagine an ancient sunlit estuary, rich in clay deposits. Countless aggregations of organic molecules stick to the clay. At first there are quantities of an amino acid called *D*. All over the estuary, *D* is incorporated into protein molecules—until the supply dwindles. Suppose a certain enzyme can promote the formation of *D* from a simpler substance (*C*)—which is still abundant. Aggregations having that enzyme will be favored in the chemical competition for starting materials. Now suppose *C* becomes scarce also. At that point, the advantage tilts to aggregations that can promote formation of *C* from the even simpler organic substances *B* and *A*—say, carbon dioxide and water. These substances are present in essentially unlimited amounts in the atmosphere and the sea. Selection has favored a synthetic pathway:

$$A + B \longrightarrow C \longrightarrow D$$

Suppose, finally, that some aggregations in the estuary are better than others at harnessing and using energy. What kind of molecules could give them this advantage? Think of photosynthesis, the energy-trapping pathway that now dominates the world of life. This metabolic route starts at light-trapping pigment molecules called chlorophylls. The part of chlorophyll that absorbs sunlight energy and gives up electrons is a porphyrin ring structure. Porphyrins also are part of cytochromes. And cytochromes are part of electron transport systems in every photosynthetic and aerobically respiring cell. As

Figure 21.4 shows, porphyrins can spontaneously assemble from formaldehyde—one of the legacies of cosmic clouds. Was porphyrin an electron transporter of the first metabolic pathways? Perhaps.

Origin of Self-Replicating Systems

Another defining characteristic of life is a capacity for reproduction, and reproduction starts with DNA's protein-building instructions. DNA is a fairly stable molecule, and it is easily replicated before each cell division. Enzymes and RNA carry out its encoded instructions.

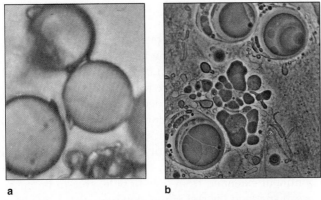

a　　　　　　　**b**

Figure 21.5　Microscopic spheres of (**a**) proteins and (**b**) lipids that self-assembled under abiotic conditions.

Small molecules called coenzymes assist most of the existing enzymes. Intriguingly, the structure of some coenzymes is the same as that of RNA nucleotides. Another clue: If you heat nucleotide precursors and short chains of phosphate groups together, they will assemble into RNA strands. On the early earth, sunlight energy alone could have driven the spontaneous formation of RNA molecules.

Simple, self-replicating systems of RNA, enzymes, and coenzymes have been created in the laboratory. Experiments with these systems suggest that RNA could have formed on clay templates. Did RNA later replace clay as information-storing templates for protein synthesis? Perhaps, although existing RNA is too chemically inept to serve this role. And we still don't know how DNA entered the picture.

Until we identify the likely chemical ancestors of RNA and DNA, the story of life's origin will be incomplete. Filling in the details will require imaginative sleuthing. For example, researchers ran a simple program on one of the most rapid, advanced computers. The program contained information on simple compounds, of the sort that are believed to have been present on the early earth. It asked the computer to subject the compounds to random chemical competition and natural selection. The program ran repeatedly, and the outcome was always the same. Simple precursors inevitably evolved into interacting systems of large, complex molecules.

Origin of the First Plasma Membranes

Experiments are more revealing of the origin of the plasma membrane, which surrounds every living cell. The membrane is a lipid bilayer, studded with proteins that carry out diverse functions. The membrane's primary function is to control which substances move into and out of the cell. Without this control, cells cannot exist.

Before cells, there must have been "proto-cells." These may have been little more than membrane sacs,

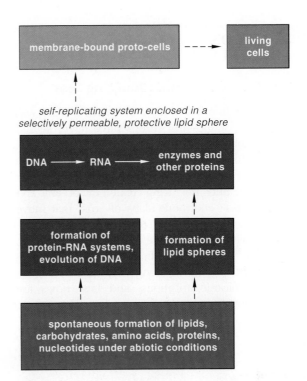

Figure 21.6　One possible sequence of events that led to the first self-replicating systems, then to the first living cells.

protecting information-storing templates and various metabolic agents from the environment. We know that simple membrane sacs can form spontaneously.

In one experiment, Sidney Fox heated amino acids until they formed protein chains, which he put in hot water. After cooling, the chains assembled into small, stable spheres (Figure 21.5a). Like cell membranes, the spheres were selectively permeable to different substances. They picked up free lipids, and a lipid-protein film formed at their surface.

In other experiments, fatty acids and glycerol combined to form long-tail lipids under conditions that simulated evaporating tidepools. The lipids self-assembled into small, water-filled sacs, which in many were like cell membranes (Figure 21.5b).

In short, there are major gaps in the story of life's origins. But there also is strong experimental evidence that chemical evolution could have led to the molecules and structures characteristic of life. Figure 21.6 summarizes the key events in that chemical evolution, which preceded the first cells.

Although the story is far from complete, laboratory experiments and computer simulations indirectly suggest that chemical and molecular evolution gave rise to proto-cells.

21.3 LIFE ON A CHANGING GEOLOGIC STAGE

Drifting Continents and Changing Seas

By about 3.8 billion years ago, the earth's crust was becoming more stable—but things have never settled down completely.

According to **plate tectonic theory**, plumes of molten rock from the mantle spread out beneath the crust or broke through it. Such fracturing created the oceanic and continental plates—enormous slabs of crust that move apart and crunch together at their margins (Figure 21.7). The plates move no more than a few centimeters a year, on the average. Yet that small displacement causes pressure to decrease or increase along vast crustal margins. Earthquakes, volcanic eruptions, and lava flows are among the short-term consequences. Crustal uplifting

and subsidence, mountain building, seafloor spreading, and continental drift are long-term consequences.

Take a look at Figure 21.8. In the remote past, land masses collided together to form supercontinents, which later split open at deep rifts that became new ocean basins. One early continent, Gondwana, drifted southward from the tropics, across the south polar region, then northward until it crunched together with other land masses to form Pangea. This single world continent extended from pole to pole, and an immense ocean spanned the rest of the globe.

Besides this, the erosive forces of wind and water have continually resculpted the earth's surface. Asteroids and meteorites have continued to bombard the crust, and huge impacts have had long-term effects on global temperature and climate.

All such changes in land masses, the oceans, and the atmosphere profoundly influenced the evolution of life.

Figure 21.7 Some forces of geologic change. (**a**) Plate tectonics. The crust is fractured in rigid plates that split, drift, and collide. (**b**) Plumes of molten material drive the crustal movements. They well up from the earth's interior and spread out beneath the crust. Such plumes have ruptured the crust at vast ridges on the ocean floor. Here, molten material escapes, cools, and forces the seafloor away from the ridge. Seafloor spreading displaces plates away from the ridge. Often the opposite plate margin is thrust under another plate and contributes to mountain building. Plumes also cause deep rifts and splitting in the interior of continents. Such rifting is taking place in Missouri, at Lake Baikal in Russia, and in eastern Africa. In the past, superplumes violently ruptured the Earth's crust. Volcanoes still form at the ruptured "hot spots." The Hawaiian Islands formed this way.

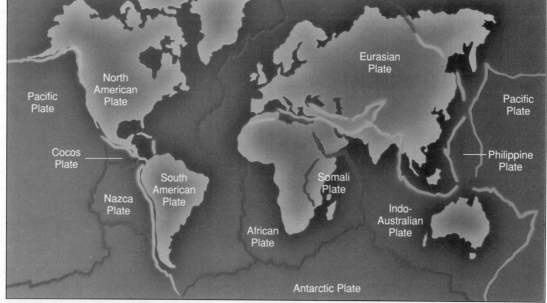

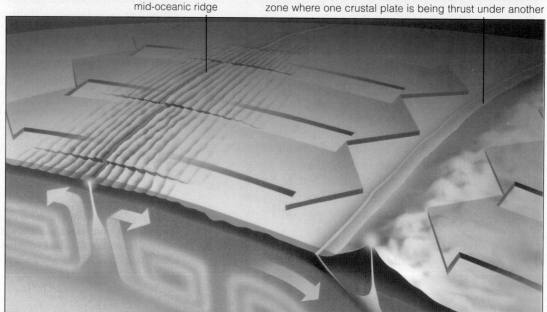

mid-oceanic ridge   zone where one crustal plate is being thrust under another

For example, life flourished in warm, shallow waters along the vast shorelines of the early continents. When continents collided, the shorelines vanished. These and other changes had dire consequences for many lineages. Yet even as old habitats were lost, new ones opened up for the survivors—and evolution took off in new directions.

The Geologic Time Scale

As you read in the preceding chapter, early naturalists had no way of dating the earth's rocks or the fossils that were contained in sedimentary beds. But there were obviously abrupt transitions in the kinds of fossils represented. They interpreted these as boundaries between four great eras in time. The most ancient fossils were from the first geologic era, the **Proterozoic**. They were followed by fossils from the **Paleozoic**, the **Mesozoic**, and finally the "modern" era, the **Cenozoic**. The boundaries between eras still serve as the basis of a **geologic time scale**, which is shown in Figure 21.8. Today we know the boundaries mark the times of mass extinctions. Thanks to radioisotope dating work (page 20), actual dates have been assigned to the time scale. The work also revealed that the Proterozoic lasted for an astoundingly immense time. The Proterozoic has since been subdivided. The first era of the geologic time scale is now called the **Archean** ("the beginning").

Over the past 3.8 billion years, changes in the earth's crust, the atmosphere, and the oceans profoundly affected the evolution of life. The boundaries of the geologic time scale mark abrupt shifts in the direction of evolution.

Figure 21.8 The geologic time scale. The time spans of the different eras are not to scale. If they were, the Archean and Paleozoic portions of the chart would run off the page.

Think of the spans of time as minutes on a clock. Life originated at midnight. The Paleozoic began at 10:04 A.M., the Mesozoic at 11:09 A.M., and the Cenozoic at 11:47 A.M. The recent epoch of the Cenozoic started during the last 0.1 second before noon.

Era	Period	Epoch	Millions of Years Ago (mya)
CENOZOIC	Quaternary	Recent	0.01–
		Pleistocene	1.65
	Tertiary	Pliocene	5
		Miocene	25
		Oligocene	38
		Eocene	54
		Paleocene	65
MEZOZOIC	Cretaceous	Late	100
		Early	138
	Jurassic		205
	Triassic		240
PALEOZOIC	Permian		290
	Carboniferous		360
	Devonian		410
	Silurian		435
	Ordovician		505
	Cambrian		550
PROTEROZOIC			2,500
ARCHEAN			4,600

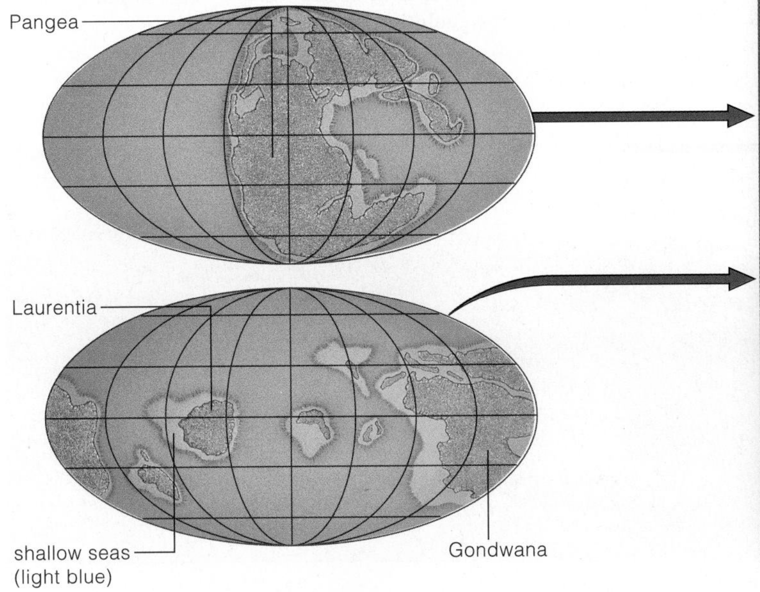

Pangea

Laurentia

shallow seas (light blue)

Gondwana

21.4 LIFE IN THE ARCHEAN AND PROTEROZOIC ERAS

The first cells emerged as molecular extensions of the evolving universe, the solar system, and the planet earth. They may have originated in tidal flats or on the seafloor, in muddy sediments warmed by heated water from volcanic vents. (Such *hydrothermal* vents are described on page 884.) Like existing bacteria, the first were **prokaryotic cells**, without a nucleus. Possibly they were little more than self-replicating, membranous sacs of DNA and other complex organic molecules. Given the absence of free oxygen, they must have secured energy through anaerobic routes—fermentation pathways, most likely. Energy and carbon sources were plentiful; dissolved organic compounds had accumulated in the seas by natural geologic processes. Thus, "food" was available, predators were absent, and biological molecules were free from oxygen attacks.

Early in the Archean era, the original prokaryotic lineage diverged in three evolutionary directions. Two of the branchings led to the domains of **archaebacteria** and **eubacteria**. You will read about these bacteria in the chapter to follow. As described in the *Focus* essay, the third branching led to the prokaryotic ancestors of **eukaryotic cells**.

Between 3.5 and 3.2 billion years ago, the cyclic pathway of photosynthesis evolved in some species of eubacteria. This pathway is described on page 112. Those eubacteria were anaerobic, but components of their electron transport systems had become modified in ways that allowed them to tap an unlimited energy source—sunlight. For nearly 2 billion years, their photosynthetic descendants dominated the world of life. Their populations formed large mats in which sediments collected. Mats accumulated one atop the other, becoming structures called **stromatolites** (Figure 21.9).

By the dawn of the Proterozoic, 2.5 billion years ago, the noncyclic pathway of photosynthesis had evolved in

some eubacterial species. Oxygen, one of its by-products, started to accumulate—imperceptibly at first, but in time it had two irreversible effects. First, *an oxygen-rich atmosphere stopped the further chemical origin of living cells*. Except in very few anaerobic habitats, spontaneously formed organic compounds would not survive attacks by free oxygen. Second, *aerobic respiration became the dominant energy-releasing pathway*. In many prokaryotic lineages, selection now favored metabolic equipment that could "neutralize" oxygen by using it as an electron acceptor. This innovation foreshadowed the evolution of multicelled eukaryotes and their invasion of far-flung environments.

Eukaryotes originated during the Proterozoic, by 1.2 billion years ago and possibly earlier. We have fossils, 900 million years old, of well-developed red algae, green algae, plant spores, and fungi. Diverse internal membranous compartments—organelles—are the hallmark of eukaryotic cells. Where did they come from? The *Focus* essay describes a few of the most plausible hypotheses.

About 800 million years ago, stromatolites began a dramatic decline. They had become an untapped food source for newly evolved, tiny, bacteria-eating animals. About the same time, small, soft-bodied animals were leaving tracks and burrows on the seafloor. They lived near the shores of a supercontinent, Laurentia. About 570 million years ago, in "Precambrian" times, their descendants may have started the first adaptive radiation of animals.

The first living cells had evolved by about 3.8 billion years ago. About 600 million years later, oxygen released by some of their photosynthetic descendants was starting to change the atmosphere.

In time, an oxygen-rich atmosphere precluded further spontaneous chemical evolution of life—and it was a key selection pressure in the evolution of eukaryotic cells.

Figure 21.9 (a) One of the oldest known fossils: a strand of walled cells, 3.5 billion years old. The strand resembles certain modern-day filamentous bacteria. (b) From Western Australia, stromatolites that formed between 2,000 and 1,000 years ago in shallow seawater. Calcium deposits preserved their structure. They are identical to stromatolites more than 3 billion years old.

10 μm

a   b

The Rise of Eukaryotic Cells

The key defining feature of eukaryotic cells is the profusion of membrane-bound organelles in the cytoplasm. Where did they come from? Speculations abound. Some organelles probably evolved gradually, through gene mutations and natural selection. For others, researchers make a good case for evolution by way of endosymbiosis.

Origin of the Nucleus and ER   Prokaryotic cells do not have a profusion of organelles. Yet some species do have infoldings of the plasma membrane (Figure *a*). Enzymes and other metabolic agents are embedded in that membrane. In ancient forerunners of eukaryotic cells, similar infoldings may have extended far enough into the cell interior to serve as a channel to the surface. Perhaps they evolved into ER channels and into an envelope around the DNA.

What would be the advantage of such membranous enclosures? Maybe they protected genes and protein products from "foreigners." Remember how bacterial species can transfer plasmid DNA among themselves? Yeasts, which are simple eukaryotic cells, also make such transfers. Some yeast species contain up to fifty plasmids. Initially, a nuclear envelope may have been favored because it got the cell's genes, replication enzymes, and transcription enzymes out of the cytoplasm. It would have allowed vital genetic messages to be produced free of metabolic competition from an unmanageable hodgepodge of foreign genes. Similarly, ER channels might have kept vital proteins and other organic compounds away from metabolically hungry "guests"—foreign cells that became permanent residents inside the host cell.

A Theory of Endosymbiosis   Accidental partnerships between different prokaryotic species must have formed countless times on the evolutionary road to eukaryotes. Some partnerships resulted in the origin of mitochondria, chloroplasts, and other organelles. This is a **theory of endosymbiosis**, developed in greatest detail by Lynn Margulis. *Endo-* means within, and *symbiosis* means living together. In endosymbiosis, one cell (a guest species) lives permanently inside another kind of cell (the host species), and the interaction benefits both. By this theory, eukaryotes arose after the noncyclic pathway of photosynthesis emerged and oxygen accumulated to significant levels in the atmosphere.

In some bacterial groups, electron transport systems had already become expanded to include "extra" cytochromes. As it happened, those cytochromes were able to donate electrons to oxygen. In other words, the bacteria could extract energy from organic compounds by way of aerobic respiration. By 1.2 billion years ago, and possibly much earlier, the forerunners of eukaryotes were engulfing aerobic bacteria. They may have been like existing amoebalike cells that weakly tolerate free oxygen. If so, they trapped food by sending out cytoplasmic extensions of the cell body. Endocytic vesicles formed around the food and delivered it to the cytoplasm for digestion.

Some aerobic bacteria resisted digestion. They actually thrived in the new, protected, nutrient-rich environment. In time they were releasing extra ATP—which the hosts came to depend on for growth, greater activity, and assembly of more structures, such as hard body parts. The guests were no longer duplicating metabolic functions that the hosts were performing for them. The anaerobic and aerobic cells were now incapable of independent existence. The guests had become mitochondria, supreme suppliers of ATP.

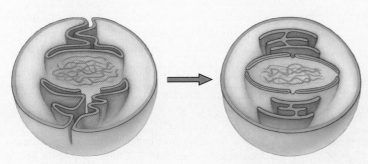

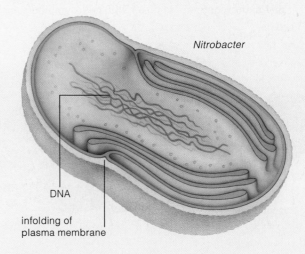

Nitrobacter

DNA

infolding of
plasma membrane

a   Over evolutionary time, infoldings of the plasma membrane may have given rise to the nuclear envelope and endoplasmic reticulum now present in eukaryotic cells. Such infoldings are present in the cytoplasm of many existing bacteria, including *Nitrobacter*, sketched here in cutaway view.

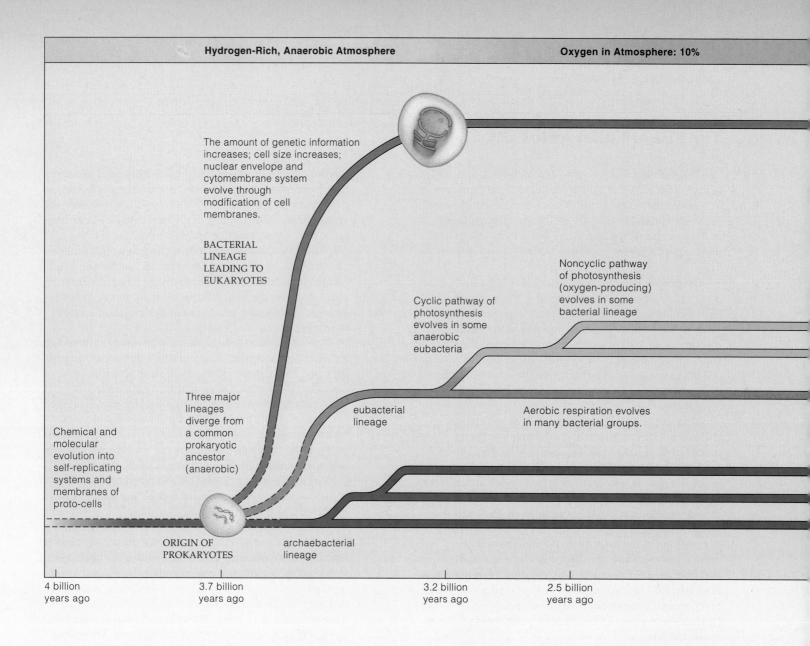

The amount of genetic information increases; cell size increases; nuclear envelope and cytomembrane system evolve through modification of cell membranes.

BACTERIAL LINEAGE LEADING TO EUKARYOTES

Noncyclic pathway of photosynthesis (oxygen-producing) evolves in some bacterial lineage

Cyclic pathway of photosynthesis evolves in some anaerobic eubacteria

Three major lineages diverge from a common prokaryotic ancestor (anaerobic)

eubacterial lineage

Aerobic respiration evolves in many bacterial groups.

Chemical and molecular evolution into self-replicating systems and membranes of proto-cells

ORIGIN OF PROKARYOTES

archaebacterial lineage

| 4 billion years ago | 3.7 billion years ago | 3.2 billion years ago | 2.5 billion years ago |

Evidence of Endosymbiosis

Strong evidence supports Margulis's theory. There are plenty of examples that nature continues to tinker with endosymbionts, including the one shown in the micrograph below. In such eukaryotic cells, mitochondria are like bacteria in size and structure. The inner mitochondrial membrane is like a bacterial plasma membrane. A mitochondrion replicates its own DNA and divides independently of the host cell's division. Its DNA contains instructions for building a few proteins required for specialized mitochondrial tasks.

A few of the genetic code words have uniquely mitochondrial meanings. That is, the "mitochondrial code" has a few slight variations, compared to the genetic code of cells.

As another example, consider the food-producing factories called chloroplasts. Chloroplasts may be descended from aerobic eubacteria that engaged in oxygen-producing photosynthesis. Such cells may have been engulfed, and they may have resisted digestion. By providing their respiring host cell with

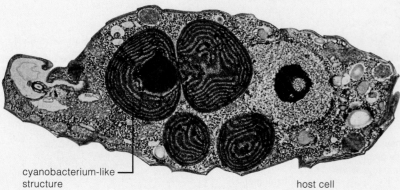

cyanobacterium-like structure

host cell

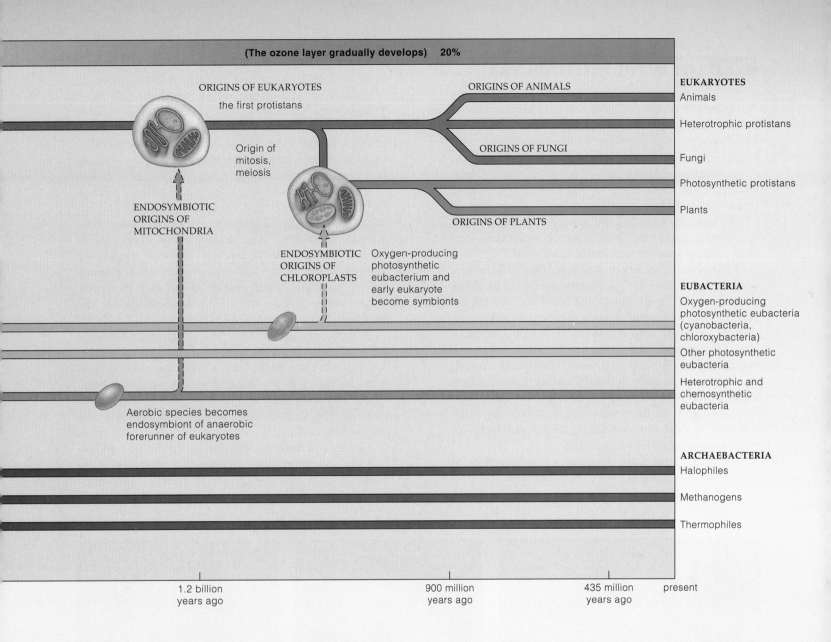

(The ozone layer gradually develops)   20%

ORIGINS OF EUKARYOTES
the first protistans

ORIGINS OF ANIMALS

ORIGINS OF FUNGI

ORIGINS OF PLANTS

Origin of
mitosis,
meiosis

ENDOSYMBIOTIC
ORIGINS OF
MITOCHONDRIA

ENDOSYMBIOTIC
ORIGINS OF
CHLOROPLASTS

Oxygen-producing
photosynthetic
eubacterium and
early eukaryote
become symbionts

Aerobic species becomes
endosymbiont of anaerobic
forerunner of eukaryotes

EUKARYOTES
Animals

Heterotrophic protistans

Fungi

Photosynthetic protistans

Plants

EUBACTERIA
Oxygen-producing
photosynthetic eubacteria
(cyanobacteria,
chloroxybacteria)

Other photosynthetic
eubacteria

Heterotrophic and
chemosynthetic
eubacteria

ARCHAEBACTERIA
Halophiles

Methanogens

Thermophiles

| 1.2 billion years ago | 900 million years ago | 435 million years ago | present |

oxygen, their endosymbiotic existence would have been favored. And they could have evolved into chloroplasts.

In metabolism and overall nucleic acid sequence, chloroplasts resemble some eubacteria. Their DNA is self-replicating, and they divide independently of the cell's division. Chloroplasts vary in shape and in their array of light-absorbing pigments, just as different photosynthetic eubacteria do. They may have originated a number of times, in a number of different lineages.

An Evolutionary Tree for Eukaryotes Whatever the routes, new kinds of cells appeared on the evolutionary stage. They were equipped with a nucleus, ER membranes, and mitochondria or chloroplasts (or both). They were the first eukaryotes, the first **protistans**.

b An evolutionary tree of life that reflects mainstream thinking about the connections among major lineages. The diagram incorporates a few ideas about the origins of some of the key eukaryotic organelles: the nucleus, ER membrane, mitochondria, and chloroplasts.

With their efficient metabolic strategies, the early protistans underwent rapid divergences and adaptive radiations. In no time at all, evolutionarily speaking, some of their descendants gave rise to the kingdoms of animals, fungi, and plants. Pulling this all together, we can put together a plausible evolutionary tree for eukaryotes. Take a moment to study Figure *b*. You can use it as a road map for chapters to follow.

21.5 LIFE IN THE PALEOZOIC ERA

As Figure 21.8 shows, we subdivide the Paleozoic into six periods: the Cambrian, Ordovician, Silurian, Devonian, Carboniferous, and Permian. Tectonic movements split the supercontinent Laurentia before this era. By Cambrian times, fragmented land masses were straddling the equator, and warm, shallow seas lapped their margins. Global conditions restricted pronounced seasonal changes in winds, ocean currents, and the upward churning of nutrients from deep waters. Thus, along the equatorial shorelines, nutrient supplies were stable but limited.

Early Cambrian organisms had flattened bodies, with a good surface-to-volume ratio for taking up nutrients (Figure 21.10a–c). Most lived on or just beneath the seafloor, where dead organisms and organic debris settled. Most major animal phyla evolved in Cambrian seas. They ranged from sponges to simple vertebrates— and they were exuberantly diverse. How could so much diversity arise so early in time? Possibly, genes governing embryonic development were less intertwined than they are today, so there may not have been as much selection against mutated alleles and novel forms of traits among the Cambrian animals.

Also, the long, warm shorelines offered plenty of vacant adaptive zones, with opportunities for new ways of securing food. Now we start seeing fossilized organisms with missing chunks, punctures, and healed-over wounds. These are not artifacts of fossilization; the organisms were damaged while they were still alive. Things were starting to get lively! In short order, diverse predators and prey with armorlike shells, spines, mouths, and novel feeding structures evolved.

Late in the Cambrian, temperatures in the shallow seas changed drastically. Trilobites (Figure 21.10d), one of the most common animals, nearly became extinct. During the Ordovician, a supercontinent called Gondwana had been drifting south, and parts became submerged in shallow seas. Vast marine environments opened up, and they favored adaptive radiations. Species of reef organisms flourished. Fast-swimming, shelled predators, the nautiloids, dominated the evolutionary stage. Their only living descendants are the chambered nautiluses (page 887).

During the late Ordovician, Gondwana straddled the South Pole. Immense glaciers formed on its surface. As water became locked up as ice, the shallow seas were drained. This was the first ice age we know about, and it may have triggered the first global mass extinction. At the Ordovician-Silurian boundary, reef life everywhere collapsed.

Gondwana drifted north during the Silurian and on into the Devonian. Reef organisms recovered, and an adaptive radiation produced formidable predators: armor-plated, massive-jawed fishes. Plants, fungi, and a

Figure 21.10 Representative Cambrian animals. (**a,b**) From Australia's Ediacara Hills, fossils of two animals, about 600 million years old. (**c**) From the Burgess Shale of British Columbia, a fossilized marine worm. (**d**) A beautifully preserved fossil of one of the earliest trilobites.

a   **b**

Figure 21.11 (**a**) Fossils of a Devonian plant (*Psilophyton*), possibly one of the earliest ancestors of seed-bearing plants. Compare this to Figure 20.2*c*, which shows some fossilized stems of the oldest known genus of land plants, *Cooksonia*. (**b**) Fossilized leaves of *Archaeopteris*, a land plant of the upper Devonian that may have belonged to the lineage leading to gymnosperms (such as pine trees) and angiosperms (the flowering plants). Some *Archaeopteris* trees were over three stories tall.

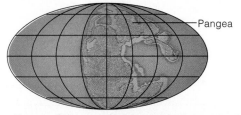

c

Figure 21.12 Lobe-finned fishes venturing onto a Devonian shoreline. They probably gulped air at the surface of shallow, stagnant water. In some lineages, lobed fins evolved into the limbs of amphibians, reptiles, birds, and mammals. Do fish out of water seem farfetched? A catfish thrashes sideways and uses its fins to pull its body forward across land (*see the photograph*).

variety of invertebrates, including segmented worms, became established in the wet lowlands. The small stalked plants foreshadowed a major radiation of terrestrial plants (Figure 21.11). Later in the Devonian, lobe-finned fishes that were ancestral to amphibians invaded the land (Figure 21.12).

At the Devonian-Carboniferous boundary, an unknown catastrophic event caused sea levels to swing dramatically, and another mass extinction occurred. Afterward, land plants and insects embarked on major adaptive radiations. Throughout the Carboniferous, land masses were gradually submerged and drained many times. Organic debris accumulated, became compacted, and was converted to coal, in the manner described on page 396.

In Permian times, insects, amphibians, and early reptiles flourished in vast swamp forests, composed in part of the ancestors of modern-day cycads, ginkgos, and conifers. As the Permian drew to a close, the greatest of all mass extinctions occurred. Nearly all known

species on land and in the seas perished. At that time, all land masses were colliding to form Pangea. This vast supercontinent extended from pole to pole. A single world ocean lapped its margins:

As you will see, the new distribution of oceans, land masses, and land elevations had catastrophic effects on global climates and on the course of evolution.

Early in the Paleozoic era, organisms of all five kingdoms were flourishing in the seas. By the end of the era, many lineages had successfully invaded land environments, including the wet lowlands of the supercontinent Pangea.

21.6 LIFE DURING THE MESOZOIC ERA

The Mesozoic lasted about 175 million years. It is divided into the Triassic, Jurassic, and Cretaceous periods. Adaptive radiations that began during this era greatly expanded the range of global diversity. For example, early in the Mesozoic, divergences in a few lineages gave rise to mammals and wonderful reptilian "monsters," including the dinosaurs. At the Triassic-Jurassic boundary, many marine organisms perished during a mass extinction.

Early in the Cretaceous, the supercontinent Pangea started to break up. As its huge fragments gradually drifted apart, the resulting geographic isolation favored divergences and speciation on a grand scale:

The Mesozoic was a time when dinosaurs and angiosperms (flowering plants) were the most visibly dominant lineages. The flowering plants emerged about 127 million years ago, and they began a spectacular radiation. Within 30 to 40 million years, they would displace the already declining conifers and related plants in nearly all environments (Figure 21.13). Marine invertebrates, fishes, and insects underwent equally spectacular radiations.

About 120 million years ago, global temperatures shot up by 25 degrees. By one theory, a superplume (or a rash of them) originated deep in the earth, then broke through the crust. Huge plumes spreading out beneath the crust "greased" the earth's plates into moving twice as fast. Around the globe, volcanoes spewed nutrient-rich ashes. Basalt and lava poured from huge fissures. The South Atlantic crust opened up like a zipper. Simultaneously, the plumes released huge amounts of carbon dioxide. This is one of the "greenhouse" gases that absorb heat radiating from the earth before it can escape into space. The nutrient-enriched planet warmed up—and stayed warm for 20 million years. On land and in shallow seas, photosynthetic organisms flourished. Their remains were slowly buried and converted into the world's oil reserves.

About 65 million years ago, the last dinosaurs and many marine organisms disappeared in a mass extinction. As the *Focus* essay indicates, their disappearance coincided with an asteroid impact, at a site that eventually drifted into a position that we now call the northern Yucatán peninsula.

The Mesozoic was a time of major adaptive radiations and of a mass extinction in which the last dinosaurs and many marine organisms disappeared.

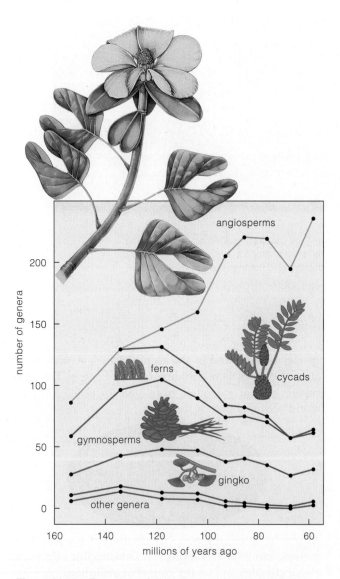

Figure 21.13 The range of diversity among vascular plants during the Mesozoic era. Accompanying the spectacular adaptive radiation of flowering plants were significant declines in the other groups shown. The painting depicts a floral shoot of *Archaeanthus linnenbergeri*, a flowering plant that lived during the Cretaceous. This now-extinct species resembled living magnolias in many respects.

Plumes, Global Impacts, and the Dinosaurs

Rise of the Ruling Reptiles   Early in the Mesozoic era, a reptilian lineage evolved into the first dinosaurs. They weren't much larger than a wild turkey. Many sprinted about on two legs. Most of the early dinosaurs had high metabolic rates, and they may have been warm-blooded. They were not the dominant land animals during the Triassic period. Center stage belonged to *Lystrosaurus* and other plant-eating, mammal-like reptiles that were too large to be bothered by most predators.

Then *Lystrosaurus* ran out of luck, and adaptive zones opened up for the dinosaurs, possibly when an asteroid struck the earth. There is a crater about the size of Rhode Island in central Quebec (Figure *a*). Even if it is not *the* impact crater, it certainly shows how huge such an impact could have been. The blast wave, firestorm, earthquakes, and lava flows would have been stupendous. The atmospheric distribution of rocks and water must have vaporized upon impact, which could have darkened the skies. If so, months of acid rain followed. What kinds of animals survived this time of mass extinction (and later ones)? Most of them were small, metabolically active, and less vulnerable than others to drastic swings in climate.

The surviving dinosaurs underwent a major adaptive radiation. For 140 million years, their descendants were the ruling reptiles. Some, including the ultrasaurs, reached monstrous proportions. These were 15 meters tall.

Many of the dinosaur lineages perished in another mass extinction at the end of the Jurassic, then in a pulse of extinctions during the Cretaceous. Perhaps plumes of molten material from deep in the earth ruptured the crust, releasing enough carbon dioxide in the atmosphere to change the global temperature and climate. Or perhaps major environmental disturbances followed a swarm of asteroid bombardments (Figure *b*). Whatever the causes, conditions changed for the dinosaurs, and not all of them lived through it.

Yet some lineages recovered and new forms replaced them. By the late Cretaceous, there were perhaps a hundred different genera of dinosaurs. Duckbilled dinosaurs appeared in forests and swamps. Tanklike *Triceratops* and other plant eaters flourished in more open regions. They were prey for the swift, agile, and fearsomely toothed *Velociraptor* of motion picture fame.

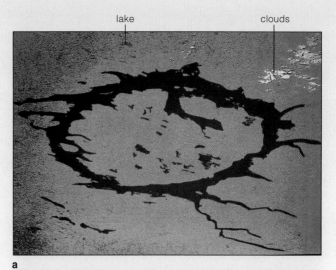

lake   clouds

a

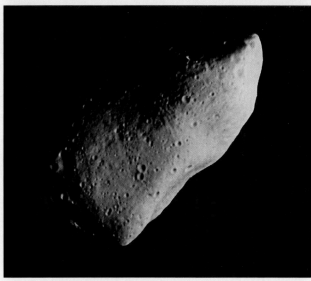

b

a   What an impact crater looks like. Aerial view of Manicougan Crater, Quebec, where an asteroid struck the earth approximately 210 million years ago. Asteroids are rocky, metallic bodies with diameters ranging from 1,000 kilometers (600 miles) to a few meters. Most were swept from the sky when the planets were forming. About 6,000 known asteroids are still orbiting the sun in a belt between Mars and Jupiter. Unfortunately, the orbits of many dozens of others take them across the earth's orbit. Think of it as Russian roulette on a cosmic scale.

b   What one asteroid looks like, courtesy of the Galileo spacecraft that flew past on its way to Jupiter. This is one of the smaller asteroids; it would only extend from Washington, D.C., halfway to Baltimore.

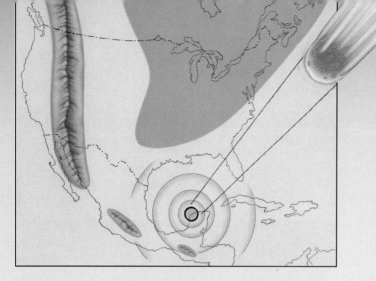

c Artist's interpretation of what might have happened during the last few minutes of the Cretaceous.

Horrendous End to Dominance   The final blow came at the Cretaceous-Tertiary (K-T) boundary. Then, all (or nearly all) of the remaining dinosaurs perished when an asteroid apparently hit the earth. A thin layer of iridium-rich rock distributed around the world dates precisely to the K-T boundary. Iridium is rare on the earth's surface but common in asteroids.

By analyzing gravity maps, iridium levels in soils, and other evidence, researchers identified what appears to be the impact site. Massive movements in the crust transported the site to what is now the northern Yucatán peninsula of Mexico (Figure *c*). The crater itself is 9.6 kilometers deep and 300 kilometers across—wider than the state of Connecticut. This means the asteroid hit the earth at 160,000+ kilometers (100,000 miles) per hour. At least 200,000 cubic kilometers of debris and dense gases were blasted skyward—enough to shut out sunlight for months. Monstrous, 120-meter waves raced across the oceans; the entire crust heaved with earthquakes.

Things haven't settled down much since. About 2.3 million years ago, for example, a huge object from space hit the Pacific Ocean. About the same time, vast ice sheets started forming abruptly in the Northern Hemisphere. Long-term shifts in climate may have been ushering in this most recent ice age, but a global impact might have accelerated the process. Water vaporized during the impact would have formed a cloud cover that prevented sunlight from reaching the earth's surface. As you will see from the next chapter, the early ancestors of humans were around when all of this happened. The extreme shift in climate surely put their adaptability to the test.

In short, the formation of ice sheets following the global impact is one more bit of information that compels us to look skyward, also, in our attempts to piece together the environments in which life evolved.

d If dinosaurs of this sort had not disappeared by the dawn of the Cenozoic, would the then-tiny mammals ever have ventured out from under the shrubbery? Would you even be here today?

Figure 21.14 A few representatives of the Cenozoic. (**a**) From the Eocene, small, four-toed horses (*Orohippus*) and a much larger herbivore (*Uintatherium*). (**b**) From the Pleistocene, the saber-tooth cat (*Smilodon*), a large bird (*Teratornis*), and an extinct horse. This reconstruction is based on fossils recovered from a pitch pool at Rancho LaBrea, California.

a

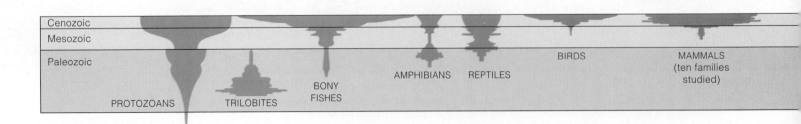

b

Cenozoic

Mesozoic

Paleozoic

PROTOZOANS TRILOBITES BONY FISHES AMPHIBIANS REPTILES BIRDS MAMMALS (ten families studied)

21.7 LIFE DURING THE CENOZOIC ERA

The breakup of Pangea set events in motion that have continued to the present. At the dawn of the Cenozoic, land masses were on collision courses:

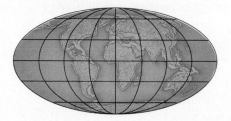

Coastlines fractured. Intense volcanic activity and uplifting produced mountains along the margins of massive rifts and plate boundaries. The Alps, Andes, Himalayas, and Cascades were born through these upheavals. The geologic changes caused major shifts in climate that affected the further evolution of life. As you will read in later chapters, vast, semiarid, cooler grasslands emerged. Diverse plant-eating mammals and their predators radiated into these new adaptive zones (Figure 21.14).

Today the distribution of land masses has favored unparalleled richness in species diversity. The tropical forests of South America, Madagascar, and Southeast Asia may well be the richest ecosystems ever to appear on earth. Marine ecosystems of the island chains of the tropical Pacific are probably not far behind. Yet we are in the middle of what may turn out to be one of the greatest of all mass extinctions. About 50,000 years ago, early humans started following migrating herds of wild animals around the Northern Hemisphere. Within a few thousand years, major groups of large mammals disappeared. The pace of extinction has been accelerating ever since, as humans hunt animals for food, fur, feath-ers, or fun, and as they destroy habitats to clear land for farm animals or crops. Chapter 49 focuses on the repercussions, which are global in scope.

The Cenozoic era has been a time of unparalleled richness in species diversity. It also is a time when the human species may be causing one of the greatest of all mass extinctions.

On the next two pages, we will be concluding our overview of life's history with a summary illustration that correlates milestones in the evolution of the earth and life. As you study this figure, keep in mind that it is only a generalized summary. It shows the five greatest mass extinctions—but there were many others in between. It shows the shrinking and expanding range of species diversity—but the range is for all the major groups combined. As you can see from the sampling of Figure 21.15, each major group has its own distinctive history of persistences, radiations, and extinctions.

With these qualifications in mind, we are ready to turn to the next chapters in this unit. They will provide you with richer detail of the history and current range of diversity for all five kingdoms of organisms.

Figure 21.15 A few representative evolutionary histories. Notice the differences in the patterns of adaptive radiations and mass extinction for this limited sampling. Notice also the spectacular current success of the insects. (For plants and animals, the widths of the lineages, shown in blue, represent the approximate numbers of families.)

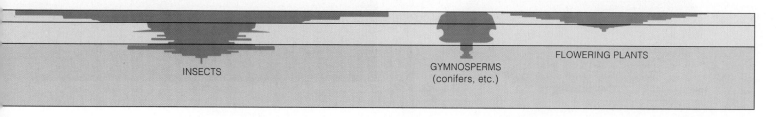

INSECTS

GYMNOSPERMS (conifers, etc.)

FLOWERING PLANTS

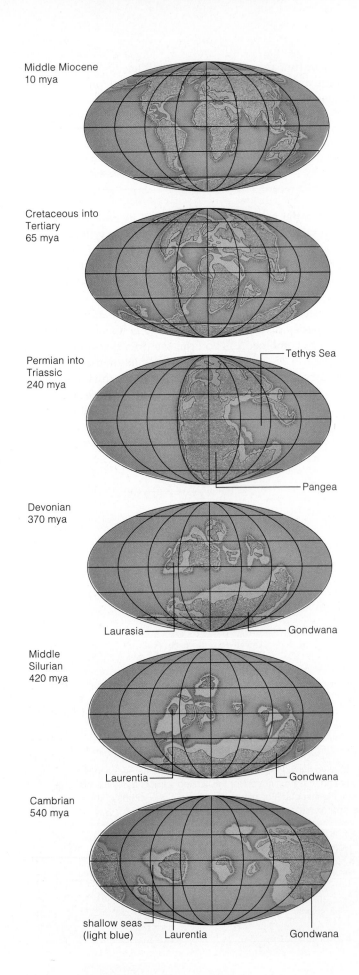

Middle Miocene
10 mya

Cretaceous into
Tertiary
65 mya

Permian into
Triassic
240 mya

Tethys Sea

Pangea

Devonian
370 mya

Laurasia

Gondwana

Middle
Silurian
420 mya

Laurentia

Gondwana

Cambrian
540 mya

shallow seas
(light blue)

Laurentia

Gondwana

Era	Period	Epoch	Millions of Years Ago (mya)	
CENOZOIC	Quaternary	Recent	0.01-	
		Pleistocene	1.65	
	Tertiary	Pliocene	5	
		Miocene	25	
		Oligocene	38	
		Eocene	54	
		Paleocene	65	
MESOZOIC	Cretaceous	Late		
			100	
		Early		
			138	
	Jurassic			
			205	
	Triassic			
			240	
PALEOZOIC	Permian			
			290	
	Carboniferous			
			360	
	Devonian			
			410	
	Silurian			
			435	
	Ordovician			
			505	
	Cambrian			
			550	
PROTEROZOIC				
			2,500	
ARCHEAN				

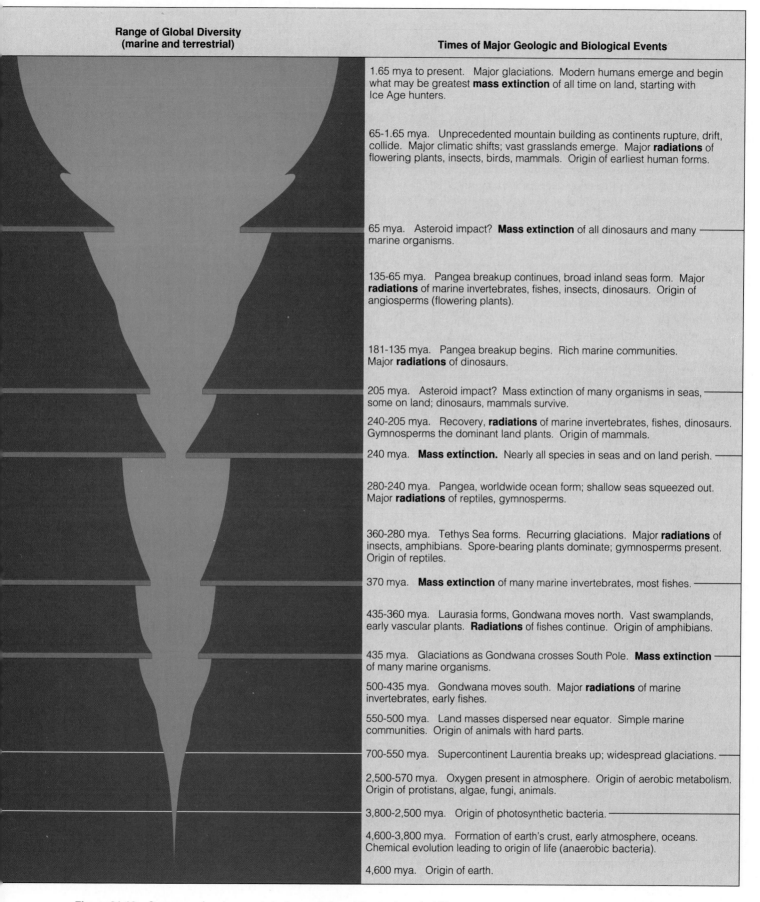

Range of Global Diversity
(marine and terrestrial)

Times of Major Geologic and Biological Events

1.65 mya to present. Major glaciations. Modern humans emerge and begin what may be greatest **mass extinction** of all time on land, starting with Ice Age hunters.

65-1.65 mya. Unprecedented mountain building as continents rupture, drift, collide. Major climatic shifts; vast grasslands emerge. Major **radiations** of flowering plants, insects, birds, mammals. Origin of earliest human forms.

65 mya. Asteroid impact? **Mass extinction** of all dinosaurs and many marine organisms.

135-65 mya. Pangea breakup continues, broad inland seas form. Major **radiations** of marine invertebrates, fishes, insects, dinosaurs. Origin of angiosperms (flowering plants).

181-135 mya. Pangea breakup begins. Rich marine communities. Major **radiations** of dinosaurs.

205 mya. Asteroid impact? Mass extinction of many organisms in seas, some on land; dinosaurs, mammals survive.

240-205 mya. Recovery, **radiations** of marine invertebrates, fishes, dinosaurs. Gymnosperms the dominant land plants. Origin of mammals.

240 mya. **Mass extinction.** Nearly all species in seas and on land perish.

280-240 mya. Pangea, worldwide ocean form; shallow seas squeezed out. Major **radiations** of reptiles, gymnosperms.

360-280 mya. Tethys Sea forms. Recurring glaciations. Major **radiations** of insects, amphibians. Spore-bearing plants dominate; gymnosperms present. Origin of reptiles.

370 mya. **Mass extinction** of many marine invertebrates, most fishes.

435-360 mya. Laurasia forms, Gondwana moves north. Vast swamplands, early vascular plants. **Radiations** of fishes continue. Origin of amphibians.

435 mya. Glaciations as Gondwana crosses South Pole. **Mass extinction** of many marine organisms.

500-435 mya. Gondwana moves south. Major **radiations** of marine invertebrates, early fishes.

550-500 mya. Land masses dispersed near equator. Simple marine communities. Origin of animals with hard parts.

700-550 mya. Supercontinent Laurentia breaks up; widespread glaciations.

2,500-570 mya. Oxygen present in atmosphere. Origin of aerobic metabolism. Origin of protistans, algae, fungi, animals.

3,800-2,500 mya. Origin of photosynthetic bacteria.

4,600-3,800 mya. Formation of earth's crust, early atmosphere, oceans. Chemical evolution leading to origin of life (anaerobic bacteria).

4,600 mya. Origin of earth.

Figure 21.16 Summary of major events in the evolution of the earth and of life.

SUMMARY

1. The full story of life begins with the "big bang," a model of the origin of the universe.

a. By this model, all matter and all of space were once compressed in a fleeting state of enormous heat and density. Time began with the near-instantaneous distribution of all matter and energy throughout the known universe, which has been expanding ever since.

b. Nearly all of the helium and other light elements, the most abundant elements of the universe, were produced immediately after the big bang. The heavier elements originated during the formation, evolution, and death of stars.

c. Every element of the solar system, the planet earth, and life itself are products of the physical and chemical evolution of the universe and its stars.

2. Four billion years ago, the earth had a high-density core, a mantle of intermediate density, and a thin, extremely unstable crust of low-density rocks. Probably gaseous hydrogen, nitrogen, carbon monoxide, and carbon dioxide made up the first atmosphere. Free oxygen and water could not have accumulated under the prevailing conditions.

3. After the crust cooled, water accumulated and rains began. Runoff from rains carried mineral salts and other compounds from the crustal rocks to depressions in the crust and formed the early seas. Life could not have originated without this salty liquid water.

4. Many studies and experiments provide indirect evidence that life originated under conditions that presumably existed on the early earth.

a. Comparative analyses of the composition of cosmic clouds and of rocks from other planets and the earth's moon suggest that precursors of the complex molecules associated with life were available.

b. In laboratory tests that simulated primordial conditions, including the absence of free oxygen, the precursors spontaneously assembled into glucose, sugars, amino acids, and other organic compounds.

c. Known chemical principles as well as computer simulations indicate that metabolic pathways could have evolved as a result of chemical competition for the limited supplies of organic molecules that had accumulated in the seas.

d. Self-replicating systems of RNA, enzymes, and coenzymes have been synthesized in the laboratory. How DNA entered the picture is not yet understood.

e. Lipid and lipid-protein membranes that show properties of cell membranes form spontaneously under the prescribed conditions.

5. Life originated by 3.8 billion years ago. Since then, life's history has been influenced profoundly by major changes in the earth's crust, atmosphere, and oceans. Forces of change have included tectonic movements, asteroid impacts, and activities of organisms—such as oxygen-producing photosynthesizers and, currently, the human species.

6. Abrupt discontinuities in the fossil record mark the times of global mass extinctions. They serve as boundary markers for five great eras in the geologic time scale. Radiometric dating has allowed us to assign absolute dates to this time scale:

a. Archean: 4.6 to 2.5 billion years ago
b. Proterozoic: 2.5 billion to 550 million years ago
c. Paleozoic: 550 to 240 million years ago
d. Mesozoic: 240 to 65 million years ago
e. Cenozoic: 65 million years ago to the present

7. The first living cells were prokaryotic. Not long after they appeared, divergences began that led to three great lineages: the archaebacteria, the eubacteria, and the prokaryotic ancestors of all eukaryotes. Among the eubacteria were species that used a cyclic pathway of photosynthesis.

8. In the Proterozoic, the noncyclic pathway of photosynthesis had evolved in some lineages, and oxygen started accumulating in the atmosphere.

a. In time, the atmospheric concentration of free oxygen prevented further spontaneous formation of large organic molecules. The spontaneous origin of life was no longer possible on earth.

b. The abundance of atmospheric oxygen was a selective pressure that brought about the evolution of aerobic respiration. Aerobic respiration was a key innovation in the origin of the first eukaryotic cells.

c. Mitochondria and chloroplasts, two key eukaryotic organelles, may have evolved following endosymbiosis between aerobic bacteria and the anaerobic forerunners of eukaryotes.

d. The oxygen-rich atmosphere promoted the formation of a layer of ozone (O_3). In time, that shield against destructive ultraviolet radiation allowed some lineages to move out of the seas, into low wetlands.

9. Early in the Paleozoic, diverse organisms of all five lineages had become established in the seas. By the end of the era, the invasion of land was under way. From that time forward, there have been pulses of mass extinctions and adaptive radiations. Asteroid bombardments and other catastrophes triggered many of these events. So did tectonic movements that changed the distribution of oceans and land as well as the prevailing global and regional climates.

Review Questions

1. Describe the chemical and physical characteristics of the earth four billion years ago. How do we know what it may have been like? *324*

2. Describe the experimental evidence for the spontaneous origin of large organic molecules, the self-assembly of proteins, and the formation of organic membranes and spheres, under conditions similar to those of the early earth. *324–327*

3. How does continental drift occur, and in what ways does this process influence changes in biological communities? *328–329*

4. Summarize the theory of endosymbiosis. Cite evidence in favor of this theory. *331–332*

5. When did plants, insects, and vertebrates invade the land? *342–343*

6. The Atlantic Ocean is widening, and the Pacific Ocean and Indian Ocean are closing. Many millions of years from now, the continents will collide and form a second Pangea. Write a short essay on what conditions might be like on that future supercontinent and what types of organisms might survive there.

Self-Quiz *(Answers in Appendix IV)*

1. Through study of the geologic record, we know that the evolution of life is linked to _____ .
   a. tectonic movements of the earth's crust
   b. bombardment of the earth by celestial objects
   c. profound shifts in land masses, shorelines, and oceans
   d. physical and chemical evolution of the earth
   e. all of the above

2. The geologic time scale begins with the _____ era.
   a. Paleozoic   d. Proterozoic
   b. Mesozoic   e. Cenozoic
   c. Archean

3. The first scientist to find indirect evidence that organic molecules could have been formed on the early earth was _____ .
   a. Darwin   c. Fox
   b. Miller   d. Margulis

4. One type of "proto-cell," the microsphere, was first produced in a laboratory by _____ .
   a. Darwin   c. Fox
   b. Miller   d. Margulis

5. An abundance of _____ was conspicuously absent from the earth's atmosphere four billion years ago.
   a. hydrogen   c. carbon monoxide
   b. nitrogen   d. free oxygen

6. Life originated by _____ .
   a. 4.6 billion years ago   c. 3.8 billion years ago
   b. 2.8 million years ago   d. 3.8 million years ago

7. Abrupt discontinuities in the fossil record indicate _____ .
   a. the death of all organisms when the atmospheric oxygen levels dropped
   b. the times of global mass extinctions
   c. boundary markers for five great eras of the geologic time scale
   d. both b and c

8. Which of the following statements is false?
   a. The first living cells were prokaryotes.
   b. The cyclic pathway of photosynthesis first appeared in early eubacterial species.
   c. Oxygen began accumulating in the atmosphere after the noncyclic pathway of photosynthesis evolved.
   d. During the Proterozoic, an increasing amount of atmospheric oxygen promoted further spontaneous generation of organic molecules.

9. Match the era with the events listed.
   ____ Archean   a. major radiations of dinosaurs, origin
   ____ Proterozoic   of flowering plants and mammals
   ____ Paleozoic   b. formation of earth's crusts, oceans,
   ____ Mesozoic   early atmosphere, chemical evolution,
   ____ Cenozoic   origin of life
                       c. major radiations of flowering plants, insects, birds, mammals, emergence of human forms
                       d. oxygen present, origin of aerobic metabolism, protistans, algae, fungi, animals
                       e. origin of amphibians, origin of reptiles, spread of early plants

Selected Key Terms

archaebacterium *330*   Paleozoic *329*
Archean *329*   plate tectonic theory *328*
Cenozoic *329*   prokaryotic cell *330*
eubacterium *330*   Proterozoic *329*
eukaryotic cell *330*   protistan *333*
geologic time scale *329*   radiation *343*
mass extinction *343*   stromatolite *330*
Mesozoic *329*   theory of endosymbiosis *331*

Readings

Bambach, R., C. Scotese, and A. Ziegler. 1980. "Before Pangea: The Geographies of the Paleozoic World." *American Scientist* 68(1): 26–38.

Dobb, E. February 1992. "Hot Times in the Cretaceous." *Discover* 13: 11–13.

Dott, R., Jr., and R. Batten. 1988. *Evolution of the Earth.* Fourth edition. New York: McGraw-Hill.

Hartman, W., and Chris Impey. 1994. *Astronomy: The Cosmic Journey.* Fifth edition. Belmont, California: Wadsworth.

Horgan, J. February 1991. "Trends in Evolution: In the Beginning . . . " *Scientific American* 264(2): 116–125.

22 BACTERIA AND VIRUSES

The Unseen Multitudes

Did a friend ever mention that you are nearly 1/1,000 of a mile tall? Probably not. What would be the point of measuring people in units as big as miles? Yet we think this way, in reverse, when we measure **microorganisms**. These are mostly single-celled organisms too small to be seen without a microscope.

The bacteria in Figure 22.1 are a case in point. To measure them, you'd have to divide a meter into a thousand units (millimeters). Then you'd have to divide one millimeter into a thousand smaller units (micrometers). One millimeter would be as small as the dot of this "i." A thousand bacteria would fit side by side across the dot!

Of all organisms, bacteria generally are the smallest. To be sure, viruses are smaller. We measure them in nanometers (billionths of a meter). But viruses are not living things. We consider them here because they infect just about every kind of organism in the great spectrum of life.

Bacteria vastly outnumber all other organisms combined. Their reproductive potential is staggering. Under ideal conditions, some bacteria can divide about every twenty minutes. At that rate, a single bacterium could have nearly a billion descendants in ten hours! So why don't bacteria take over the world? Sooner or later, their activities typically ruin the conditions favoring their reproduction. Also, bacteria eat one another, other organisms attack them, and viruses and seasonal changes in living conditions help keep them in check.

Many kinds of bacteria are **pathogens** (infectious, disease-causing agents). They invade and multiply inside other organisms, and disease follows when their

Figure 22.1 How small are bacteria? Shown here, *Bacillus* cells on the tip of a pin. The cells in (**c**) are magnified 14,000 times. (**d**) How small are viruses? Shown here, bacteriophage particles, each about 225 nanometers tall, infecting a bacterial cell.

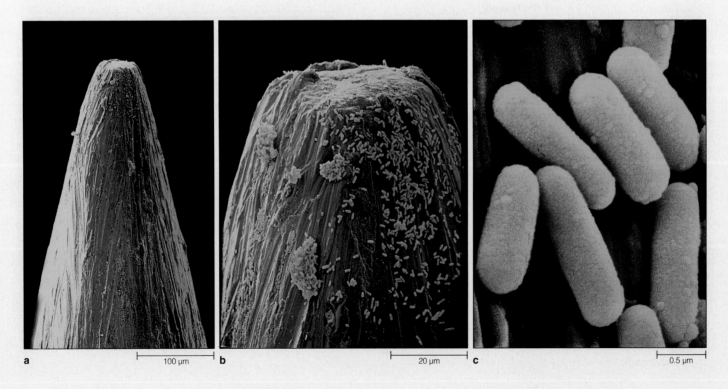

a |———| 100 µm   b |———| 20 µm   c |———| 0.5 µm

activities damage tissues and interfere with normal body functions. However, the pathogenic types should not give the whole kingdom a bad name.

Think back on the vast numbers of photosynthetic bacterial cells in the seas (page 117). They provide food and oxygen for entire communities, and they have major roles in the global cycling of carbon. Other bacteria decompose organic debris and so help cycle nutrients that sustain entire communities. Trace the lineages of *any* organism back far enough in time, and you discover bacterial ancestors.

And so, from *Escherichia coli* to elephants, clams, and coast redwoods, all organisms interconnect with bacteria—regardless of differences in size, numbers, and evolutionary distance.

one virus particle

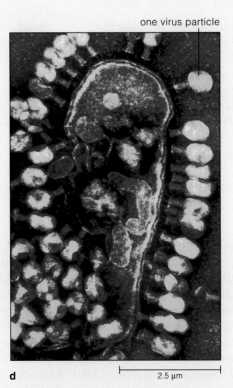

d

2.5 µm

KEY CONCEPTS

1. Bacteria are microscopically small, single-celled, prokaryotic organisms. They do not have a profusion of internal, membrane-bound organelles in the cytoplasm, as eukaryotic cells do. Yet bacteria show great metabolic diversity, and many make complex behavioral responses to the environment.

2. Bacteria reproduce by binary fission, a cell division mechanism that immediately follows DNA replication and that divides the parent cell into two genetically equivalent daughter cells.

3. Today there are two great bacterial lineages, the archaebacteria and eubacteria. Archaebacteria live in such inhospitable, anaerobic habitats that they may resemble the first living cells on earth. Most bacterial species are eubacteria, which inhabit nearly all existing environments.

4. From the human perspective, bacteria are good, bad, or dangerous. Basically, however, they simply are busy at surviving and reproducing like the rest of us.

5. A virus is a nonliving infectious particle that consists of nucleic acid and a protein coat (and sometimes an outer envelope). Viruses cannot be replicated without pirating the metabolic machinery of a specific type of host cell.

6. Nearly all viral multiplication cycles include five steps: attachment to a suitable host cell, penetration, viral DNA or RNA replication and protein synthesis, assembly of new viral particles, and release from the infected cell.

22.1 CHARACTERISTICS OF BACTERIA

Of all organisms, bacteria are the most abundant and far-flung. Thousands of species live in places ranging from deserts to hot springs, snow, the seafloor, and the ocean. Billions of bacterial cells may be present in a handful of rich soil. The ones in your gut and on your skin outnumber your body cells! (Your cells are larger, however, so you are only "a few percent bacterial" by weight.) And of all organisms, bacteria have the longest evolutionary histories. Two great lineages, **archaebacteria** and **eubacteria**, have endured from the time of life's origin to the present. A third lineage gave rise to eukaryotic cells (page 330). Table 22.1 and Figure 22.2 introduce the main features that help characterize these remarkable organisms.

Splendid Metabolic Diversity

All organisms must take in energy and carbon to meet their nutritional requirements. However, compared to other organisms, bacteria show the greatest diversity in their modes of securing these resources.

Like plants, some bacteria are **photoautotrophs**. These "self-feeders" can make their own biological molecules by using sunlight as an energy source and carbon dioxide as a carbon source for photosynthesis. Some species use a noncyclic pathway, based on light-trapping pigments, transport systems, and ATP synthases embedded in their plasma membrane. In this pathway, water molecules give up electrons and hydrogen that are used for the reactions, and oxygen is released as a by-product (page 112). Other photoautotrophic species use a cyclic pathway. They either cannot use oxygen or die in its presence. These anaerobes strip electrons and hydrogen from gaseous hydrogen (H_2), sulfur, hydrogen sulfide (H_2S), and other inorganic compounds in their environment.

Other bacteria are **photoheterotrophs**. As their name suggests, they are not self-feeders. They *can* use sunlight as an energy source for a cyclic pathway of photosynthesis—but they can't harness carbon dioxide from the surroundings. Instead, they get carbon from fatty acids, carbohydrates, and other organic compounds that other organisms produced.

Then there are the self-feeding **chemoautotrophs**. These self-feeders harness carbon dioxide as the main carbon source. They get the required electrons and hydrogen from a variety of inorganic substances, including gaseous hydrogen, sulfur, sulfur compounds, nitrogen compounds, and ferrous iron (Fe^{++}).

And then there are the **chemoheterotrophs**. These are parasites or saprobes, not self-feeders. The *parasitic* types live on or in a living host and draw glucose and other nutrients from it. The *saprobic* types obtain nutri-

Table 22.1	Characteristics of Bacterial Cells

1. Bacterial cells are prokaryotic (they have no membrane-bound nucleus or other organelles in the cytoplasm).

2. Bacterial cells have a single chromosome (a circular DNA molecule); many species also have plasmids (page 248).

3. Most bacteria have a cell wall composed of peptidoglycan.

4. Most bacteria reproduce by binary fission.

5. Collectively, bacteria show great diversity in their modes of metabolism.

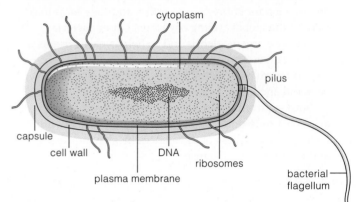

Figure 22.2 Generalized body plan of a bacterium.

ents from the organic products, wastes, or remains of other organisms.

Bacterial Sizes and Shapes

From the chapter introduction, you already have a sense of the microscopically small sizes of bacteria. Typically, the width or length of these cells falls between 1 and 10 micrometers.

Three basic shapes are common among bacteria. A spherical shape is a **coccus** (plural, cocci; from a word that really means berries). A rod shape is a **bacillus** (plural, bacilli, meaning small staffs). A bacterium with one or more twists in the cell body is a **spiral**:

coccus   bacillus   spiral

Don't let these simple categories fool you. Cocci may also be oval or partly flattened, and bacilli may be skinny (like straws) or tapered (like cigars). After they divide, daughter cells remain stuck together in various ways. For example, chains of daughter cells are called *strepto*cocci, and sheets of them are called *staphylo*cocci. Some spiral bacteria (vibrios) are curved, like a comma, and others (spirilla) are stiff, flagellated, and twisted helically, like a corkscrew. Other helically twisted

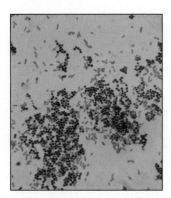

Figure 22.3 Example of Gram staining. Gram-positive *Staphylococcus aureus* remains purple when washed with organic solvents. Gram-negative *Escherichia coli* loses color easily. *E. coli* cells shown here appear red because they have been treated with a light-red dye (safranin) after being washed. Without this "counter-stain," they would be colorless.

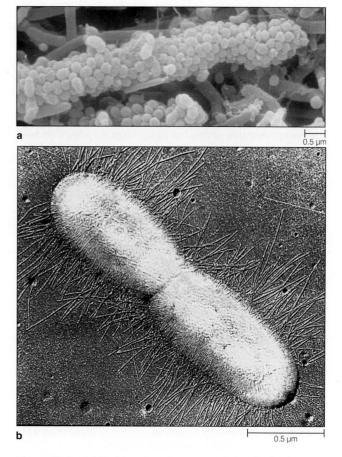

Figure 22.4 (**a**) Surface view of numerous bacteria that have become attached, by means of their sticky glycocalyx, to the surface of a human tooth—and doesn't this micrograph make you want to grab a toothbrush? (**b**) An abundance of the filamentous structures called pili, projecting from the surface of a cell (*Escherichia coli*) that is dividing in two.

forms, the spirochetes, have a flexible body with a sheathed filament attached.

Structural Features

Bacterial cells alone are **prokaryotic**, meaning they were around before the origin of the nucleus and of eukaryotic cells. (*Pro-*, before; *karyon*, nucleus.) Most bacteria are not internally subdivided into metabolic compartments. Reactions proceed in the cytoplasm or at the plasma membrane. (For example, proteins are synthesized at ribosomes distributed through the cytoplasm or attached to the plasma membrane.) This doesn't mean bacteria are somehow inferior to eukaryotic cells. Being tiny, fast reproducers, they do very well without great internal complexity.

Nearly all bacteria have a semirigid, permeable **cell wall** around the plasma membrane (Figure 22.2). The wall helps the cell maintain its shape and resist rupturing when internal fluid pressure increases. Eubacterial cell walls consist of **peptidoglycan**, a type of molecule in which peptides crosslink many polysaccharide strands to one another.

The precise structure and composition of the wall helps clinicians identify many bacterial species. In one staining reaction, the Gram stain, cells are exposed to a purple dye, then washed with alcohol, then exposed to a pink dye. Walls of *Gram-positive* bacteria retain the purple stain after the alcohol wash. Walls of *Gram-negative* bacteria lose color after the wash and become pink (Figure 22.3).

Surrounding the cell wall is a **glycocalyx**, a sticky mesh of polysaccharides, polypeptides, or both. When the mesh is highly organized and attached firmly to the wall, it is called a capsule. When less organized and loosely attached, it's called a slime layer. The glycocalyx helps bacterial cells attach to teeth, assorted membranes, rocks in streambeds, and other surfaces (Figure 22.4*a*). In some encapsulated types, it deters a host's infection-fighting cells.

Some bacteria have one or more motile structures called **bacterial flagella** (singular, flagellum). These don't have the same structure as eukaryotic flagella, and they don't operate the same way. They move the cell by rotating like a propeller. Many bacteria have **pili** (singular, pilus), short, filamentous proteins that project above the cell wall. Figure 22.4*b* shows an example. Some pili help cells adhere to surfaces. Others help them attach to one another as a prelude to conjugation, described next.

Bacteria are microscopic, prokaryotic cells having one bacterial chromosome and, often, a number of plasmids. The cells of nearly all species have a wall around the plasma membrane, and a capsule or slime layer around the cell wall.

22.2 BACTERIAL REPRODUCTION

Most bacterial cells reproduce by **binary fission**, a cell division mechanism. Each cell has only one bacterial chromosome, which is a circular, double-stranded DNA molecule with only a few proteins attached to it. Just after a parent cell replicates its DNA and starts moving the two molecules apart, the cytoplasm divides into two genetically equivalent daughter cells. As Figure 22.5 shows, the division requires suitable membrane growth and deposition of wall material at the midsection of the dividing cell.

In many species, daughter cells also inherit one or more plasmids. A **plasmid** is a small, self-replicating circle of extra DNA with only a few genes (page 248). Usually the genes confer a survival advantage, as when they specify an enzyme that allows a cell to synthesize or use some nutrient. Some plasmid genes confer resistance to antibiotics, substances that kill or inhibit the growth of bacteria. Others confer the means to conjugate. In **bacterial conjugation**, one bacterial cell transfers plasmid DNA to another cell, even of a different species (page 350 and Figure 22.6).

Bacteria use binary fission, a cell division mechanism that immediately follows DNA replication and divides the parent cell into two genetically equivalent daughter cells.

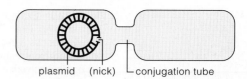

plasmid (nick) — conjugation tube

a *Conjugation tube unites a donor and a recipient cell*

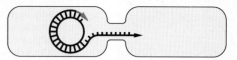

b *Replication starts on plasmid DNA in donor cell; displaced single DNA strand enters recipient cell*

c *Replication starts on transferred DNA strand*

d *Cells separate; plasmids circularize*

Figure 22.6 Bacterial conjugation—transfer of a plasmid from a donor to a recipient cell. A long pilus brings the two cells into close contact so that a conjugation tube can form between them. For clarity, the bacterial chromosome is not shown and the plasmid in the diagrams is enormously enlarged (*compare* Figure 16.2).

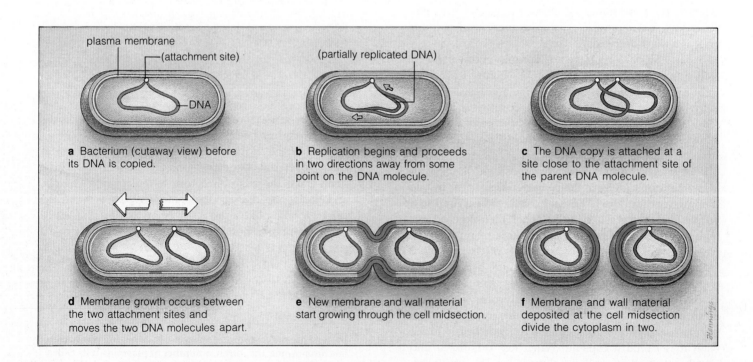

a Bacterium (cutaway view) before its DNA is copied.

b Replication begins and proceeds in two directions away from some point on the DNA molecule.

c The DNA copy is attached at a site close to the attachment site of the parent DNA molecule.

d Membrane growth occurs between the two attachment sites and moves the two DNA molecules apart.

e New membrane and wall material start growing through the cell midsection.

f Membrane and wall material deposited at the cell midsection divide the cytoplasm in two.

Figure 22.5 Bacterial reproduction by binary fission, a cell division mechanism.

22.3 BACTERIAL CLASSIFICATION

A few decades ago, reconstructing the phylogeny of bacteria seemed to be an impossible task. Except for stromatolites, early bacteria are not spectacularly represented in the fossil record. (Figure 21.9 shows some stromatolites—the fossilized, stacked mats of bacterial populations and sediments.) Most bacterial groups are not represented at all. Given their elusive evolutionary histories, the many thousands of known bacterial species traditionally have been grouped together on the basis of cell shape, modes of nutrition, metabolic patterns, staining attributes of their cell wall, and other traits. Table 22.2 lists representatives from some of the major groupings. Comparative biochemistry, including nucleic acid hybridization studies, is now providing insight into bacterial phylogeny. Such studies provide strong evidence that three great bacterial lineages diverged from a common ancestor not long after living cells originated. These led to the ancestors of eukaryotes, and to the archaebacteria and eubacteria.

Table 22.2 Some Major Groups of Bacteria

Group	Main Habitats	Characteristics	Representatives
Archaebacteria			
Methanogens	Anaerobic sediments of lakes, swamps; also animal gut	Chemosynthetic; methane producers; used in sewage treatment facilities	*Methanobacterium*
Halophiles	Brines (extremely salty water)	Heterotrophic; also have photosynthetic machinery of a unique sort	*Halobacterium*
Extreme thermophiles	Acidic soil, hot springs, hydrothermal vents on seafloor	Heterotrophic or chemosynthetic; use inorganic substances as electron donors	*Sulfolobus, Thermoplasma*
Photoautotrophic eubacteria			
Cyanobacteria, green sulfur bacteria, and purple sulfur bacteria	Mostly lakes, ponds; some marine, terrestrial	Photosynthetic; use sunlight energy and carbon dioxide (carbon source); cyanobacteria produce oxygen as by-product	*Anabaena, Nostoc, Chloroflexus, Rhodopseudomonas*
Photoheterotrophic eubacteria			
Purple nonsulfur and green nonsulfur bacteria	Anaerobic, organically rich muddy soils, and sediments of aquatic habitats	Use sunlight, organic compounds (electron donors); some purple nonsulfur may also grow chemotrophically	*Rhodospirillum, Chlorobium*
Chemoautotrophic eubacteria			
Nitrifying, sulfur-oxidizing, and iron-oxidizing bacteria	Soil; freshwater, marine habitats	Use carbon dioxide; electrons from inorganic compounds; some have roles in agriculture, cycling of nutrients in ecosystems	*Nitrosomonas, Nitrobacter, Thiobacillus*
Chemoheterotrophic eubacteria			
Spirochetes	Aquatic habitats; parasites of animals	Helically coiled, motile; free-living and parasitic species; some major pathogens	*Spirochaeta, Treponema*
Gram-negative, aerobic rods and cocci	Soil, aquatic habitats; parasites of animals, plants	Some major pathogens; some (e.g., *Rhizobium*) fix nitrogen	*Pseudomonas, Neisseria, Rhizobium, Agrobacterium*
Gram-negative, facultative anaerobic rods	Soil, plants, animal gut	Many are major pathogens; one (*Photobacterium*) is bioluminescent	*Salmonella, Proteus, Escherichia, Photobacterium*
Rickettsias and chlamydias	Host cells of animals	Intracellular parasites; many pathogens	*Rickettsia, Chlamydia*
Myxobacteria	Decaying plant, animal matter; bark of living trees	Gliding, rod-shaped; cells aggregate and migrate together	*Myxococcus*
Gram-positive cocci	Soil; skin and mucous membranes of animals	Some major pathogens	*Staphylococcus, Streptococcus*
Endospore-forming rods and cocci	Soil; animal gut	Some major pathogens	*Bacillus, Clostridium*
Gram-positive nonsporulating rods	Fermenting plant, animal material; gut, vaginal tract	Some important in dairy industry, others major contaminators of milk, cheese	*Lactobacillus, Listeria*
Actinomycetes	Soil; some aquatic habitats	Include anaerobes and strict aerobes; major producers of antibiotics	*Actinomyces, Streptomyces*

22.4 ARCHAEBACTERIA

There are three intriguing groups of archaebacteria—methanogens, halophiles, and extreme thermophiles. In many respects, their cell structure, metabolism, and nucleic acid sequences are unique. (For example, none has a cell wall of peptidoglycan.) They differ as much from other bacteria as they do from eukaryotes. They probably resemble the first living cells, which must have originated in extremely hot, acidic, and salty habitats. Archaebacteria live in such habitats—hence their name (*archae-* means beginning).

The **methanogens**, or "methane-makers," live in the muck of swamps, sewage, the animal gut, and other anaerobic habitats. They make ATP by converting carbon dioxide and hydrogen gases to methane (CH_4). Pungent stockyard fumes testify to their presence. So does the "marsh gas" that hangs over swamps and sewage treatment plants. As a group, methanogens produce about 2 billion tons of methane gas each year. As described on page 857, they affect atmospheric levels of carbon dioxide and the global carbon cycle.

Halophiles, or "salt-lovers," live in brackish ponds, salt lakes, near volcanic vents on the seafloor, and other high-salinity habitats (Figure 22.7). They can contaminate salted fish and animal hides as well as commercially produced sea salt.

Most halophiles are heterotrophs that form ATP by aerobic pathways. When oxygen levels decline, however, some strains produce ATP by photosynthesis. Patches of bacteriorhodopsin, a light-trapping pigment, form in the plasma membrane. Absorption of light energy triggers an increase in the hydrogen ion gradient across the membrane. When the ions flow down the gradient, through other membrane proteins, ATP forms (compare the discussion of chemiosmosis on page 114).

Extreme thermophiles (heat lovers) live in seemingly inhospitable places as hot springs, highly acidic soils, and sediments near hydrothermal vents. Vents are volcanic fissures in the ocean floor, where water temperature can exceed 250°C. They spew hydrogen sulfide, which thermophiles use as a source of electrons for ATP formation. Thermophiles inhabiting such places are the basis of remarkable food webs (page 885). They are cited as evidence that life could have originated deep in the early oceans. Waste heaps of coal mines are the only known habitat of *Thermoplasma*. Because this is a habitat of recent origin—coal mines have been around for only a few hundred years—*Thermoplasma* must have evolved in places we don't know about.

22.5 EUBACTERIA

The *eu-* in eubacteria implies "typical." Eubacteria are far more common than the three bacterial groups just described. Let's look at a few types, using modes of nutrition as a conceptual framework.

Photoautotrophic Eubacteria

Cyanobacteria (also called blue-green algae) are the most common photoautotrophic bacteria. Most species live in freshwater ponds, where they often grow as chains of cells, sheathed in mucus. The chains form dense, slimy mats near the surface of nutrient-enriched water (Figure 22.8). *Anabaena* and other types produce oxygen during photosynthesis. They also can convert nitrogen to ammonia, a nitrogen source for biosynthesis. When nitrogen-containing compounds are in short supply, some cells develop into **heterocysts**. These modified cells can make a nitrogen-fixing enzyme. Heterocysts share nitrogen compounds with the photosynthetic cells and get carbohydrates in return. Substances move freely through junctions that connect the cytoplasm of neighboring cells.

a

b

5 μm

Figure 22.7 Archaebacteria. (**a**) Pinkish, saline water in Great Salt Lake, Utah—a sign of colonies of halophilic bacteria and certain algae that contain pinkish to red-orange carotenoid pigments. (**b**) A colony of methanogenic cells (*Methanosarcina*).

Photoautotrophs that do not produce oxygen include green bacteria and purple bacteria. They cannot use water as a source of electrons. Green bacteria strip electrons from hydrogen sulfide or hydrogen gas. They may resemble the ancient anaerobic bacteria in which the cyclic pathway of photosynthesis first emerged.

Chemoautotrophic Eubacteria

Chemoautotrophs, recall, get their carbon from carbon dioxide and electrons from inorganic substances. Many eubacteria in this category affect the global cycling of nitrogen, sulfur, phosphorus, and other nutrients. Consider nitrogen, a building block for amino acids and proteins. Without it, there would be no life. Nitrifying bacteria, such as the one shown in Figure 22.9, strip electrons from ammonia. Plants can use the end product, nitrate, as a nitrogen source.

Chemoheterotrophic Eubacteria—
Wonderfully Beneficial Types

Most bacteria are chemoheterotrophs, and many are beneficial. Most species are major decomposers. Their enzymes break down organic compounds, even pesticides in soil. Pseudomonads are an example (Figure 22.10). We also use species of *Lactobacillus* when manufacturing pickles, sauerkraut, buttermilk, and yogurt. We use actinomycetes as sources of antibiotics. *Escherichia coli*, a gut dweller, produces vitamin K and compounds useful in fat digestion, and it helps newborns digest milk. Its activities keep many food-borne pathogens from colonizing the gut. Sugarcane and corn plants benefit from a symbiotic association with a nitrogen-fixing spirochete, *Azospirillum*. Plants use some of the nitrogen and the spirochete uses some of the plant's carbohydrates. Beans and other legumes benefit from the nitrogen-fixing activities of *Rhizobium*, which dwells in their roots.

a

b

resting spore   heterocyst

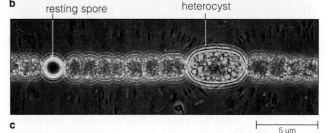

c   5 µm

Figure 22.8 *Cyanobacterium*—a premier photoautotroph. (**a**) A cyanobacterial population near the surface of a nutrient-enriched pond. (**b,c**) Resting spores form when conditions do not favor growth. A nitrogen-fixing heterocyst is shown in (**c**).

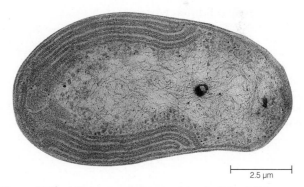

2.5 µm

Figure 22.9 *Nitrobacter*, a chemoautotroph. This nitrifying bacterium has a role in the global cycling of nitrogen.

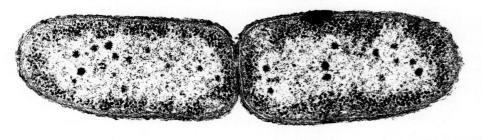

1.2 µm

Figure 22.10 One of the diverse chemoheterotrophs: a *Pseudomonas* cell, nearly completing binary fission.

22.6 REGARDING THE PATHOGENIC EUBACTERIA

Also in the category of chemoheterotrophs are most of the pathogenic bacteria. We admire pseudomonads when they grow in soil, not when they grow on "our" carbon-containing materials, including antiseptics and bits of soap. Pseudomonad infections can be serious, because they transmit plasmids that carry antibiotic-resistant genes (see the *Focus* essay). *Listeria monocytogenes*, a relative of *Lactobacillus*, can contaminate even refrigerated foods. Some *E. coli* strains cause a form of diarrhea that is the leading cause of infant death in developing countries.

Or consider *Clostridium tetani* (Figure 22.11). It causes *tetanus*, a terrible disease described on page 569. As part of its life cycle, *C. tetani* forms an **endospore** around its DNA and a bit of cytoplasm. Endospores are structures that resist heat, drying, boiling, and radiation. New bacterial cells form after the endospore germinates. A relative, *C. botulinum*, can taint fermented grain as well as food in improperly sterilized or sealed cans and jars. Its toxins cause *botulism*, a form of poisoning that can lead to respiratory failure and death.

Also consider *Borrelia burgdorferi*, which taxis from host to host inside the gut of blood-sucking ticks. Tick bites transmit this spirochete from deer, field mice, and some other wild animals to humans, who develop *Lyme disease*. The first sign is a circular "bull's-eye" rash around the tick bite (Figure 22.12). Severe headaches, backaches, chills, and fatigue follow. Without prompt treatment, the condition worsens.

Viewed through the prism of human interests, bacteria are good, bad, or dangerous. Yet their lineages are the most ancient, their adaptations are breathtakingly diverse, and they simply are surviving and reproducing like the rest of us.

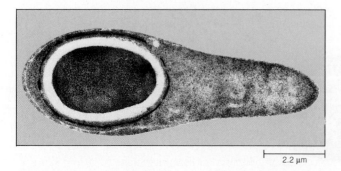

Figure 22.11 An endospore developing in *Clostridium tetani*. This chemoheterotroph is a dangerous pathogen.

2.2 μm

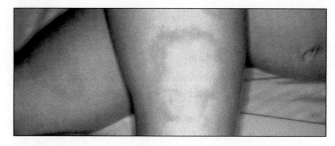

actual size:

Figure 22.12 An extreme reaction to an infection by *Borrelia burgdorferi*, a spirochete. The "bull's-eye" rash is a classic symptom of what may now be the most common tickborne disease in the United States: Lyme disease. Tick bites deliver the spirochete from one host to another. Statewide incidences in 1989 are mapped (*beige*, less than 10; *yellow*, between 15 and 90; *gold*, more than 90).

Focus on Health

Antibiotics

When your grandparents were children, bacterial agents of tuberculosis, pneumonia, and scarlet fever may have caused 25 percent of all deaths each year in the United States. Bacterial agents of dysentery, diphtheria, and whooping cough also were common killers. Bacterial infections during childbirth killed or maimed women by the thousands. From the 1940s onward, we started treating these and other diseases with antibiotics.

An **antibiotic** is a normal metabolic product of actinomycetes and other microorganisms, and it kills or inhibits the growth of other microbial species. For example, streptomycins block protein synthesis in their targets. Penicillins disrupt formation of covalent bonds that hold bacterial cell walls together. Penicillin derivatives cause the wall to weaken until it ruptures. The known antibiotics

22.7 REGARDING THE "SIMPLE" BACTERIA

Bacteria are small. Their insides are not elaborate. *But bacteria are not simple.* A brief look at their behavior will reinforce this point. Bacteria move toward regions with more nutrients. Aerobes move toward oxygen; anaerobes move away from it. Photosynthetic species move toward more intense light (or away from light if it is too intense). Many species tumble away from toxins.

Many bacterial behaviors involve membrane receptors that change shape when they absorb light or encounter chemical compounds. When a bacterium changes direction, its receptors are stimulated in a different way. This triggers a fleeting "memory," a changing biochemical condition that can be compared against that of the immediate past.

Magnetotactic bacteria contain a chain of magnetite particles that serves as a tiny compass (Figure 22.13). The compass helps them sense which way is north and also down. These bacteria swim toward the bottom of a body of water, where oxygen concentrations are lower and therefore more suitable for their growth.

In a jarring imitation of multicellularity, some species even show collective behavior. Millions of *Myxococcus xanthus* cells form "predatory" colonies. Their enzyme secretions digest "prey"—cyanobacteria and other microorganisms—that become stuck to the colony.

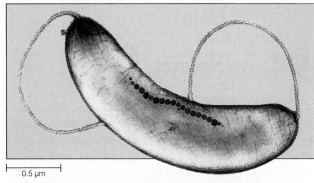

0.5 µm

Figure 22.13 Transmission electron micrograph of a magnetotactic bacterium, showing a chain of magnetite particles that acts like a tiny compass.

don't have comparable effects on protein coats of viruses. If you have a viral infection, antibiotics won't help.

Antibiotics must be carefully prescribed. Besides performing their intended function of counterattacking pathogens, some can disrupt the normal populations of bacteria in the intestines and of yeast cells in the vaginal canal. Such disruptions can lead to secondary infections.

Overprescribed antibiotics have lost their punch. Over time, they destroyed the most susceptible cells of target populations—and favored their replacement by more resistant ones. Antibiotic-resistant strains have made tuberculosis, typhoid, gonorrhea, "staph" infections, and other bacterial diseases more difficult to treat. In a few cases, "superbugs" that cause tuberculosis cannot be treated successfully.

Figure 22.14 *Chondromyces crocatus*, a myxobacterium. Cells of this species aggregate and form a fruiting body.

The cells absorb the breakdown products. What's more, the cells migrate, change direction, and move as a single unit toward what may be food!

Many myxobacterial species also form **fruiting bodies**, which are a type of spore-bearing structure (Figure 22.14). Under appropriate conditions, some cells in the colony differentiate and produce a slime stalk, other cells form branches, and still others form clusters of single-celled spores. When the clusters burst open, the spores are dispersed; each may form a new colony. As you will see in the next chapter, certain eukaryotic organisms also form such structures.

22.8 THE VIRUSES

Defining Characteristics

In ancient Rome, *virus* meant "poison" or "venomous secretion." In the late 1800s, this rather nasty word was bestowed on newly discovered pathogens, smaller than the bacteria being studied by Louis Pasteur and others. Many viruses deserve the name. They attack humans, cats, cattle, insects, crop plants, fungi, bacteria, even other viruses. You name it, and there probably are viruses that can infect it.

Today we define a **virus** as a noncellular infectious agent having two characteristics. First, each viral particle consists of a protein coat around a nucleic acid core—that is, its genetic material. Second, a virus cannot reproduce itself. It can be reproduced only after its genetic material and maybe a few enzymes enter a host cell and subvert the cell's biosynthetic machinery. Thus a virus is no more alive than a chromosome is alive.

The genetic material of a virus is either DNA or RNA. A viral coat consists of one or more types of protein subunits, organized into a rodlike or many-sided shape (Figure 22.15). It protects the genetic material during the journey from one host cell to the next. It also contains proteins that are capable of binding with specific receptors on host cells. Some coats are enclosed in an envelope of mostly membrane remnants from the previously infected cell. Protein-carbohydrate structures spike out from some of these envelopes. Coats of complex viruses have sheaths, tail fibers, and other accessory structures attached.

The immune system of vertebrates can detect certain viral proteins. The problem is, genes for many viral proteins mutate frequently, so a virus may elude the immune fighters. People who are vulnerable to lung infections get new "flu shots" each year because envelope spikes on influenza viruses keep changing.

Examples of Viruses

Each virus can multiply only in cells of particular species. Unless researchers culture living host cells, they cannot easily study viruses. This is why much of our knowledge of viruses comes from **bacteriophages**, which infect bacterial cells. Unlike the cells of humans and other complex, multicelled organisms, bacterial cells can be cultured easily and rapidly. This also is why bacteriophages and bacteria were used in the experiments to determine DNA function (page 210). They are still used as research tools in genetic engineering.

Table 22.3 lists the main groups of animal viruses, which cause herpes, influenza, warts, the common cold, some forms of cancer, and many other diseases. Figure 22.16 shows examples.

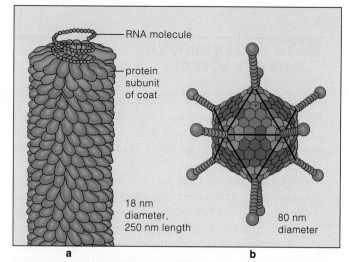

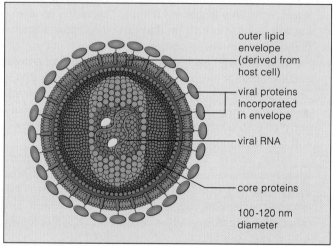

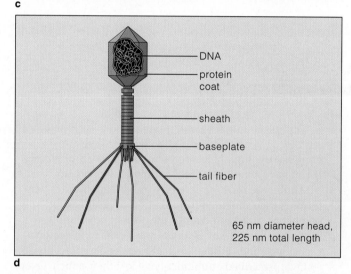

Figure 22.15 Body plans of viruses. (**a**) *Helical* viruses have a rod-shaped coat of protein subunits, coiled helically around the nucleic acid. For this portion of a tobacco mosaic virus, the upper subunits have been removed to reveal the RNA. (**b**) *Polyhedral* viruses, such as this adenovirus, have a many-sided coat. (**c**) *Enveloped* viruses, such as HIV, have an envelope around a helical or polyhedral coat. (**d**) *Complex* viruses, such as the T-even bacteriophages, have additional structures attached to the coat.

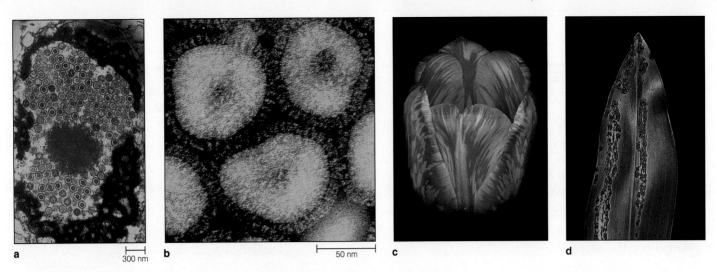

a

300 nm

b

50 nm

c

d

Figure 22.16 (**a**) Particles of a DNA virus that causes a herpes infection. This virus has a many-sided protein coat. (**b**) Particles of an RNA virus that causes influenza. Spikes project from the lipid envelope. Visible effects of two plant viruses: (**c**) Streaking of a tulip blossom. A harmless virus infected pigment-forming cells in the colorless parts. (**d**) An orchid leaf infected by a rhabdovirus.

Table 22.3 Classification of Animal Viruses	
DNA Viruses	**Some Diseases**
Adenoviruses	Respiratory infections
Hepatitis virus	Liver diseases
Herpesviruses:	
H. simplex type I	Oral herpes, cold sores
H. simplex type II	Genital herpes
Varicella-zoster	Chicken pox, shingles
Epstein-Barr	Infectious mononucleosis, implicated in some cancers
Papovaviruses	Benign and malignant warts
Parvoviruses	Roseola (fever, rash) in small children; aggravation of sickle-cell anemia
Poxviruses	Smallpox, cowpox
RNA Viruses	**Some Diseases**
Picornaviruses:	
Enteroviruses	Polio, hemorrhagic eye disease; hepatitis A (infectious hepatitis)
Rhinoviruses	Common cold
Togaviruses	Encephalitis, yellow fever, dengue fever
Paramyxoviruses	Measles, mumps
Rhabdoviruses	Rabies
Coronaviruses	Respiratory infections
Orthomyxoviruses	Influenza
Arenaviruses	Hemorrhagic fevers
Reoviruses	Respiratory, intestinal infections
Retroviruses:	
HTLV I, II	Associated with cancer (leukemia)
HIV	AIDS

Animal viruses range in size from parvoviruses (18 nanometers) to poxviruses (350 nanometers across). Their genetic material, double- or single-stranded DNA or RNA, is replicated in different ways. HIV, one of the RNA viruses, is the trigger for AIDS. As described in Chapter 40, it attacks certain white blood cells and so weakens the human immune system. Infections that otherwise might not kill the individual end up doing so. Most often, researchers attempting to find cures for AIDS and other diseases must use living animals for their experiments. For some viruses, they can use HeLa cells (page 145) and other immortal cell lines instead.

Plant viruses must breach plant cell walls to cause diseases. They typically hitch rides on the piercing or sucking devices of insects that feed on plant juices. RNA viruses infect tobacco plants (tobamovirus, or the tobacco mosaic virus), barley (hordeivirus), potatoes (potax-virus), and many other major crops. So do DNA viruses (caulimovirus infects cauliflower, and geminivirus infects maize). Figures 22.16*c* and *d* show the visible effects of other viral infections.

Some pathogens are even more stripped down than viruses. Prions, which are protein particles, cause rare, fatal degeneration of the nervous system in humans, sheep, and other animals. Viroids are strands or circles of RNA, smaller than any viral DNA or RNA molecule, and they have no protein coat. Viroids are plant pathogens that can destroy entire fields of citrus, potatoes, and other crop plants.

A virus is a nonliving infectious particle that consists of nucleic acid enclosed in a protein coat and sometimes an outer envelope. It cannot be replicated without pirating the metabolic machinery of a specific type of host cell.

22.9 VIRAL MULTIPLICATION CYCLES

Viruses multiply in a variety of ways. But nearly all multiplication cycles include five basic steps:

1. The virus attaches to a host cell. Any cell is a suitable "host" if a virus can chemically recognize and lock onto specific molecular groups at its surface.

2. The whole virus or its genetic material alone penetrates the cell's cytoplasm.

3. In an act of molecular piracy, the viral DNA or RNA directs the host cell into producing many copies of viral nucleic acids and proteins, including enzymes.

4. The viral nucleic acids and proteins are put together to form new virus particles.

5. Newly formed virus particles are released from the infected cell.

Two pathways are common among bacteriophages. In a **lytic pathway**, steps 1 through 4 proceed rapidly, with new particles released by lysis (Figure 22.17). "Lysis" means the cell's plasma membrane is damaged and its cytoplasm leaks out. Cell death is quick. In **lysogenic pathways**, an infection enters a latent period; the host cell isn't killed outright. Genetic recombination may take place during latency. A viral enzyme cuts the host chromosome and integrates viral genes into it. The recombinant DNA is replicated and passed on to all of the infected cell's descendants. Later, if viral genes move out of the chromosome, the multiplication cycle will resume.

Multiplication cycles differ among animal viruses. Figure 22.18 shows an example of an enveloped DNA animal virus. It penetrates a cell by endocytosis, becomes uncoated, and undergoes replication. Enzymes and other proteins are assembled from viral gene products. Virus particles are released by exocytosis.

By contrast, RNA viruses multiply only in the cytoplasm. Their RNA is used as a template either for mRNA or protein synthesis. Some animal viruses also enter latency. Herpesviruses that cause recurring cold sores in just about everybody are like this. So is HIV, a retrovirus. It carries along its own enzymes, which assemble DNA on viral RNA by the process of reverse transcription, as described on page 690.

Nearly all viral multiplication cycles include five steps: attachment to a suitable host cell, penetration, viral DNA or RNA replication and protein synthesis, assembly of new viral particles, and release from the infected cell.

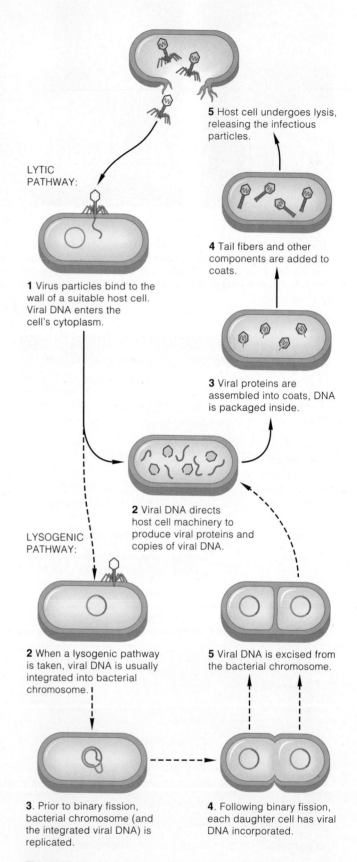

LYTIC PATHWAY:

1 Virus particles bind to the wall of a suitable host cell. Viral DNA enters the cell's cytoplasm.

2 Viral DNA directs host cell machinery to produce viral proteins and copies of viral DNA.

3 Viral proteins are assembled into coats, DNA is packaged inside.

4 Tail fibers and other components are added to coats.

5 Host cell undergoes lysis, releasing the infectious particles.

LYSOGENIC PATHWAY:

2 When a lysogenic pathway is taken, viral DNA is usually integrated into bacterial chromosome.

3. Prior to binary fission, bacterial chromosome (and the integrated viral DNA) is replicated.

4. Following binary fission, each daughter cell has viral DNA incorporated.

5 Viral DNA is excised from the bacterial chromosome.

Figure 22.17 Multiplication cycle for some of the bacteriophages. New viral particles may be produced and released by a lytic pathway alone. For certain viruses, the lytic pathway may be expanded to include a lysogenic pathway.

Human Disease—Modes and Patterns of Infection

Humans are hosts for many viruses and other pathogens, including bacteria, fungi, protozoans, and parasitic worms. **Infection** means a pathogen has invaded the body and is multiplying in cells and tissues. Its outcome, **disease**, results when defenses cannot be mobilized fast enough to prevent the pathogen's activities from interfering with normal body functions. Pathogens reach hosts in several ways:

1. *Direct contact*, as by coming into contact with open sores caused by an infection. ("Contagious disease" comes from the Latin *contagio*, meaning touch or contact.) Gonorrhea, syphilis, and other sexually transmitted diseases are spread primarily by direct contact.

2. *Indirect contact*, as by touching doorknobs, food, diapers, hypodermic needles, or other objects that were previously in contact with an infected person. Food that is moist, not refrigerated, and not too acidic can be contaminated by various pathogens, including the ones responsible for amoebic dysentery and typhoid fever.

3. *Inhaling airborne pathogens*, which may be spread by coughing and sneezing.

4. *Transmission by biological vectors*. Mosquitoes, flies, fleas, ticks, and other arthropods transport pathogens from infected people or contaminated material to new hosts. Many pathogens use vectors only as taxis. Others use vectors as intermediate hosts, in that part of the infectious cycle must proceed inside an organism other than the primary host. Mosquitoes, for instance, are intermediate hosts for the causative agent of malaria (page 369).

Some diseases occur in patterns. Whooping cough and other *sporadic* diseases break out irregularly and affect few people. Tuberculosis and other *endemic* diseases occur more or less continuously, but they do not spread far in large populations. Impetigo, a highly contagious bacterial infection, is like this; it is often confined to day-care centers. During an **epidemic**, a disease abruptly spreads through large portions of a population for a limited period. When influenza breaks out along the east coast of North America, this is an epidemic. When epidemics break out in several countries around the world in a given period, as is happening in the case of AIDS, this is a **pandemic**.

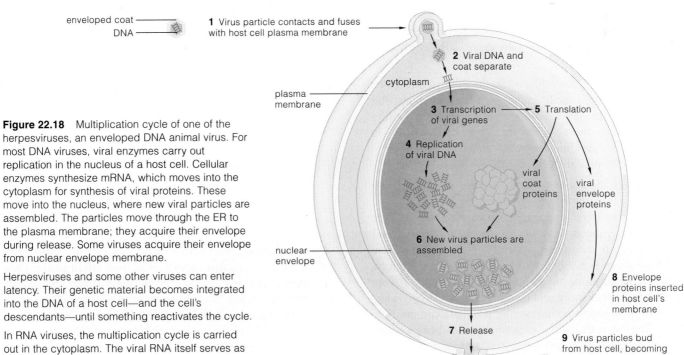

Figure 22.18 Multiplication cycle of one of the herpesviruses, an enveloped DNA animal virus. For most DNA viruses, viral enzymes carry out replication in the nucleus of a host cell. Cellular enzymes synthesize mRNA, which moves into the cytoplasm for synthesis of viral proteins. These move into the nucleus, where new viral particles are assembled. The particles move through the ER to the plasma membrane; they acquire their envelope during release. Some viruses acquire their envelope from nuclear envelope membrane.

Herpesviruses and some other viruses can enter latency. Their genetic material becomes integrated into the DNA of a host cell—and the cell's descendants—until something reactivates the cycle.

In RNA viruses, the multiplication cycle is carried out in the cytoplasm. The viral RNA itself serves as a template either for mRNA synthesis or for protein synthesis.

SUMMARY

1. Microorganisms are mostly single-celled species of bacteria and protistans that are too small to be seen without the aid of a microscope.

2. Soon after the origin of life, divergences led to three great prokaryotic lineages: archaebacteria, eubacteria, and the forerunners of eukaryotes.

   a. Archaebacteria (methanogens, halophiles, and extreme thermophiles) live in extreme environments, like those in which life probably originated.

   b. All other existing bacteria are eubacteria. These more common types differ from archaebacteria in wall structure and other features.

3. All bacteria are single-celled prokaryotes. Some have membrane infoldings and other structures in the cytoplasm, but none has a profusion of organelles.

   a. Three basic bacterial shapes are common: cocci (spheres), bacilli (rods), and spirals.

   b. Nearly all eubacteria have a cell wall composed of peptidoglycan. It protects the plasma membrane and helps it resist rupturing. Its composition and structure help clinicians identify particular bacterial species.

   c. A sticky mesh of polysaccharides (glycocalyx) surrounds the wall as a capsule or slime layer. It helps bacteria attach to substrates and sometimes to resist a host organism's infection-fighting mechanisms.

   d. Some species have bacterial flagella, unique motile structures that rotate like a propeller. Many species have pili, filamentous proteins that help cells adhere to a surface or facilitate conjugation.

4. Bacteria reproduce by binary fission, a cell division mechanism that follows DNA replication and that divides a parent cell into two genetically equivalent daughter cells.

5. Many species have plasmids, small circles of DNA that are replicated independently of the single, circular bacterial chromosome. Some genes on plasmids confer antibiotic resistance. Plasmids are transmitted to daughter cells and may be transferred by conjugation to cells of the same species or a different species.

6. Bacteria show great metabolic diversity, as in their modes of acquiring energy and carbon.

   a. *Photoautotrophs* use sunlight and carbon dioxide for photosynthesis. They include cyanobacteria and other oxygen-producing species of eubacteria. They include green nonsulfur and purple nonsulfur bacteria that do not produce oxygen; these obtain electrons from sulfur and other inorganic compounds, not from water.

   b. *Photoheterotrophs* use sunlight but not carbon dioxide. They use organic compounds as carbon sources. They include certain archaebacteria and eubacteria.

   c. *Chemoautotrophs* such as the nitrifying bacteria use carbon dioxide but not sunlight. They get energy by stripping electrons from a variety of inorganic substances, such as nitrogen compounds.

   d. *Chemoheterotrophs* are parasites (which draw carbon and energy from living hosts) or saprobes (which feed on the products, wastes, or remains of other organisms). Most bacteria are in this category. They include major decomposers and pathogens.

7. Bacteria make behavioral responses to stimuli, as when they move toward regions with more nutrients. The responses depend on activation of membrane receptors. Some bacteria have a magnetic compass in the cytoplasm that assists in directional behavior. Some species show collective behavior, as when they aggregate and move as a unit toward prey or form fruiting bodies by which spores are dispersed.

8. Viruses are nonliving, noncellular agents that infect particular species of nearly all organisms. They have two defining traits:

   a. Each virus particle consists of a nucleic acid core and a protein coat that sometimes is enclosed in a lipid envelope. Spikes of proteins and carbohydrates project from these envelopes. The coats of complex viruses have sheaths, tail fibers, and other accessory structures attached.

   b. A virus particle cannot reproduce itself. Its genetic material must enter a host cell and direct the cellular machinery to synthesize the materials necessary to produce new virus particles.

9. Nearly all viral multiplication cycles include five steps: attachment to a suitable host cell, penetration, viral DNA or RNA replication and protein synthesis, assembly of new viral particles, and release from the infected cell.

10. Two pathways are common in the multiplication cycle of bacteriophages, which infect bacteria. In a lytic pathway, multiplication proceeds rapidly and new viral particles are released by lysis. In a lysogenic pathway, the infection enters a latent period so that the host cell is not killed outright. The viral nucleic acid may undergo genetic recombination with a host cell chromosome.

11. Multiplication cycles of animal viruses are diverse. They may proceed rapidly or enter a latent phase. Penetration and release may be by way of endocytosis and exocytosis. For DNA viruses, part of the cycle proceeds in the nucleus of a host cell. For RNA viruses, it proceeds exclusively in the cytoplasm. The viral RNA serves as a template for mRNA and for protein synthesis.

Review Questions

1. Label the major structures on the generalized bacterial cell below. *348*

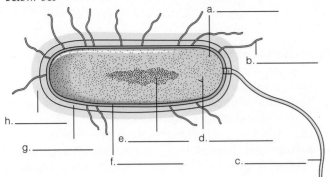

a. _____
b. _____
c. _____
d. _____
e. _____
f. _____
g. _____
h. _____

2. Describe the key characteristics of bacteria. What are some differences between archaebacteria and eubacteria? *348–355*

3. Name a few photoautotrophic, chemoautotrophic, and chemoheterotrophic eubacteria. Describe some that are likely to give humans the most trouble medically. *351–355*

4. What is a virus? Why is a virus considered to be no more alive than a chromosome? *356*

Self-Quiz *(Answers in Appendix IV)*

1. Nondividing bacteria have _____ chromosome(s) and may have extra circles of _____ called plasmids.
   a. one; RNA
   b. two; RNA
   c. one; DNA
   d. two; DNA

2. _____ live in extreme environments, much like those of the early earth.
   a. Cyanobacteria
   b. Eubacteria
   c. Archaebacteria
   d. Protozoans

3. Which of the following does not belong in the Archaebacteria?
   a. Halophiles
   b. Cyanobacteria
   c. Thermophiles
   d. Methanogens

4. Bacteria reproduce by _____ .
   a. mitosis
   b. meiosis
   c. binary fission
   d. longitudinal fission

5. Eubacterial cell walls are composed of _____ ; a sticky mesh of polysaccharides, the _____ , surrounds the wall as a capsule or slime layer.
   a. peptidoglycan; plasma membrane
   b. cellulose; glycocalyx
   c. cellulose; plasma membrane
   d. peptidoglycan; glycocalyx

6. Most bacteria are _____ and include major decomposers and pathogens.
   a. photoautotrophs
   b. photoheterotrophs
   c. chemoautotrophs
   d. chemoheterotrophs

7. Viruses are _____ .
   a. the simplest living organisms
   b. infectious particles
   c. nonliving
   d. both a and b
   e. both b and c

8. Viruses have _____ .
   a. DNA cores and carbohydrate coats
   b. DNA or RNA cores; and plasma membranes
   c. DNA-containing nuclei; and lipid envelopes
   d. DNA or RNA cores; and protein coats

9. Which of the following is *not* characteristic of a viral lysogenic pathway?
   a. bacteria are infected
   b. the infection enters a latent phase
   c. multiplication of virus particles proceeds rapidly
   d. the host cell is not killed outright

10. Match each term appropriately.
   ____ archaebacteria
   ____ eubacteria
   ____ viruses
   ____ cyanobacteria
   ____ plasmids

   a. bacteria other than archaebacteria
   b. nonliving infectious particle, nucleic acid core, protein coat
   c. small circles of bacterial DNA
   d. methanogens, halophiles, extreme thermophiles
   e. photoautotrophs

Selected Key Terms

antibiotic *354*
archaebacteria *348*
bacillus *348*
bacterial conjugation *350*
bacterial flagella *349*
bacteriophage *356*
binary fission *350*
cell wall *349*
chemoautotroph *348*
chemoheterotroph *348*
coccus *348*
disease *359*
endospore *354*
epidemic *359*
eubacteria *348*
extreme thermophiles *352*
fruiting body *355*
glycocalyx *349*

halophiles *352*
heterocyst *352*
infection *359*
lysogenic pathway *358*
lytic pathway *358*
methanogens *352*
microorganism *346*
pandemic *359*
pathogen *346*
peptidoglycan *349*
photoautotroph *348*
photoheterotroph *348*
pili *349*
plasmid *350*
prokaryotic *349*
spiral *348*
virus *356*

Readings

Brock, T., and M. Madigan. 1988. *Biology of Microorganisms.* Fifth edition. Englewood Cliffs, New Jersey: Prentice-Hall.

Frankel-Conrat, H., P. Kimball, and J. Levy. 1988. *Virology.* Second edition. Englewood Cliffs, New Jersey: Prentice-Hall.

Frazier, W., and D. Westoff. 1988. *Food Microbiology.* Fourth edition. New York: McGraw-Hill. Good reference on the microbes that have major effects on our food supplies.

Stanier, R., et al. 1986. *The Microbial World.* Fifth edition. Englewood Cliffs, New Jersey: Prentice-Hall.

Woese, C. 1981. "Archaebacteria." *Scientific American* 244(6): 98–125.

23 PROTISTANS

Kingdom at the Crossroads

More than 2 billion years ago, in tidal flats and soils, in estuaries, streams, lakes, and lagoons, prokaryotic cells were inconspicuously changing the world. Ever since the time of life's origin, the earth's atmosphere had been free of oxygen, and anaerobic bacteria had reigned supreme. Now their realm was about to shrink, drastically. An oxygen-producing pathway of photosynthesis had been operating in vast populations of cells. Gaseous oxygen had been dissolving in the surrounding waters and escaping into the air—and its atmospheric concentration was beginning to approach its current level.

The oxygen-enriched atmosphere was a selection pressure of global dimensions. Anaerobic species that could not neutralize the highly reactive, potentially lethal gas were thereafter restricted to black muds, stagnant waters, and other anaerobic habitats. Oxygen tolerance developed in other species. And it must have been a short evolutionary step from having the capacity to neutralize oxygen to using it in metabolic reactions, for at least some aerobic species emerged in nearly all bacterial groups.

This was the start of rampant competition for resources. In the presence of so much oxygen, energy-rich organic compounds could no longer accumulate at the earth's surface through geochemical processes. Now the organic compounds formed by living cells became the premier source of carbon and energy. Through evolutionary experimentation, novel ways of acquiring and using those compounds developed. Diverse groups of bacteria engaged in new kinds of partnerships, in predatory and parasitic interactions. Within 500 million years, some experiments led to the first eukaryotic cells. Such cells were alike in having a nucleus and other organelles, some of which resulted from the coevolution of bacterial symbionts (page 331). Even so, the eukaryotic lineages soon branched in confounding directions.

Of all existing organisms, protistans are most like those early eukaryotes. Figure 23.1 will give you an idea of the range of diversity in this kingdom. Many are microscopic members of aquatic communities. The photosynthetic types range from single cells to giant seaweeds. Saprobic types resemble certain bacteria and fungi. Some predatory and parasitic types resemble animals.

As far as we can determine, ancestral forms very much like them were once poised at the evolutionary crossroads leading from prokaryotic lineages to the kingdoms of plants, fungi, and animals.

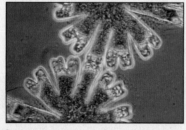

a

b

Figure 23.1 A sampling of the confounding diversity among protistans. (**a**) *Micrasterias*, a single-celled freshwater green alga, here dividing in two. (**b**) *Postelsia palmaeformis*, a multicelled brown alga. (**c**) *Physarum*, a plasmodial slime mold. This aggregation of cells is migrating along a rotting log. (**d**) Mealtime for *Didinium*, a ciliated protozoan with a big mouth. *Paramecium*, a different ciliated protozoan, is poised at the mouth (*left*) and swallowed (*right*).

c

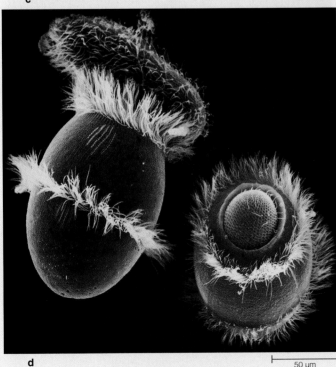

d

50 µm

1. The kingdom Protista includes all of the single-celled eukaryotes and some of the structurally simple multicelled forms. Many species live in aquatic habitats, where they are key components of phytoplankton and zooplankton.

2. Protistans are easily distinguishable from bacteria but are difficult to classify with respect to other eukaryotes. They also differ enormously from one another in their morphology and life-styles.

23.1 WHAT IS A PROTISTAN?

Through no fault of their own, protistans are looked upon as the rag-bag of classification schemes, a confusing bunch of eukaryotes that are assigned their own kingdom because they don't seem to fit anywhere else. Most protistans are single celled. Yet nearly every lineage includes multicelled forms, and some have close evolutionary ties to other kingdoms.

It's been said that it is easier to classify protistans by what they are *not*—as in, "They are not bacteria, fungi, plants, or animals." Remember, though, *we* impose the boundaries of kingdoms. *We make our cuts across continuous, unbroken lines of descent that extend from the origin of life to the present.* Some boundaries, such as the one between bacteria and protistans, are obvious; protistans are the simplest eukaryotes. Other boundaries are not as obvious, although they are now coming into focus through comparative biochemistry and other investigations.

Recall that protistans and other eukaryotes differ from bacteria in several respects. At the least, eukaryotic cells have a double-membraned nucleus, mitochondria, endoplasmic reticulum, and large ribosomes. They have microtubules, which have uses as cytoskeletal elements, as spindles for chromosome movements, and in the 9 + 2 core of flagella and cilia. These cells contain two or more chromosomes, which consist of DNA complexed with many histones and other proteins. Eukaryotic cells alone engage in mitosis and meiosis. Many types have chloroplasts and other plastids.

Later chapters will describe structural and functional traits that distinguish plants, fungi, and animals from protistans. By a process of elimination, "protistans" are organisms that do not show these traits and that are not bacteria. The ones called chytrids, water molds, slime molds, protozoans, and sporozoans are diverse heterotrophs. Euglenoids are photosynthetic, heterotrophic, or both. Nearly all chrysophytes, the dinoflagellates, and the red, brown, and green algae are evolutionarily committed to photosynthesis.

23.2 CHYTRIDS AND WATER MOLDS

Most of the **chytrids** (phylum Chytridiomycota) live in freshwater and marine habitats. Most of the 575 species are saprobic decomposers; some are parasites. Like fungi, all chytrids secrete enzymes that digest organic compounds of other organisms, such as marsh grasses, then they absorb the breakdown products.

Single-celled species produce flagellated asexual spores. The spores settle onto a host cell, germinate, then develop into globe-shaped cells with rhizoids, which are rootlike absorptive structures (Figure 23.2). At maturity, the globe-shaped cells become spore-producing structures. The cells of more complex species grow and develop into a mesh of absorptive filaments, called a **mycelium** (plural mycelia). Cell walls usually do not cut across the filaments, so there is an uninterrupted flow of cytoplasm that distributes enzymes and nutrients throughout the mycelium.

Chitin reinforces the cell wall of some chytrids. This, together with biochemical evidence, suggests an evolutionary link between chytrids and fungi.

The 580 species of **water molds** (phylum Oomycota) may be distantly related to red algae. Most produce an extensive mycelium, and some of the filaments develop into gamete-producing structures. At fertilization, a male and female gamete fuse to form a diploid zygote, which develops into a thick-walled resting spore. Following spore germination, a new mycelium develops. Water molds also produce flagellated, asexual spores.

Most water molds are key decomposers of aquatic habitats. Some parasitize aquatic animals and land plants (Figure 23.3). The cottony growths you may have seen on goldfish or tropical fish are mycelia of a parasitic type, *Saprolegnia*.

Some water molds have influenced human affairs. More than a century ago, Irish peasants cultivated potatoes as their main food source. Between 1845 and 1860, growing seasons were cool and damp, year after year. The cool conditions encouraged the rapid spread of *Phytophthora infestans*, a water mold that causes *late blight*—a rotting of potato and tomato plants. The mold produced abundant spores, spore dispersal through the watery film on plants went unimpeded, and destruction was rampant. During a fifteen-year period, a third of Ireland's population starved to death, died in the outbreak of typhoid fever that followed as a secondary effect, or fled to the United States and other countries.

Most chytrids and water molds are saprobic decomposers of aquatic habitats. Some are single cells. Others form a mesh of absorptive filaments (a mycelium) during the life cycle.

23.3 SLIME MOLDS

Call someone a slimy scum and you may get punched in the nose. Yet that is an apt description of the **slime molds**. During the life cycle of these heterotrophic protistans, free-living, amoebalike cells are produced. The cells crawl about on rotting plant parts, such as decaying leaves and bark. Like true amoebas, they are phagocytes; they engulf bacteria, spores, and organic compounds. When cells are starving, they aggregate into a slimy mass that may migrate to a new place. It moves through contractions of myosin molecules, present in secretions that sheath the mass. Later, the amoebalike cells develop into a few cell types that

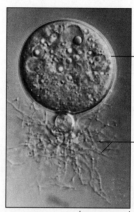

— globe-shaped cell which, when mature, is a spore-producing structure

— rhizoids, rootlike absorptive structures that grow through organic material

25 μm

Figure 23.2 A common chytrid (*Chytridium confervaie*).

a

b

Figure 23.3 Effects of two water molds. (**a**) A parasitic water mold (*Saprolegnia*) has destroyed tissues of this aquarium fish. (**b**) *Plasmopara viticola* causes downy mildew in grapes. At times it has threatened large vineyards in Europe and North America. Since the late 1800s, mixtures of copper sulfate, lime, and other chemicals have been applied to vines to control the disease.

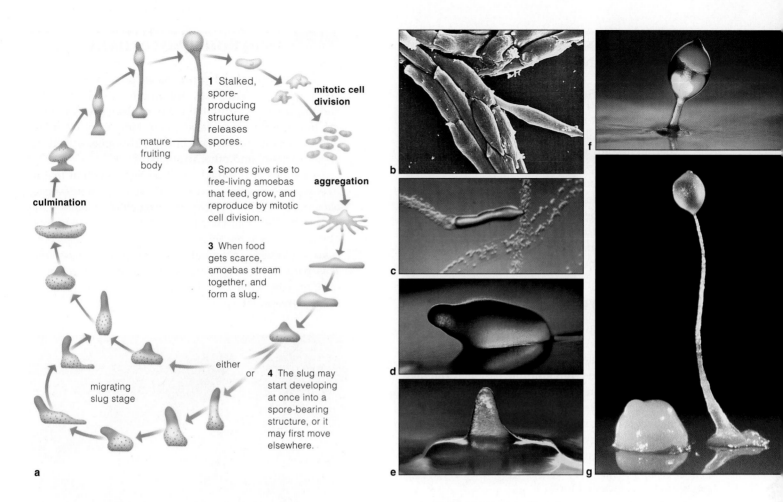

a

1 Stalked, spore-producing structure releases spores.

mature fruiting body

mitotic cell division

culmination

2 Spores give rise to free-living amoebas that feed, grow, and reproduce by mitotic cell division.

aggregation

3 When food gets scarce, amoebas stream together, and form a slug.

either or

4 The slug may start developing at once into a spore-bearing structure, or it may first move elsewhere.

migrating slug stage

b

c

d

e

f

g

differentiate and form a stalked, spore-bearing reproductive structure. After a spore lands on a suitably warm, damp surface, it germinates and gives rise to an amoebalike cell. Sexual reproduction (by gametes) also is common among slime molds.

Figure 23.4 provides a close look at *Dictyostelium discoideum* at different stages of the life cycle. This is the best known of seventy species of *cellular* slime molds (phylum Acrasiomycota).

There are about 500 species of *plasmodial* slime molds (phylum Myxomycota), of obscure ancestry. When their amoebalike cells aggregate, the plasma membranes and cell walls break down. The cytoplasm flows as a unit, distributing nutrients and oxygen through the entire mass. This streaming mass (the plasmodium) may grow over several square meters. When food runs out, it migrates. You may have seen one crossing lawns or roads, even climbing trees. During dry seasons, the mass becomes encysted and often turns bright yellow and orange (Figure 23.1*c*).

During a slime mold life cycle, amoeboid cells aggregate to form a migrating mass. Cells in the mass differentiate, forming reproductive structures and spores or gametes.

Figure 23.4 *Dictyostelium discoideum*, a cellular slime mold. (**a**) The life cycle includes a spore-producing stage. Spores give rise to free-living amoebas, which grow and divide until food (soil bacteria) dwindles. Then, in response to a chemical signal (cyclic AMP) that they themselves secrete, amoebas stream toward one another. As amoebas aggregate, their plasma membranes become sticky, and they adhere to one another. A cellulose sheath forms around the aggregation, and it starts crawling, like a slug (**b**–**d**). Some of these slugs may consist of 100,000 or more amoebas.

As a slug migrates, the amoebas develop into prestalk, prespore, and anteriorlike cells. Prestalk cells (shaded *red* in **a**) form an anterior tissue. Prespore cells (*white*) and anteriorlike cells (*scattered brown dots*) form a posterior tissue. Prestalk cells secrete ammonia, in amounts that vary with temperature and light intensity. The slug moves most rapidly in response to intermediate amounts of ammonia (not too little, not too much), which correspond to warm, moist conditions (not too cold or hot, not soppy or dry).

Where such conditions exist at the surface of a substrate, prestalk cells and prespore cells differentiate and form a stalked, spore-bearing structure (**e**–**g**). Anteriorlike cells sort into two groups at each end of the posterior tissue. Possibly they function in elevating the nonmotile spores for dispersal from the top of the spore-bearing structure.

23.4 CONCERNING THE ANIMAL-LIKE PROTISTANS

Protozoan Classification

Many chytrids, water molds, and slime molds are free-living predators or parasites some of the time and simple experiments in multicellularity at other times. About 65,000 other species of protistans are unequivocally single-celled predators and parasites. Collectively, they are called **protozoans** ("first animals") because they may resemble the single-celled, heterotrophic protistans that gave rise to animals. By one scheme, they are classified as the amoeboid, ciliated, and flagellated protozoans and the sporozoans.

Protozoan Life Cycles

Asexual reproduction dominates protozoan life cycles. Following increases in size, DNA replication, and multiplication of organelles, cells undergo fission or budding. Some species undergo multiple fission. More than two nuclei form, then each nucleus and a bit of cytoplasm around it separate to form a daughter cell. Sexual reproduction also occurs.

During the life cycle, many parasitic types form **cysts**, protective capsules that allow them to survive adverse conditions. Many species spread through host populations in encysted form.

Fewer than two dozen protozoans cause diseases in humans, but they receive our serious attention because there are no effective vaccines against them. In any given year, hundreds of millions of humans are affected by protozoan-caused diseases.

23.5 AMOEBOID PROTOZOANS

Among the **amoeboid protozoans** (phylum Sarcodina) are phagocytic amoebas, foraminiferans, heliozoans, and radiolarians. Adults move or capture prey by sending out pseudopods ("false feet"), temporary cytoplasmic extensions of the cell body. Most species feed on algae, bacteria, and other protozoans.

Amoebas are "naked," soft-bodied cells that live in freshwater, seawater, and soil. There they engulf algae, bacteria, and other protozoans. Amoebas are among the simplest protistans. They include *Amoeba proteus* of biology laboratory fame (Figure 23.5*a*). *Entamoeba histolytica* causes a severe intestinal disorder, *amoebic dysentery*. This parasite travels in cysts within feces, which may contaminate water and soil in regions with inadequate sewage treatment.

a

c

e

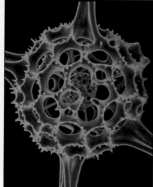
f

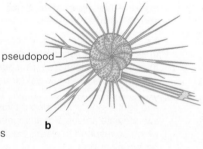
pseudopod
b

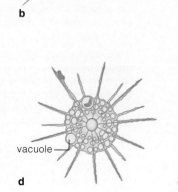
vacuole
d

Figure 23.5 Amoeboid protozoans. (**a**) *Amoeba proteus*. The earliest eukaryotes may have been as soft-bodied as this. (**b**,**c**) Shells and body plan of foraminiferans. Needlelike spines support the pseudopods. (**d**,**e**) A living heliozoan. Its pseudopods have a supportive core of microtubules. Fluid-filled vacuoles make it neutrally bouyant so that it can remain suspended in the water. (**f**) A radiolarian shell.

Foraminiferans live mostly in the seas. Their hardened shells are often peppered with hundreds of thousands of tiny holes through which sticky, threadlike pseudopods extend. Figures 23.5*b* and *c* show some foraminiferan shells. Often the shells bear spines, which in some species are long enough that the shell can be seen with the naked eye.

The **heliozoans**, or "sun animals," have fine, needlelike pseudopods that radiate from the body like sun rays (Figure 23.5*e*). These largely freshwater protozoans are generally floaters or bottom-dwellers. Part of the cytoplasm forms an outer sphere around a core composed of denser cytoplasm and the bases of microtubular rods.

Among the structurally stunning protozoans are the **radiolarians**, which are found mostly in marine plankton. The cells basically resemble the heliozoans, but most also have a skeleton of silica (Figure 23.5*f*). Some radiolarian species form colonies in which many individual cells are cemented together. Accumulated shells of radiolarians and foraminiferans are key components of many ocean sediments, and are testimony to the abundance of these organisms in the past.

Figure 23.6 Ciliated protozoans. (**a**) Body plan of *Paramecium* and (**b**) surface view. (**c**) From Bimini, the Bahamas, a hypotrich with long, stiff, leglike tufts of cilia.

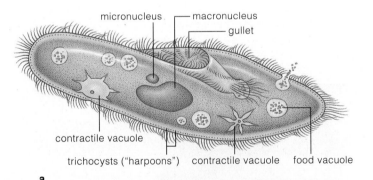

micronucleus — macronucleus — gullet

contractile vacuole

trichocysts ("harpoons")   contractile vacuole   food vacuole

a

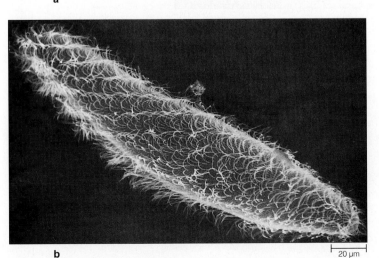

b   20 µm

23.6 CILIATED PROTOZOANS

About 8,000 species of **ciliated protozoans** (phylum Ciliophora) have distinctive arrays of cilia, which are used as motile structures (Figure 23.1*d* and 23.6). Various organelles extrude mucus, toxins, and other materials at the cell surface.

The ciliates abound in freshwater and marine habitats, where they feed on bacteria, algae, and one another. *Paramecium* is a typical member of the group. Its rows of cilia beat in synchrony, sweeping water laden with bacteria and food particles into a gullet. The gullet is a cavity that leads into the cell body. There, food is enclosed in enzyme-filled vesicles and digested. Wastes move to an "anal pore" and are eliminated. Like the amoeboid protozoans, *Paramecium* has **contractile vacuoles**. Excess water that has entered the cytoplasm by osmosis collects in these organelles. A filled vacuole contracts, forcing water through a small pore to the outside (page 84).

When under attack or otherwise stressed, *Paramecium* discharges many sticky protein threads from organelles called trichocysts. This does not notably reduce stress or deter predators. Perhaps the threads function as anchors at feeding time.

Ciliates reproduce sexually as well as asexually. They have a large macronucleus and one or more haploid micronuclei. During **protozoan conjugation**, two cells exchange micronuclei, which fuse to form a diploid macronucleus in each cell. When the cells later divide by fission, the daughter cells are diploid, with genetic instructions from two parents.

In their movements and behavior, the hypotrichs are the most animal-like ciliates. They run around on leglike tufts of cilia. Some have a distinct "head" end equipped with modified, sensory cilia (Figure 23.6*c*). Hypotrichs prey on bacteria, but they also scavenge bits of plant and animal material.

Amoeboid and ciliated protozoans are single-celled predators and parasites, some of which have intriguing animal-like traits.

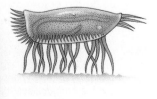

c

23.7 FLAGELLATED PROTOZOANS

The **flagellated protozoans** (phylum Mastigophora) are equipped with one to several flagella. Many species are internal parasites; others are free-living members of either freshwater or marine habitats.

The parasitic types include the trypanosomes. *Trypanosoma brucei* (Figure 23.7a) causes a dangerous disease called *African sleeping sickness*. Bites of the tsetse fly transmit the parasite from one host to another. Early symptoms include fever, headaches, rashes, and anemia. Later, the central nervous system becomes damaged. Without treatment, death follows. Another trypanosome (*Trypanosoma cruzi*) is prevalent in Mexico and South America, where it causes *Chagas disease*. A variety of bugs pick up the parasite when they feed on infected humans, armadillos, opossums, and other animals. The parasite multiplies in the insect gut, then the insect may excrete it onto the skin of a host animal. Scratched or abraded skin regions are open doors for infection. A terrible disease follows infection. The liver and spleen enlarge, the face and eyelids swell, then the brain and heart become severely damaged. There is no specific treatment or cure.

Flagellated protozoans also include trichomonads, of the sort shown in Figure 23.7b. *Trichomonas vaginalis*, a worldwide nuisance, is transferred to new human hosts during sexual intercourse. Without treatment, trichomonad infection damages the membranes in the urinary and reproductive tracts. This sexually transmitted disease and others are described on page 791.

By some estimates, about 10 percent of the human population in the United States is infected with another flagellated protozoan, *Giardia lamblia* (Figure 23.7c). Infection leads to mild intestinal disturbances, including diarrhea, but it has severe and sometimes fatal consequences in a few susceptible people. This flagellated protozoan also is common among wild animals and foraging cattle. It forms cysts that leave the body in feces. It infects a new host who drinks water or ingests food that has become contaminated with the feces. Even the water of remote mountain streams may contain the cysts and should be boiled before drinking.

Figure 23.7 A few flagellated protozoans. (**a**) *Trypanosoma brucei* causes African sleeping sickness. (**b**) *Giardia lamblia* causes intestinal disturbances. (**c**) *Trichomonas vaginalis* causes a sexually transmitted disease, trichomoniasis.

23.8 SPOROZOANS

Sporozoan is an informal designation for parasitic protistans that must live part of the time inside specific cells of host species. All of these intracellular parasites produce infective, motile stages called sporozoites. Many sporozoans also become encysted during some phase of the life cycle.

Trout, salmon, and other commercially important freshwater fishes are hosts for some sporozoans. The infective stage develops into an amoebalike form that may become quite massive.

The life cycle of *Plasmodium*, a sporozoan that causes the disease malaria, is described in the *Focus* essay. *Toxoplasma*, another sporozoan, completes the sexual phase of its life cycle in cats and asexual phases in humans as well as cattle, pigs, and other animals. It may enter new hosts by traveling along in infected meat that is either raw or undercooked. Sporozoite-containing cysts in the feces of infected cats are spread by houseflies, cockroaches, and other insects. The resulting disease, *toxoplasmosis*, is not prevalent in the population but is a major cause of birth defects. A woman who becomes infected during pregnancy may transmit the disease to her fetus, which may suffer brain damage or die. A pregnant woman should never empty litterboxes or otherwise clean up after any cat.

Certain parasitic species of flagellated protozoans and sporozoans cause dangerous human diseases.

a

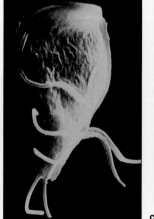
b

c

Malaria and the Night-Feeding Mosquitoes

Each year, about 150 million people come down with *malaria*, and 1 to 2 million die from it. At one time, this long-lasting disease was prevalent mostly in tropical and subtropical parts of Africa. However, the number of cases in North America and elsewhere is increasing dramatically, owing to globe-hopping travelers and unprecedented levels of immigration.

A sporozoan (*Plasmodium*) causes malaria. The mosquito *Anopheles* transmits its sporozoites to human (or bird) hosts. The sporozoites repeatedly engage in multiple fission, producing thousands of progeny that infect and multiply inside of red blood cells. The cells rupture and release the parasite's metabolic wastes, which cause fever and chills.

The *Plasmodium* life cycle (Figure *a*) is attuned to the host's body temperature, which normally rises and falls rhythmically every 24 hours. Gametocytes, the infective stage that is transmitted to new hosts, are ready for travel at night—when mosquitoes feed.

Travelers in countries with high rates of malaria are advised to use antimalarial drugs such as chloroquine. Drug resistance is common, however. And a vaccine has been difficult to develop. Vaccines induce the body to build up its resistance to a specific pathogen. Experimental vaccines for malaria have not been equally effective against all the different stages that develop during sporozoan life cycles. This is true of most parasites with complex life cycles.

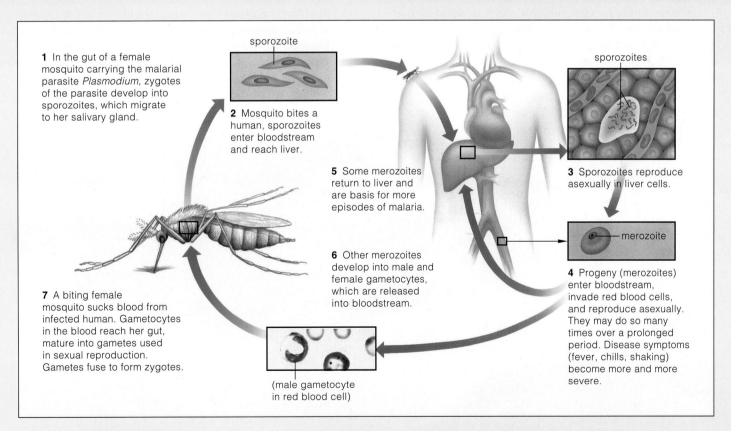

1 In the gut of a female mosquito carrying the malarial parasite *Plasmodium*, zygotes of the parasite develop into sporozoites, which migrate to her salivary gland.

sporozoite

2 Mosquito bites a human, sporozoites enter bloodstream and reach liver.

sporozoites

3 Sporozoites reproduce asexually in liver cells.

merozoite

4 Progeny (merozoites) enter bloodstream, invade red blood cells, and reproduce asexually. They may do so many times over a prolonged period. Disease symptoms (fever, chills, shaking) become more and more severe.

5 Some merozoites return to liver and are basis for more episodes of malaria.

6 Other merozoites develop into male and female gametocytes, which are released into bloodstream.

7 A biting female mosquito sucks blood from infected human. Gametocytes in the blood reach her gut, mature into gametes used in sexual reproduction. Gametes fuse to form zygotes.

(male gametocyte in red blood cell)

a *Plasmodium*, a sporozoan that causes malaria. Its life cycle requires a human host and an insect intermediate host.

23.9 EUGLENOIDS

The **euglenoids** (phylum Euglenophyta) are a classic example of evolutionary experimentation. About 1,000 species abound in freshwater or stagnant ponds and lakes. Like the flagellated protozoans, many are heterotrophs. Most, however, are photosynthetic.

Figure 23.8 shows one *Euglena* cell, which is 40,000 times larger and more complex than a bacterium. Notice the profusion of organelles, including the large chloroplasts and contractile vacuole. Beneath the plasma membrane are spiral strips of a translucent, protein-rich material. The strips help form a pellicle, a firm yet flexible layer. Light readily passes through the pellicle. An "eyespot" of carotenoid pigment granules partly shields a light-sensitive receptor. By moving a long flagellum, the cell keeps the receptor exposed to light and so keeps itself where light is most suitable for its activities.

Yet *Euglena* also can subsist heterotrophically, on dissolved organic compounds. Were its ancestors heterotrophs that acquired chloroplasts by way of endosymbiosis? Probably. The chloroplasts of green algae and plants may be stripped-down descendants of the same kind of endosymbionts. Their chlorophyll pigments are like the ones in *Euglena*.

23.10 CONCERNING "THE ALGAE"

Algae is a term that originally was used to define simple aquatic "plants." It no longer has formal meaning in classification schemes. The phyla once grouped under the term are now classified as follows:

Monera	Cyanobacteria	*(blue-green algae)*
Protista	Chrysophyta	*(golden algae, diatoms, yellow-green algae)*
	Pyrrhophyta	*(dinoflagellates)*
	Rhodophyta	*(red algae)*
	Phaeophyta	*(brown algae)*
	Chlorophyta	*(green algae)*

Many of the single-celled species are suspended in vast numbers in the ocean, seas, lakes, and smaller bodies of water. Collectively, the photosynthetic types are **phytoplankton**, the food producers that are the basis of nearly all food webs in aquatic habitats. (The word is derived from the Greek *planktos*, meaning to wander.) The food producers provide organic compounds *and* dissolved oxygen for **zooplankton**—mostly microscopic heterotrophs, drifting or weakly swimming through the water. These include bacteria, protozoans, tiny crustaceans, and animal larvae, which in turn are food for fishes and, in the oceans, most of the great whales. All of these communities are hurt by pollution.

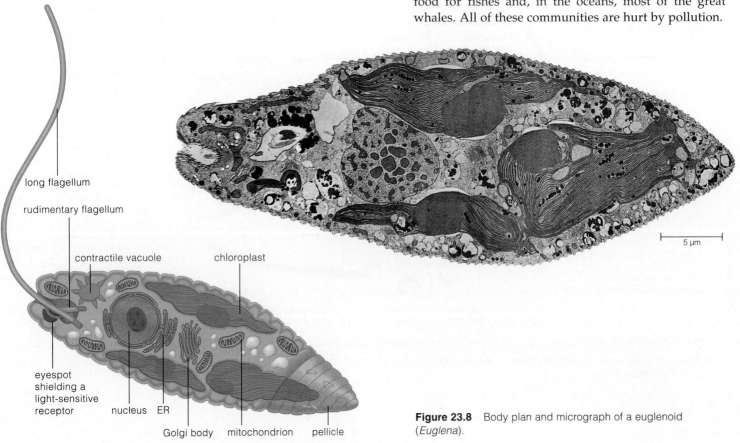

long flagellum

rudimentary flagellum

contractile vacuole

chloroplast

eyespot shielding a light-sensitive receptor

nucleus　ER

Golgi body　mitochondrion　pellicle

5 μm

Figure 23.8 Body plan and micrograph of a euglenoid (*Euglena*).

23.11 CHRYSOPHYTES AND DINOFLAGELLATES

Chrysophytes are mostly photosynthetic single cells with chlorophylls *a* and *c*. Some members have a cellulose wall, others have no wall at all, and still others have scales or mineral-impregnated shells.

Many of the 500 known species of **golden algae** have scales or skeletal elements of silica. Nearly all are photosynthetic. Fucoxanthin, a golden-brown carotenoid pigment, largely masks the chlorophylls. A few amoeboid types engulf bacteria. Golden algae are important in freshwater habitats and in "nanoplankton," marine communities composed of tremendous numbers of extremely tiny cells.

Similarly, the 5,600 existing species of **diatoms** are abundant in aquatic habitats. Although most are photosynthetic, many are temporary or permanent heterotrophs. Each diatom has an outer "shell"—two perforated, glasslike structures that overlap like a pillbox. Substances move to and from the plasma membrane through perforations in the shell.

We know of 35,000 species of extinct diatoms. For about 100 million years, their silica shells have been accumulating at the bottom of lakes and seas. Many sediments contain deposits of finely crumbled diatom shells, which we use in abrasives, filters, and insulating materials. More than 270,000 metric tons of this crumbly material are quarried annually near Lompoc, California.

Fucoxanthin is absent from the 600 species of **yellow-green algae**. One genus, *Vaucheria*, is common in aquatic habitats; it even grows as continuous filaments over wet soils (Figure 23.9*d*).

Of 1,200 species of **dinoflagellates**, most are members of marine phytoplankton. A few photosynthesizers live in freshwater, and a few heterotrophs live in the seas. Some have flagella that fit in grooves between stiff cellulose plates at the body surface (Figure 23.10*a*).

Dinoflagellates appear yellow-green, green, brown, blue, or red, depending on the photosynthetic pigments. Every so often, the red types undergo population explosions and color the seas red or brown. Because some forms produce a neurotoxin, the resulting **red tides** can have devastating effects (Figure 23.10*b*). Hundreds of thousands of fish that feed on plankton may be poisoned and wash up along the coasts. The neurotoxin does not affect clams, oysters, and other mollusks, but it builds up in their tissues. Humans who eat the tainted mollusks may die.

The single-celled photosynthetic protistans, including most euglenoids, chrysophytes, and dinoflagellates, are members of phytoplankton—the "pastures" of most aquatic habitats.

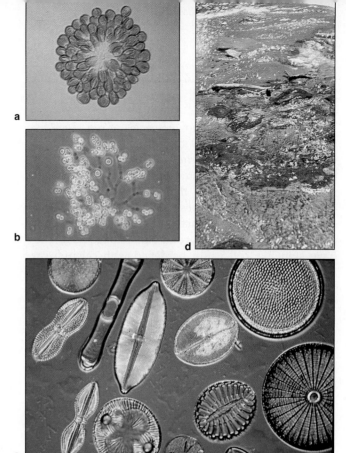

Figure 23.9 Chrysophytes. (**a**) *Synura*, a golden alga that grows as colonies in phytoplankton. It has a fishy odor. (**b**) *Mischococcus*, a yellow-green alga that forms colonies in phytoplankton. (**c**) Diatom shells. (**d**) *Vaucheria*, a yellow-green alga growing over red sandstone in Arizona.

Figure 23.10　(**a**) *Gymnodinium breve*, a dinoflagellate that causes red tides along Florida's coast. (**b**) Part of a fish kill resulting from a dinoflagellate "bloom."

23.12 RED ALGAE

Nearly all of the 4,000 known species of **red algae** live in the seas; fewer than 100 species are found in freshwater lakes, streams, and springs. Red algae are especially abundant in warm currents or tropical seas, often at surprising depths—more than 265 meters—when the water is clear. A few types are planktonic, and a few form colonies. Some colonial types contribute to the formation of coral reefs and coral banks (page 886). They have stonelike cell walls, hardened with calcium carbonate deposits.

The name of their phylum (Rhodophyta) is taken from the Greek *rhodon*, meaning rose, and *phyton*, meaning plant. Figure 23.11 shows two of the "red" species, but others appear green, purple, or greenish-black. They differ in the types and amounts of accessory pigments—mainly phycobilins—which tend to mask their chlorophyll *a*. Phycobilins are good at absorbing green and blue-green wavelengths, which can penetrate deep waters. (Chlorophylls are more efficient at absorbing red and blue wavelengths that may not reach as far below the water's surface.) The chloroplasts bear strong biochemical and structural resemblances to cyanobacteria. The resemblance suggests that these organelles had endosymbiotic origins.

Red algae have complex reproductive modes. For example, a typical life cycle includes haploid and diploid phases. In haploid individuals, structures near the tips of some filaments produce gametes. The male gametes are unflagellated. When released, currents carry them to egglike structures on the same haploid individual or a different one. There, after fertilization, a zygote develops into a diploid, spore-producing structure. Spores develop and are released into the water. After germinating, a spore grows by mitosis into a diploid, spore-producing individual that looks very much like the haploid ones. Some of its cells undergo meiosis, giving rise to *another* set of spores. After release and germination, a spore grows to become a new haploid individual.

Most species are multicelled, with a filamentous, often branching body plan. Unlike plants, the cells do not form tissues and organs. The cell walls incorporate mucous material, which gives red algae a flexible, slippery texture. **Agar** is made from extracts of wall material from a number of species. We use this inert, gelatinous substance as a moisture-preserving agent in cosmetics and bakery goods, as a setting agent for jellies and desserts, and as a culture medium. We also shape it into soft capsules (for packaging drugs and food supplements). We use **carrageenan**, extracted from the red alga *Euchema*, to stabilize paints, dairy products, and many other emulsions.

23.13 BROWN ALGAE

Walk along a rocky coast at low tide and you are likely to come across brown-tinged seaweeds. They are examples of **brown algae**. Nearly all of the 1,500 species inhabit temperate or cool marine waters throughout the world. Figure 23.12 shows examples. Depending on their photosynthetic pigments, brown algae appear olive-green, golden, or dark brown. Like the chrysophytes to which they may be related, they contain carotenoids, especially fucoxanthin (a xanthophyll), as well as chlorophylls *a* and *b*.

Brown algae range from microscopically small, filamentous *Ectocarpus* to giant kelps, 20 to 30 meters tall. The giant kelps, including *Macrocystis* and *Laminaria*, are the largest and most complex of all protistans. Dur-

a

b

Figure 23.11 Red algae. (**a**) *Bonnemaisonia hamifera*, showing the most common growth pattern among red algae: filamentous and branched. This alga reproduces asexually when the hooklike branchlets break off and grow into new plants after they become caught on branches of other algae. (**b**) From a tropical reef, a red alga showing sheetlike growth.

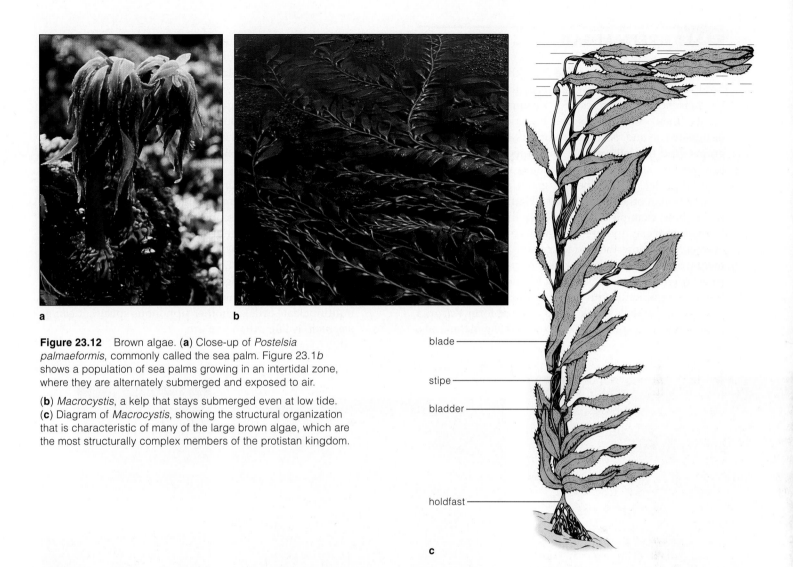

blade

stipe

bladder

holdfast

c

Figure 23.12 Brown algae. (**a**) Close-up of *Postelsia palmaeformis*, commonly called the sea palm. Figure 23.1*b* shows a population of sea palms growing in an intertidal zone, where they are alternately submerged and exposed to air.

(**b**) *Macrocystis*, a kelp that stays submerged even at low tide. (**c**) Diagram of *Macrocystis*, showing the structural organization that is characteristic of many of the large brown algae, which are the most structurally complex members of the protistan kingdom.

ing the life cycle, the formation of gametophytes (gamete-producing bodies) alternates with the formation of sporophytes (spore-producing bodies). The large, multicelled sporophytes have tissues that are differentiated into **holdfasts** (anchoring structures), **stipes** (stemlike parts), and **blades** (photosynthetic, leaf-shaped parts). Hollow, gas-filled bladders impart buoyancy to stipes and blades and help keep them upright. The stipes contain tubelike arrays of elongated cells. Like the food-conducting tissues of plants, the tubes rapidly distribute dissolved sugars and other products of photosynthesis through the individual.

Large populations of giant kelps are productive ecosystems. Think of them as underwater forests in which great numbers of diverse bacteria and protistans, as well as fishes and other animals, carry out their lives. Extensive masses of another brown alga, *Sargassum*,

serve as floating ecosystems in the vast Sargasso Sea, which lies between the Azores and the Bahamas.

If you enjoy ice cream, pudding, salad dressing, canned and frozen foods, jellybeans, or beer, if you use cough syrup, toothpaste, cosmetics, paper, or floor polish, thank the brown algae. The cell walls of some species contain **algin**, which is used as a thickening, emulsifying, and suspension agent. Especially in the Far East, kelps are harvested as sources of mineral salts and as a fertilizer for crops.

Red algae and brown algae are conspicuous photosynthetic members of aquatic habitats, especially the seas. Some species are microscopically small; others are among the largest of the multicelled protistans.

23.14 GREEN ALGAE

Of all protistans, **green algae** bear the greatest structural and biochemical resemblances to plants and may be their nearest relatives. For example, as is true of plants, their chlorophylls are the molecular versions designated *a* and *b*. Their chloroplasts contain starch grains. And the cell walls of some species contain cellulose, pectins, and other polysaccharides typical of plants.

With at least 7,000 species, green algae show more diversity than other algal groups. Figures 23.13 through 23.15 show examples. Most species live in freshwater. But you also can find green algae at the surface of open oceans, just below the surface of soil and marine sediments, on rocks, on tree bark as well as on other organisms, even on snow. Look closely and you find some living as symbionts with fungi, protozoans, and even a few marine animals. Look at the colonial form *Volvox*, a hollow whirling sphere of 500 to 60,000 flagellated cells.

Sink your feet in a white, powdery beach in the tropics; it is largely the work of countless green algal cells (*Halimeda*) that formed calcified cell walls, then went on to die and disintegrate. Green algae may even accompany astronauts on long journeys. They could grow in small spaces on light, carbon dioxide, and some minerals. Besides giving off oxygen, space-traveling algae would take up carbon dioxide exhaled by the aerobically respiring crew.

You won't see many species without the aid of a microscope. For example, Figure 23.1*a* shows one of thousands of microscopically small, mostly single-celled species of desmids. These freshwater algae are important food producers in nutrient-poor ponds and peat bogs. Figures 23.13*a* and *b* show larger members of the same class of algae. Unlike desmids, these two species are siphonous (no crosswalls separate their multinucleate cells). Another siphonous species, *Codium magnum*, is taller than you are.

a

b

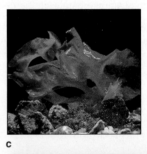

c

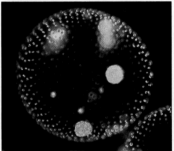

d

e   dormant cell ⎯

Figure 23.13 Green algae. (**a**) A marine species (*Codium*) that has a branched form. *Codium* from Puget Sound, Washington, was accidentally introduced into the mouth of the Connecticut River in 1956. There are no native species of *Codium* in that part of the Atlantic—and no predators that evolved with and are adapted to eating them. The introduced alga spread from the coasts of Maine to North Carolina in just over two decades.

(**b**) Also from a marine habitat, *Acetabularia*, fancifully called the mermaid's wineglass. Each individual in the cluster is a multinucleate cell mass with a rootlike structure, stalk, and cap in which gametes form. (**c**) Sea lettuce (*Ulva*), common to shallow seas throughout the world.

(**d**) *Volvox*, a colony of interdependent cells that bear resemblances to free-living, flagellated cells of the genus *Chlamydomonas*.

(**e**) One of the snow algae. "Red snow" above the summer timberline in Utah indicates the presence of dormant cells of *Chlamydomonas nivalis*. This green alga has an abundance of red accessory pigments that protect its chlorophyll.

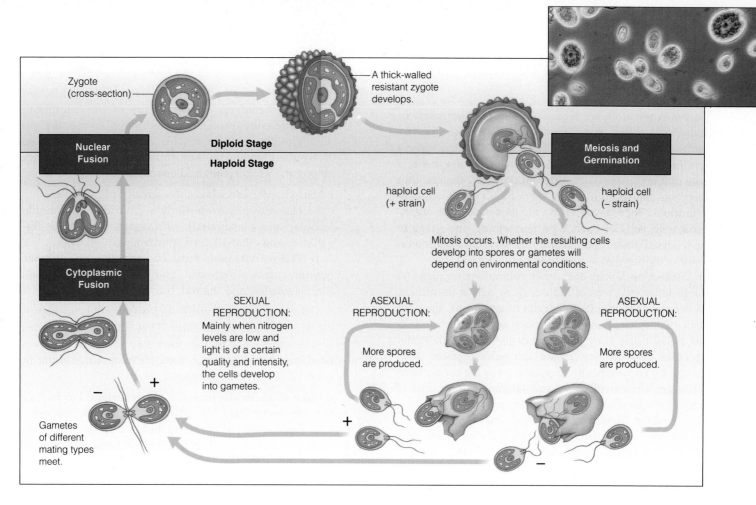

Zygote
(cross-section)

A thick-walled
resistant zygote
develops.

Nuclear Fusion

Diploid Stage

Haploid Stage

Meiosis and Germination

haploid cell
(+ strain)

haploid cell
(– strain)

Mitosis occurs. Whether the resulting cells
develop into spores or gametes will
depend on environmental conditions.

Cytoplasmic Fusion

SEXUAL
REPRODUCTION:
Mainly when nitrogen
levels are low and
light is of a certain
quality and intensity,
the cells develop
into gametes.

ASEXUAL
REPRODUCTION:

More spores
are produced.

ASEXUAL
REPRODUCTION:

More spores
are produced.

Gametes
of different
mating types
meet.

Figure 23.14 Life cycle of a single-celled species of *Chlamydomonas*, one of the most common green algae of freshwater habitats. *Chlamydomonas* reproduces asexually most of the time. It also reproduces sexually under certain environmental conditions.

Green algae employ diverse modes of reproduction. *Chlamydomonas*, shown in Figure 23.14, provides a classic example. This freshwater alga is single celled, no more than 25 micrometers across. It can reproduce sexually. But most of the time it engages in asexual reproduction, with as many as sixteen daughter cells forming (by mitotic cell division) within the confines of the parent cell wall. The daughter cells may live at home for a while. But sooner or later they leave by secreting enzymes that digest what's left of their parent. Figure 23.15 shows how a filamentous green alga, *Spirogyra*, reproduces sexually.

Green algae show great diversity in size, morphology, lifestyles, and habitats. The structure and biochemistry of some groups indicate they may be evolutionarily linked with the plant kingdom.

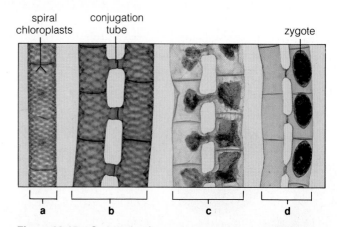

spiral chloroplasts

conjugation tube

zygote

a   b   c   d

Figure 23.15 One mode of sexual reproduction in *Spirogyra*, commonly called watersilk. (**a**) This green alga has ribbonlike, spiral chloroplasts. (**b**) A conjugation tube forms between cells of adjacent haploid filaments of different mating strains. (**c**,**d**) The cellular contents of one strain pass through the tubes into cells of the other strain, where zygotes form. The zygotes will develop thick walls. Later, as they germinate, they will undergo meiosis and give rise to new haploid filaments.

23.15 ON THE ROAD TO MULTICELLULARITY

So far in this unit, we have only hinted at the kinds of experiments that must have gone on among the first protistans, their prokaryotic ancestors, and their earliest multicellular descendants. A **multicellular organism** is one that is composed of different types of cells and that survives and reproduces by way of their combined contributions. There is a division of labor in such an organism, with cells of each type performing one or more specialized tasks even while engaging in basic metabolic activities that assure their own survival.

Reflect on *Volvox* and other colonial green algae. A large sphere has tens of thousands of cells at its surface. When the cells beat their flagella in synchrony, they can move the sphere forward. The task of reproduction falls on a few cells. They divide and give rise to daughter colonies that develop inside the parent sphere, then escape by enzymatically digesting away the connections between surface cells. In some species, certain cells of the colony produce eggs and others produce sperm.

Consider *Trichoplax*, a flattened ball of ciliated cells, half a millimeter across. It has no right side, left side, front, or back. It simply moves in any direction. And yet, as you will read in Chapter 26, *Trichoplax* is an animal. It may even bear close resemblances to one of the novel forms of ancient protozoans that gave rise to animals.

As a final example, green algae are eaten by *Placobranchus*, a marine mollusk. But the algal chloroplasts escape digestion. They accumulate in one of the mollusk's tissues—where they continue to function and provide their "host" with oxygen:

dorsal projections pushed back to show functional chloroplasts incorporated in tissues

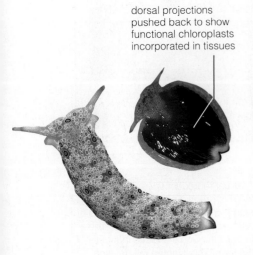

In short, the experimenting hasn't stopped.

SUMMARY

1. In this chapter and in other parts of the book, we have considered the characteristics of the "simplest" eukaryotes—the protistans—and have speculated on their evolutionary links with other kingdoms. Table 23.1 pulls together the key similarities and differences among the prokaryotes and eukaryotes.

2. The chapter described these groups of protistans:

   a. Heterotrophs: water molds, chytrids, slime molds (cellular and plasmodial), protozoans (amoeboid, flagellated, and ciliated), and sporozoans.

   b. Photoautotrophs: most euglenoids, chrysophytes (golden algae, diatoms, and yellow-green algae), dinoflagellates, and the red, brown, and green algae.

3. Like fungi, most chytrids and water molds are saprobic decomposers or parasites. They all secrete enzymes that digest organic matter, then they absorb breakdown products. Also like fungi, some develop a mycelium (a mesh of absorptive filaments).

4. Like animals, slime molds are predators. For part of the life cycle, they are phagocytic, amoebalike cells. Like fungi, their life cycle includes a spore-producing stage. Numerous single cells aggregate and differentiate into a spore-bearing structure.

5. Like animals, the 65,000 species of single-celled protozoans are predators and parasites. Fewer than two dozen cause diseases in humans, but these affect hundreds of millions of people each year.

Table 23.1	Comparison of Prokaryotes with Eukaryotes	
	Prokaryotes	Eukaryotes
Organisms represented:	Bacteria only	Protistans, fungi, plants, and animals
Ancestry:	Two major lineages (archaebacteria and eubacteria) that began more than 3.5 billion years ago	Equally ancient prokaryotic ancestors gave rise to the forerunners of eukaryotes, which emerged more than 1.2 billion years ago
Level of organization:	Single celled	Mostly multicelled, often with division of labor among differentiated cells and tissues
Typical cell size:	Small (1–10 micrometers)	Large (10–100 micrometers)
Cell wall:	Mostly distinctive sugars and peptides	Cellulose or chitin; none in animal cells
Membrane-bound organelles:	Very rarely	Typically profuse
Modes of metabolism:	Both anaerobic and aerobic	Aerobic modes predominate
Genetic material:	Single bacterial chromosome (and sometimes plasmids)	Complex chromosomes (DNA and many associated proteins) within a nucleus
Mode of cell division:	Binary fission, mostly	Mitosis, meiosis, or both

a. Amoeboid protozoans are amoebas and shelled foraminiferans, heliozoans, and radiolarians.

b. Ciliated protozoans, such as *Paramecium* and hypotrichs, use arrays of cilia as motile structures.

c. Flagellated protozoans are internal parasites or are free living in aquatic habitats. The trichomonads, trypanosomes, *Giardia*, and others cause human diseases that range from annoying to lethal.

d. Sporozoans are intracellular parasites that produce infective, motile stages and that may form cysts during the life cycle. *Plasmodium*, which causes malaria, is an example.

6. Most euglenoids and the chrysophytes and dinoflagellates are single-celled photosynthesizers. Many are members of phytoplankton.

7. Most red and brown algae are multicelled, photosynthetic members of aquatic habitats, especially the seas. Some are the largest protistans.

8. Green algae show great diversity in size, body plan, life-styles, and habitats. Plants may be descended from ancient green algae.

Review Questions

1. Generally describe the types of organisms classified in the kingdom Protista. In what habitats are they found? *363*

2. How do protistans differ from bacteria? *363*

3. Name the major categories of protistans. Think about where most of them live, then correlate some of their structural features with environmental conditions. *364–375*

Self-Quiz *(Answers in Appendix IV)*

1. Protistans and other eukaryotes differ from bacteria in which one of the following ways?
   a. Eukaryotes possess cytoplasm in their cells.
   b. Some protistans and other eukaryotes have cell walls.
   c. Protistans and other eukaryotes have DNA.
   d. Protistans and other eukaryotes engage in mitosis and meiosis.

2. Most chytrids and water molds are _____ decomposers of aquatic habitats.
   a. parasitic   c. autotrophic
   b. saprobic   d. chemosynthetic

3. Free-living, amoebalike cells crawl around on rotting plant parts as they engulf bacteria, spores, and organic compounds. This description best fits the _____ .
   a. water molds   c. sporozoans
   b. amoeboid protozoans d. slime molds

4. During a _____ life cycle, amoeboid cells aggregate and form a migrating mass. Cells in the mass then differentiate, forming reproductive structures and spores or gametes.
   a. slime mold   c. protozoan
   b. water mold   d. chytrid

5. Amoebas, foraminiferans, radiolarians, and heliozoans are all classified as _____ .
   a. ciliated protozoans   c. amoeboid protozoans
   b. flagellated protozoans d. sporozoans

6. Parasitic, flagellated protozoans of tropical regions known as trypanosomes are associated with which of the following diseases?
   a. African sleeping sickness and Chagas disease
   b. toxoplasmosis
   c. malaria
   d. amoebic dysentery

7. Euglenoids and chrysophytes are mostly _____ .
   a. photoautotrophic   c. heterotrophic
   b. chemoautotrophic   d. omnivorous

8. Single-celled photosynthetic protistans, including most euglenoids, chrysophytes, and dinoflagellates, are members of the _____ , the "pastures" of most aquatic habitats.
   a. zooplankton   c. brown algae
   b. red algae   d. phytoplankton

9. Algin is used in ice cream, pudding, salad dressing, jelly beans, beer, cough syrup, toothpaste, cosmetics, and other products. The source of algin is the _____ .
   a. green algae   c. red algae
   b. brown algae   d. dinoflagellates

Selected Key Terms

agar *372*	green algae *374*
algin *373*	heliozoan *367*
amoeba *366*	holdfast *373*
amoeboid protozoan *366*	multicellular organism *376*
blade *373*	mycelium *364*
brown algae *372*	phytoplankton *370*
carrageenan *372*	protozoan *366*
chytrid *364*	protozoan conjugation *367*
ciliated protozoan *367*	radiolarian *367*
contractile vacuole *367*	red algae *372*
cyst *366*	red tide *371*
diatom *371*	slime mold *364*
dinoflagellate *371*	sporozoan *368*
euglenoid *370*	stipe *373*
flagellated protozoan *368*	water mold *364*
foraminiferan *367*	yellow-green algae *371*
golden algae *371*	zooplankton *370*

Readings

Margulis, L. 1993. *Symbiosis in Cell Evolution.* Second edition. New York: Freeman. Paperback.

Margulis, L., and K. Schwartz. 1992. *Five Kingdoms.* Second edition. New York: Freeman. Paperback.

24 FUNGI

Dragon Run

When the first winter storm blasts through southeastern Virginia, Dragon Run—part swamp, part marsh, part old-growth woodland—pays tribute to the wind. Oaks, maples, gums, and beeches release dead leaves by the millions and these shower to earth, where they pile up as crisp, ankle-deep mounds. Branches snap off; sometimes trees crash down. Sheaves of dead grasses sink into marsh muds, and other plants buckle into the shallow, murky waters of the bottomlands. The storm kills uncounted numbers of insects, some birds, a few squirrels. By late December, Dragon Run is partially buried in organic debris.

Dig down through the debris and you will discover the accumulated litter of many past seasons—moist cushions of decayed leaves, bits of spiders and insects, mouldering branches, the carcasses of small mammals.

A new growing season begins with the warm rains of February. Buds develop on shrubs and trees, including a massive silver beech that has been budding each spring for 300 years. Woodland violets are the first to sprout in the thawing soil. By April, the resurgence of growth has imparted such a sharp, green freshness to Dragon Run that you might overlook the organisms on which the resurgence depends. On logs, under leaves, in the soil, fungi are commandeering resources and engaging in vegetative growth (Figure 24.1).

Fungi are nature's premier decomposers. Like other heterotrophs, they feast on organic compounds produced by other organisms. But few organisms besides fungi digest their dinner out on the table, so to speak. As fungi grow in or on organic matter, they secrete enzymes that digest it into bits that their individual cells can absorb. This "extracellular digestion" of organic matter liberates carbon and other nutrients *that also can be absorbed by plants*—the primary producers of Dragon Run and nearly all other ecosystems on earth.

Keep the global perspective in mind. Why? The metabolic activities of some fungi do cause diseases in humans, pets and farm animals, ornamental plants, and important crop plants. Some species are notorious spoilers of food supplies. Others have uses in the commercial production of substances ranging from antibiotics to excellent cheeses. We tend to assign "value" to fungi and other organisms in terms of their direct effect on our lives. There is nothing wrong with battling dangerous species and admiring beneficial ones—as long as we do not lose sight of the greater roles of fungi or any other kind of organism in nature.

a Sulfur shelf fungus (*Polyporus*) on a living tree

b Rubber cup fungus (*Sarcosoma*)

c Purple coral fungus (*Clavaria*)

Figure 24.1 Fungal species from southeastern Virginia. This small sampling merely hints at the rich diversity within the kingdom Fungi.

d Scarlet hood (*Hygrophorus*)

g Frost's bolete (*Boletus*)

e Yellow coral fungus (*Clavaria*)

h Trumpet chanterelle (*Craterellus*)

f Big laughing mushroom (*Gymnophilus*)

1. Fungi are heterotrophs. Together with heterotrophic bacteria, they are decomposers of the biosphere. Saprobic types obtain nutrients from nonliving organic matter. Parasitic types obtain them from tissues of living hosts.

2. Fungi secrete enzymes that digest food outside their body, then fungal cells absorb breakdown products. Their metabolic activities also release carbon dioxide to the atmosphere and return many nutrients to the soil, where they become available to producer organisms.

3. Most fungi are multicelled. They form a mycelium, which is the food-absorbing part of the fungal body. A mycelium is a mesh of hyphae, which are elongated filaments that develop by repeated mitotic cell divisions.

4. Commonly, modified hyphae form a reproductive structure in or upon which spores develop. A "mushroom" is such a structure. Germinating spores grow and develop into a new mycelium.

5. Many fungi are symbionts. Some are part of lichens. Many are part of mycorrhizae; they are locked in mutually beneficial relationships with young roots of land plants. Fungal hyphae provide the plants with nutrients, and the plants provide the fungi with carbohydrates.

24.1 MAJOR GROUPS OF FUNGI

For many of us, our thoughts about fungi are limited to deciding between whole or sliced brown mushrooms in grocery stores. Yet those drab mushrooms are produced by a fungus that belongs to a huge group of diverse species. Figure 24.1 shows just a few of its relatives. These don't begin to do justice to the 80,000 fungal species we know about. And there may be at least a million more species we don't know about!

In this chapter we consider the three major groups of fungi. These are the zygomyctes (Zygomycota), sac fungi (Ascomycota), and club fungi (Basidiomycota). We also consider some of the "imperfect fungi," the members of which cannot be assigned to any of the recognized taxonomic groups.

We know of funguslike fossils 900 million years old. Together with simple plants, species resembling the zygomycetes started invading the land about 430 million years ago. Apparently this was a trigger for major radiations. About 100 million years later, all three of the major lineages were well established, with a number of diverse species.

Nutritional Modes

Fungi are heterotrophs, meaning they require organic compounds synthesized by other organisms. Most are **saprobes**; they obtain nutrients from nonliving organic matter and so cause its decay. Others are **parasites**; they extract nutrients from tissues of a living host.

Fungal cells grow in or on organic matter. As they do, they secrete digestive enzymes, then absorb the breakdown products. This "extracellular digestion" benefits plants, which absorb some of the nutrients being released. They also release great quantities of carbon dioxide, which also benefits plants.

Together with heterotrophic bacteria, fungi are nature's decomposers. Without them, communities would become buried in their own garbage, nutrients would not be cycled, and life could not go on.

Fungal Body Plans

The vast majority of fungi are multicelled, land-dwelling species. During the life cycle, a mesh of branching filaments develops (Figure 24.2). The mesh, a **mycelium** (plural, mycelia), functions in food absorption. It rapidly grows over or into organic matter and has a good surface-to-volume ratio for absorption. Each filament in a mycelium is a **hypha** (plural, hyphae). Commonly, hyphae consist of tube-shaped cells with chitin-reinforced walls. The cytoplasm of these cells interconnects, so that nutrients flow unimpeded throughout the mycelium.

Reproductive Modes

Fungi reproduce asexually most often but, given the opportunity, also engage in sexual reproduction (Figure 24.3). They produce great numbers of nonmotile spores. Generally, **spores** are reproductive cells or multicelled structures, often walled, that germinate following dispersal from the parent body. Those of single-celled fungi simply form inside the parent cell. Most multicelled fungi produce asexual spores on or in reproductive structures called **sporangia** (singular, sporangium), which are made of interwoven hyphae. Sexual reproduction proceeds through the formation of gametes as well as spores. The gamete-producing structures are **gametangia** (singular, gametangium).

Fungi are spore-producing saprobes or parasites. Nearly all are multicelled. Together with heterotrophic bacteria, they are the decomposers of the biosphere.

There are about 765 species of **zygomycetes**. Their name refers to a thick-walled structure that forms when they undergo sexual reproduction.

Consider Figure 24.4, which shows the life cycle of the black bread mold, *Rhizopus stolonifer*. The sexual phase begins when hyphae of two different mating strains grow into each other and fuse. Two gametangia form between the hyphae, and several haploid nuclei are produced inside each. Later, their nuclei fuse, forming a zygote. A thick, protective wall forms around the zygote. This thick-walled structure is called a **zygosporangium** (plural, zygosporangia). Meiosis proceeds and spores are produced when this structure germinates. Each resulting spore can give rise to stalked structures that can produce many asexual spores, each of which

Figure 24.2 An example of the filaments (hyphae) of a mycelium, the food-absorbing portion of many fungi.

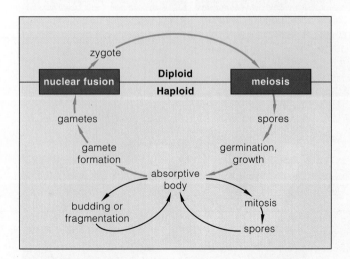

Figure 24.3 Generalized life cycle for many fungi. Asexual reproductive events (*black arrows*) are most common. Sexual reproduction (*blue arrows*) is less frequent.

can be the start of an extensive mycelium. The spores are very small, dry, and easily dispersed by winds.

R. stolonifer is a familiar saprobe with a bad reputation as a spoiler of baked goods. Most saprobic zygomycetes are at home in soil, decaying plant and animal matter, and stored food. *Pilobolus* prefers animal feces, from which it disperses its spores in a blastful way (Figure 24.5). Parasitic zygomycetes prefer living on houseflies and other insects.

Rhizopus is normally harmless to humans but, like other saprobic fungi, it is an opportunist. If its spores are inhaled or if they land on cuts in skin, they can cause diseases, especially in diabetics or individuals with weakened immune systems. Fungal infections are usually chronic (long lasting).

Zygosporangia (zygotes surrounded by a thick protective wall) are the key defining feature of the zygomycetes.

Figure 24.5 *Pilobolus*, a zygomycete. *Pilobolus* grows on animal feces and forms stalked, spore-bearing structures. Dark sacs at the end of each structure contain the spores. A stalk grows in such a direction that rays of sunlight converge at the base of its swollen portion, just below the dark sacs. Turgor pressure inside a vacuole in the swollen portion becomes so great that the spore sac can be blasted 2 meters away—a remarkable feat, considering the stalk is less than 10 millimeters tall!

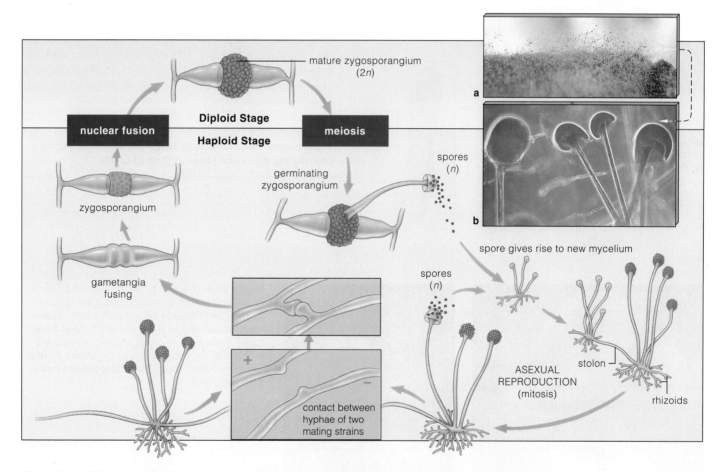

Figure 24.4 Life cycle of the black bread mold, *Rhizopus stolonifer*. As for other zygomycetes, the haploid phase dominates. Asexual reproduction is common, but different mating strains (+ and −) also reproduce sexually. Either way, haploid spores form and give rise to new mycelia. Chemical attraction between a + hypha and a − hypha causes them to fuse. Two gametangia form, each with several haploid nuclei inside. Later their nuclei fuse to form a zygote. The zygote develops a thick wall, thereby becoming a zygosporangium, and may remain dormant for several months. Meiosis occurs as the zygosporangium germinates, and new spores form.

There are more than 30,000 known species of **sac fungi**. About 500 of these are single-celled yeasts. (Many other yeasts are classified with the club fungi.) The yeasts live in the nectar of flowers and on fruits and leaves. Bakers and vintners rely on the by-products of vast populations of busily fermenting yeasts (page 133). For example, the carbon dioxide by-product of *Saccharomyces cerevisiae* leavens bread. Its ethanol end product is central to the commercial production of wine, beer, and other alcoholic beverages. Many yeast strains with desirable properties have been developed through artificial selection and genetic engineering. Yeasts also are used in genetics research.

The vast majority of sac fungi are multicelled. They include most of the red, bluish-green, and brown molds that spoil stored food. One of them, the salmon-colored *Neurospora sitophila*, causes considerable damage when it invades bakeries or research laboratories. It produces

so many easily dispersed spores, it is difficult to eradicate. One of its relatives, *N. crassa*, is an important organism in genetic research.

Multicelled sac fungi also include the edible truffles and morels, one of which is shown in Figure 24.6. Trained pigs and dogs have been used to snuffle out truffles, which grow underground as symbionts with the young roots of oak and hazelnut trees. Truffles are now cultivated commercially on the roots of inoculated seedlings in France. Even so, these fungi remain one of the most expensive luxury foods; they are priced by the gram, not by the pound.

As is true of other fungi, asexual reproduction is common among the sac fungi. Yeasts can reproduce sexually by budding. Multicelled species form specialized spores of a type called **conidia** (singular, conidium).

When sac fungi reproduce sexually, they form distinctive reproductive structures called **ascocarps**. In multicelled species, some hyphae become tightly interwoven into structures that resemble flasks, globes, and shallow cups of the sort shown in Figures 24.1*b* and 24.6. Spore-producing sacs called **asci** (singular, ascus) usually form on the inner surface of these structures. (Single-celled yeast cells simply fuse and so become the sac.) Following meiosis, haploid spores form and are dispersed. After a spore germinates, it gives rise to a mycelium that grows through soil, decaying wood, and other substrates.

The sac fungi alone form asci (distinctive spore-producing sacs) during the sexual phase of their life cycle.

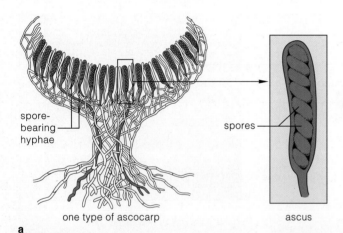

spore-bearing hyphae

spores

one type of ascocarp

ascus

a

b

c

Figure 24.6 Sac fungi. (**a**) Diagram and (**b**) photograph of ascocarps of the scarlet cup fungus, *Sarcoscypha coccinia*. This reproductive structure forms during the sexual phase of the life cycle. Saclike spore-producing structures (asci) form within the cup. (**c**) One of the choice morels (*Morchella esculenta*). This edible species has a poisonous relative.

A Few Fungi We Would Rather Do Without

You know you are a serious student of biology when you view organisms objectively in terms of their place in nature, not in terms of their impact on humans generally and yourself in particular. As a student you can indeed respect saprobic fungi as vital decomposers and salute parasitic fungi that keep populations of destructive insects and weeds in check.

The true test is when you open the refrigerator for a dish of high-priced raspberries and discover that a fungus beat you to them or when another fungus starts feeding on warm, damp tissues between your toes. Who among us praises a fungus that makes skin reddened, scaly, and cracked? Which home gardener waxes poetic about black spot or powdery mildew on roses? Which farmers happily lose millions of dollars each year to rust or smut invaders of their crops?

Who would knowingly inhale the airborne spores of *Histoplasma capsulatum*, a pathogen that causes *histoplasmosis*? This disease passes for minor respiratory infections most of the time and for tuberculosis some of the time. When the fungal spores land on soil, they give rise to mycelia. When airborne spores are inhaled, they become decidedly yeastlike. At first, the fungus takes up residence in moist lung tissues. In rare cases, it ends up damaging most organs in the body. *H. capsulatum* is common worldwide. In the United States, it is most prevalent in many regions drained by the Mississippi and Ohio rivers, where it is especially partial to the nitrogen-rich droppings of birds and bats.

And who denies that fungi have influenced the course of human history? *Claviceps purpurea* is a fungal parasite of rye and other grains. We use its metabolic by-products (alkaloids) to treat migraine headaches and to shrink the uterus (to prevent hemorrhaging) after childbirth. But large amounts of the alkaloids are toxic. Eat a lot of bread made with contaminated rye flour and you end up with *ergotism*. Disease symptoms include hysteria, hallucinations, convulsions, vomiting, diarrhea, dehydration, and gangrenous limbs. Severe cases are fatal.

Ergotism epidemics were common in Europe in the Middle Ages, when rye was a major crop. Ergotism thwarted Peter the Great, the Russian czar who was obsessed with conquering ports along the Black Sea for his vast and nearly landlocked empire. Soldiers laying siege to the ports ate mostly rye bread and fed rye to their horses. The former went into convulsions and the latter into "blind staggers." Quite possibly, outbreaks of ergotism were an excuse to launch the Salem witch-hunts in colonial Massachusetts.

Some Pathogenic and Toxic Fungi

Zygomycetes

Rhizopus	Food spoilage

Sac fungi

Ophiostoma ulmi	Dutch elm disease
Cryphonectria parasitica	Chestnut blight
Venturia inaequalis	Apple scab (Figure *a*)
Claviceps purpurea	Ergot of rye, ergotism
Monilinia fructicola	Brown rot of stone fruits

Club fungi

Puccinia graminis	Black stem wheat rust
Ustilago maydis	Smut of corn
Amanita (some)	Severe mushroom poisoning

Imperfect fungi

Verticillium	Plant wilt
Microsporum, Trichophyton, Epidermophyton	Various species cause ringworms, including athlete's foot
Candida albicans	Infection of mucous membranes
Histoplasma capsulatum*	Histoplasmosis

*Now being reassigned to the sac fungi.

a telltale evidence of *V. inaequalis*

24.5 CLUB FUNGI

A Sampling of Spectacular Diversity

You probably are familiar with some of the 25,000 or so species of **club fungi**, the largest and most diverse group of the fungal kingdom. They include the grocery-store-variety mushrooms as well as the more exotic mushrooms, shelf fungi, and coral fungi shown in Figures 24.1 and 24.7. Other splendid types include bird's nest fungi and puffballs, as well as stinkhorns of the sort shown in Figure 1.6d.

Some of the saprobic types are major decomposers of plant debris. As you will see, others are symbionts that live in association with the young roots of forest trees. Still others, including the rust and smut fungi, cause serious plant diseases that can destroy entire fields of wheat, corn, and other major crops. Then again, cultivation of the common mushroom (*Agaricus brunnescens*) is a multimillion-dollar business.

Have you ever asked which organisms are the oldest and the largest? The club fungus *Armillaria bulbosa* is among them. The mycelium of one specimen, discovered in a northern Michigan forest, extends through at least 15 hectares. A hectare is 10,000 square meters. By one estimate, this individual weighs more than 10,000 kilograms and has been spreading through the forest soil for more than 1,500 years!

Figures 24.7c and d show two species of *Amanita*. Both contain toxins and intoxicants. *A. muscaria*, the fly agaric mushroom, causes hallucinations when ingested. It was used ritualistically in ancient societies in Central America, Russia, and India. The death cap mushroom (*A. phalloides*) can kill. Within eight to twenty-four hours of ingesting even as little as 5 milligrams of its toxin, a person starts vomiting and suffering diarrhea.

Figure 24.7 Club fungi. (**a**) Bird's nest fungus. (**b**) Light-red coral fungus (*Ramaria*). (**c**) Shelf fungus (*Polyporus*) growing on a rotting log. (**d**) From California, *Amanita ocreata*, which has caused fatalities when eaten. (**e**) The death cap mushroom (*A. phalloides*) is usually fatal when ingested. (**f**) The fly agaric mushroom (*A. muscaria*) causes hallucinations when eaten.

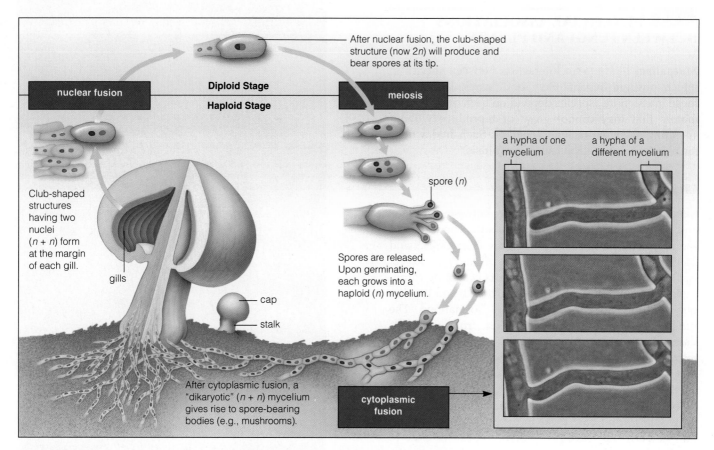

After nuclear fusion, the club-shaped structure (now 2n) will produce and bear spores at its tip.

nuclear fusion

Diploid Stage

Haploid Stage

meiosis

Club-shaped structures having two nuclei (n + n) form at the margin of each gill.

gills

cap

stalk

a hypha of one mycelium

a hypha of a different mycelium

spore (n)

Spores are released. Upon germinating, each grows into a haploid (n) mycelium.

After cytoplasmic fusion, a "dikaryotic" (n + n) mycelium gives rise to spore-bearing bodies (e.g., mushrooms).

cytoplasmic fusion

Figure 24.8 Generalized life cycle for many club fungi. When two compatible mating strains grow together, the cytoplasm (but not the nuclei) of two hyphal cells may fuse. Cell divisions produce a "dikaryotic" mycelium (its cells have two nuclei). When mushrooms develop, club-shaped structures form over the surface of the gills. Each structure contains two nuclei, which fuse to form a diploid zygote.

Later, kidney and liver cells start to degenerate; death can follow within a few days.

No general rule enables us to distinguish harmless from deadly mushrooms. *No one should eat mushrooms gathered in the wild unless they have been accurately identified as edible.* Said another way, there are old mushroom hunters and bold mushroom hunters, but no old, bold mushroom hunters.

Generalized Life Cycle

The spore-producing cells of club fungi are called **basidia** (singular, basidium). They usually are club-shaped, and they always bear the sexual spores on their outer surface (Figure 24.8). The club-shaped cells typically develop on a short-lived reproductive structure, the **basidiocarp**. Basidiocarps are merely the visible portion of the fungus; the living mycelia are buried in soil or decaying wood.

What most of us call a "mushroom" is a short-lived basidiocarp. About 10,000 species of club fungi produce them. Each consists of a stalk and a cap. Its spore-producing cells are on the sides of gills, which are sheets of tissue in the cap.

When a spore dispersed from a mushroom lands on a suitable site, it germinates and gives rise to a haploid mycelium. When hyphae of two compatible mating strains grow next to each other, they may undergo cytoplasmic fusion (Figure 24.8). Nuclear fusion does not follow at once, so the hyphae form a **dikaryotic mycelium**. Each cell in this mycelium contains one nucleus of each mating type. After an extensive mycelium develops and when conditions are favorable, mushrooms form. At first, each spore-producing cell of a mushroom is dikaryotic, but then its two nuclei fuse to form a short-lived zygote. The zygote quickly undergoes meiosis and haploid spores are produced, then dispersed by air currents.

Club fungi, the fungal group with the greatest diversity, produce club-shaped, spore-bearing structures during the sexual phase of their life cycle.

24.6 BENEFICIAL ASSOCIATIONS BETWEEN FUNGI AND PLANTS

Mutualism refers to an interaction between species in which positive benefits flow both ways. In some of these interactions, species depend on each other so intimately that they cannot grow or reproduce without spending their entire lives together. Such interactions are a form of mutualism called **symbiosis**.

Lichens

Lichens are mutualistic interactions between a fungus and cyanobacteria, green algae, or both. Thousands of sac fungi enter into such interactions. Figures 24.9 and 24.10 show examples. Lichens form after a fungal hypha penetrates a host cell and starts absorbing carbohydrates from it. If the cell survives, both it and the fungus multiply together into crusty formations. The fungus continues to draw nutrients from the cell's descendants. The photosynthetic species suffers in terms of its own growth, but it also benefits from the lichen's sheltering effect.

Sheltering is sometimes vital. Lichens colonize places that are too hostile for other organisms. They grow (slowly) on bare rocks in deserts and mountains, on tree bark and fence posts. Some types live close to the South Pole! They endure by suspending activities when it becomes extremely hot, cold, or bright. Then, their crusty part thickens and blocks sunlight, so photosynthesis stops. They become active only when moistened with raindrops, fog, or dew. Ancient lichens may have been the first invaders of land.

Lichens absorb minerals from rocks and nitrogen from the air. Their metabolic activities can slowly change the composition of their substrates. They may contribute to soil formation, thus setting the stage for colonization by different species. Lichens also are early warning signals that environmental conditions are deteriorating. They absorb but cannot rid themselves of toxins. When lichens die around cities, air pollution is getting bad. We know this from extensive studies in industrialized regions of England and in New York City.

Mycorrhizae

The small, young roots of nearly all vascular plants associate symbiotically with fungi in permanent, mutually beneficial ways. Such associations are called **mycorrhizae**, or "fungus-roots." Figure 24.11 shows an example. The fungus absorbs carbohydrates from the host plant, which absorbs mineral ions from the fungus. Collectively, fungal hyphae have an enormous surface area for taking up water and dissolved ions. The fungus takes up ions when they are abundant in soil and

Figure 24.9 Lichens. (**a**) *Usnea*, old man's beard. (**b**) *Cladonia rangiferina*, sometimes called reindeer moss.

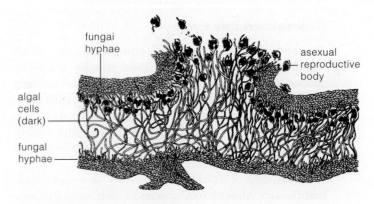

Figure 24.10 One type of lichen in cross section.

Where Have All the Fungi Gone?

Throughout European forests, wild mushrooms are getting smaller, and species are declining at an alarming rate. Mushroom gatherers aren't to blame; toxic as well as edible species may be on the road to extinction. Also, throughout Europe, driving cars, burning coal, prodding crop plants with nitrogen fertilizers, and other human practices are pumping ozone, nitrogen oxides, and sulfur oxides into the air. The decline of fungi is correlated with the rise in air pollution.

Forest trees and assorted mycorrhizal fungi are symbiotic partners. The fungi promote enhanced uptake of water and nutrients, and the trees provide the fungi with carbohydrates. Normally, as a tree ages, one fungal species gradually supplants another in predictable patterns. If fungi are indeed disappearing, trees will lose their support system and will become highly vulnerable to severe frost and drought. Entire forests may already be at risk. In Europe, collectors have been recording information about the wild mushroom populations since the 1900s. Similarly extensive records have not been kept in the United States, but comparable environmental conditions suggest that North American forests also are at risk.

releases them to the plant when ions are scarce. Many plants cannot grow as efficiently when mycorrhizae are not available to help them absorb phosphorus and other crucial ions (Figure 24.12).

The example in Figure 24.11 is an *exo*mycorrhiza, in which hyphae form a dense net around living cells in roots but do not penetrate them. Other hyphae form a velvety wrapping around the root, and the mycelium radiates outward from it. Exomycorrhizae are common in temperate regions, where beeches, oaks, willows, cottonwood, poplars, pines, and eucalyptus grow. The interaction makes the trees more resistant to adverse seasonal changes in temperature and rainfall. About 5,000 species of fungi enter these close associations. Most types are club fungi, including those truffles described earlier.

*Endo*mycorrhizae are much more common. They form on the roots of about 80 percent of all vascular plants. In this case, the fungal hyphae penetrate plant cells, as they do in lichens. Fewer than 200 species of zygomycetes serve as the fungal partner. Their hyphae branch extensively, forming tree-shaped absorptive structures within cells. They also extend for several centimeters into the surrounding soil.

As the *Focus* essay suggests, air pollution damages mycorrhizae, and this is affecting the world's forests. We return to this topic in Chapter 50.

Lichens and mycorrhizae are examples of mutualistic interactions between fungi and other organisms.

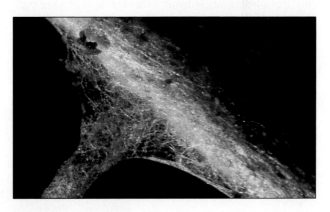

Figure 24.11 Mycorrhiza of a hemlock tree. The white threads are hyphal strands around a small, young root.

Figure 24.12 Effects of the presence or absence of mycorrhizae on plant growth. The juniper seedlings shown at the left are six months old. They were grown in sterilized, phosphorus-poor soil with a mycorrhizal fungus. The seedlings at the right were grown without the fungus.

24.7 "IMPERFECT" FUNGI

Some fungi are not easily classified. If a species has no sexual phase in its life cycle, it is called an "imperfect" fungus. If researchers later detect a sexual phase, the fungus is assigned to a recognized group—the sac or club fungi, most often.

Some imperfect fungi cause human diseases. *Candida albicans* is a notorious cause of vaginal infections (see Figure 24.13 and the *Commentary* on page 383). Other, predatory types ensnare tiny worms (Figure 24.13*b*). *Aspergillus* produces citric acid for candies and soft drinks, and it ferments soybeans for soy sauce. Certain *Penicillium* species "flavor" Camembert and Roquefort cheeses; and others produce penicillins, which we use as antibiotics.

SUMMARY

1. Fungi are heterotrophs. Many species are major decomposers of organic matter. Saprobic types feed on nonliving organic matter. Parasitic types obtain nutrients from living organisms. Mutualistic types are partners with other organisms in lichens and mycorrhizae.

2. The cells of all fungi secrete digestive enzymes that break down food into small molecules, which are absorbed across the plasma membrane.

3. Nearly all fungi are multicelled. Their food-absorbing part (mycelium) is a mesh of filaments (hyphae). Hyphae often interweave to form aboveground reproductive structures.

4. Fungi commonly reproduce asexually by spore formation or by budding from the parent body. When they reproduce sexually, distinctive spore-producing structures form.

5. The major groups of fungi are the zygomycetes, sac fungi, and club fungi. These groups are distinguished from one another largely on the basis of the reproductive structures that form during the sexual phase of the life cycle.

   a. Zygomycetes form a thick wall around the zygote, thus producing a zygosporangium.

   b. Sac fungi form saclike, spore-producing structures (asci), often in reproductive structures shaped like flasks, globes, and cups.

   c. Club fungi usually form club-shaped, spore-producing structures (basidia).

Figure 24.13 Imperfect fungi. (**a**) *Candida albicans*, cause of "yeast infections" of the vagina and mouth. In a more recent scheme, this fungal species has been assigned to ascomycetes. (**b**) Hyphae of *Arthrobotrys dactyloides*, a predatory fungus, form nooselike rings that swell rapidly with incoming water when stimulated. The "hole" in the noose shrinks and captures this worm. (**c**) Rows of conidia (asexual spores) of *Penicillium*.

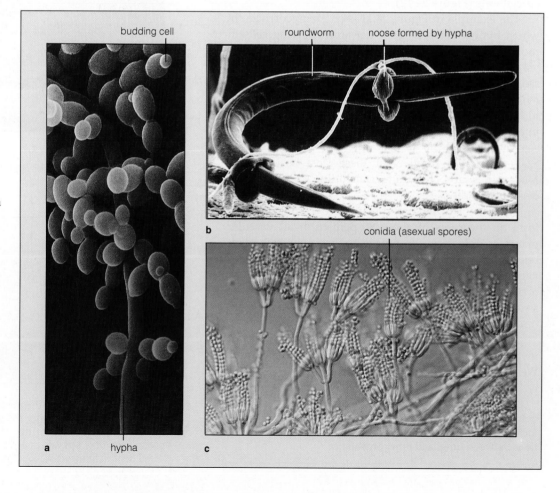

6. When a sexual phase cannot be detected or is absent from the life cycle of a fungal specimen, that specimen is assigned to an informal category called the "imperfect" fungi.

7. Fewer than 200 of the zygomycetes enter into mutualistic relationships with cyanobacteria or green algae. A lichen is an intimate association between such a fungus and its photosynthetic partner.

8. Many sac fungi and club fungi are symbiotic with young roots of shrubs and trees. In these associations, called mycorrhizae, fungal hyphae provide the plant with nutrients; the plant provides the fungus with carbohydrates.

Review Questions

1. Describe the fungal mode of nutrition. *379–380*

2. How is a mycelium constructed and what is its function? *380*

3. Describe one of the major groups of fungi. *379–385*

4. Some fungi trap living nematodes as food. Do these fungi feed as parasites or saprobes? Why? *388*

5. Define sporangium, gametangium, and basidiocarp. *380, 385*

6. How does a mycorrhiza differ from a lichen? *386–387*

Self-Quiz *(Answers in Appendix IV)*

1. The major groups of fungi are categorized mainly on the basis of their _____ .
   a. habitat
   b. color and size
   c. reproductive structures
   d. digestive enzymes

2. Fungi can reproduce asexually by _____ .
   a. forming spores
   b. budding
   c. fragmenting of parent bodies
   d. all of the above

3. New mycelia form following the germination of _____ .
   a. hyphae
   b. spores
   c. mycelia
   d. mushrooms

4. A "mushroom" is _____ .
   a. the food-absorbing part of a fungal body
   b. the part of the fungal body not constructed of hyphae
   c. a reproductive structure
   d. a nonessential part of the fungus

5. A mycorrhiza is a _____ .
   a. fungal disease of the foot
   b. fungus-plant relationship
   c. parasitic water mold
   d. fungus endemic to barnyards

6. A lichen is an intimate symbiotic association between a fungus and a _____ .
   a. mycorrhiza
   b. green alga
   c. parasitic fungus
   d. cyanobacteria
   e. both b and d

7. Parasitic fungi obtain nutrients from _____ .
   a. tissues of living host organisms
   b. nonliving organic matter
   c. only living plants
   d. only living animals
   e. none of the above

8. Saprobic fungi _____ .
   a. feed on nonliving organic matter
   b. feed on nutrients from living organisms
   c. are partners with other organisms
   d. are rarely multicelled

9. When the sexual phase is undiscovered or undetected from the life cycle of a fungus, that specimen is assigned to an informal category called _____ .
   a. zygomycetes
   b. ascomycetes
   c. imperfect fungi
   d. basidiomycetes

10. Match these terms appropriately.
   ____ zygomycetes
   ____ dikaryotic mycelium
   ____ hypha
   ____ basidiomycetes
   ____ mutualism
   ____ ascomycetes

   a. refers to an interaction between species in which positive benefits flow each way
   b. some yeasts and morels
   c. term for each filament in a mycelium
   d. black bread mold
   e. mushrooms, puffballs, shelf fungi, stinkhorns
   f. each cell contains two nuclei (one of each mating type)

Selected Key Terms

asci *382*
ascocarp *382*
basidia *385*
basidiocarp *385*
club fungi *384*
conidia *382*
dikaryotic mycelium *385*
gametangia *380*
hypha *380*
lichen *386*
mutualism *386*
mycelium *380*
mycorrhizae *386*
parasite *380*
sac fungi *382*
saprobe *380*
sporangia *380*
spore *380*
symbiosis *386*
zygomycetes *380*
zygosporangium *380*

Readings

Bold, H., and J. LaClaire. 1987. *The Plant Kingdom.* Fifth edition. Englewood Cliffs, New Jersey: Prentice-Hall. Paperback.

Kendrick, B. 1991. *The Fifth Kingdom.* Second edition. Waterloo: Mycologue Publications.

Moore-Landecker, E. 1990. *Fundamentals of the Fungi.* Third edition. Englewood Cliffs, New Jersey: Prentice-Hall. Well-written introduction to the kingdom Fungi.

25 PLANTS

Pioneers in a New World

Seven hundred million years ago, no shorebirds stirred and noisily announced the dawn of a new day. No crabs clacked their tiny claws together and skittered off to burrows. The only sounds were the rhythmic muffled thuds of waves in the distance, at the outer limits of another low tide. Nearly 3 billion years before, life had its beginnings in the waters of the earth—and now, quietly, the invasion of the land was under way.

Astronomical numbers of photosynthetic cells had come and gone, and oxygen-producing ones had slowly changed the atmosphere. High above the earth, the sun's energy had converted much of the oxygen into a dense ozone layer. This became a shield against lethal doses of ultraviolet radiation—which had kept early organisms below the water's surface.

By one scenario, cyanobacteria were the first to adapt to intertidal zones, where the land dried out with each retreating tide. They were the first to move into shallow, freshwater streams meandering down to the coasts. Later, certain green algae and fungi made the same journey together. Every land plant around you is a descendant of green algae that lived near the water's edge or made it onto land. Fungi are still associated with nearly all of them.

We have some tantalizing fossils of the pioneers. We also are learning about them through comparative biochemistry and studies of existing species. Today, as in Precambrian times, cyanobacteria and green algae grow in mats in nearshore waters and on the banks of freshwater streams (Figure 25.1). When a volcanic eruption or some other event exposes rocks for the first time, cyanobacteria colonize them. Mutually beneficial interactions of green algae and fungi follow. The biochemical activities and accumulated remains of these

Figure 25.1 (**a**) Filaments of a green alga, massed in a shallow stream. More than 400 million years ago, green algae that may have been ancestral to plants lived in similar streams that meandered down to the shore of early continents. (**b**) A land-dwelling descendant of those ancestral forms—a ponderosa pine growing on a mountaintop above Yosemite Valley, California.

a

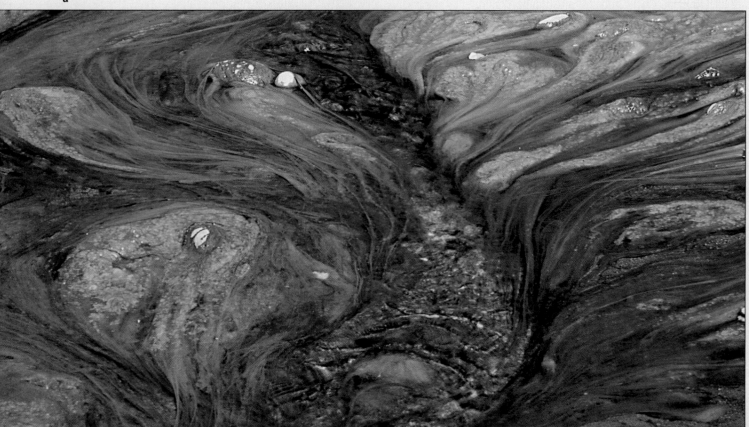

b

organisms slowly enrich the sediments around them with organic and inorganic substances. In this way they actually create soils in which mosses and other plants can take hold.

With this chapter, we turn to the plant kingdom. With few exceptions, its members are multicelled photoautotrophs. They produce their own organic compounds, using sunlight as their energy source and carbon dioxide as their carbon source. Plants have the metabolic machinery for the noncyclic as well as the cyclic pathway of photosynthesis. In other words, by splitting water molecules, they can obtain the huge numbers of electrons and hydrogen atoms required for their growth into multicellular forms as tall as giant redwoods, as vast as an aspen forest that is one continuous clone.

There are more than 275,000 known species of plants, and the vast majority reside on land. Be glad their ancient ancestors left the water. Without them, we humans and other land-dwelling animals never would have made it onto the evolutionary stage.

KEY CONCEPTS

1. Nearly all plants are multicelled photoautotrophs. Together with photosynthetic bacteria and protistans, they are the primary producers for nearly all communities.

2. Nearly all plants live on land, but their earliest ancestors were aquatic. Adaptations among different land-dwelling species include a waxy cuticle, root and shoot systems, and internal tissues for conducting water and solutes. They include special means of nourishing, protecting, and dispersing gametes and offspring.

3. Existing plants are bryophytes, seedless vascular plants, and seed-bearing vascular plants. The seed producers have been the most successful in radiating into drier environments.

25.1 CLASSIFICATION OF PLANTS

The plant kingdom includes 275,000 known species of photoautotrophs and a few heterotrophs. Most are **vascular plants**, with internal tissues that conduct and distribute water and solutes. These plants have roots, stems, and leaves, which are defined in part by the presence of vascular tissues. Fewer than 16,000 species are **bryophytes**, the "nonvascular" plants. Together with the photosynthetic bacteria and protistans, plants are the primary producers of organic compounds for nearly all communities.

This chapter surveys the major groups of existing plants. The liverworts, hornworts, and mosses are bryophytes. The whisk ferns, lycophytes, horsetails, and ferns are seedless vascular plants. The cycads, ginkgo, gnetophytes, and conifers are seed-bearing plants of a type called **gymnosperms**. A different type of seed-bearing plant, the **angiosperms**, also produces flowers. There are two classes of flowering plants, informally called the dicots and monocots.

Bryophytes, seedless vascular plants, and seed-bearing vascular plants constitute a kingdom of multicelled photoautotrophs, nearly all of which live on land.

25.2 EVOLUTIONARY TRENDS AMONG PLANTS

From the time their protistanlike ancestors arose until about 435 million years ago, plants evolved in the seas. Evolutionarily speaking, the pace picked up after that. Stalked species evolved along coasts and streams, in partnership with fungi. Within 60 million years, plants cloaked much of the land. Some long-term changes in structure and reproductive events help explain how the diversity came about.

Evolution of Roots, Stems, and Leaves

Underground structures started evolving among the first colonizers of land. In the lineages leading to vascular plants, these developed into **root systems**. Most root systems consist of underground, cylindrical absorptive structures that have a large surface area for rapidly taking up soil water and scarce mineral ions, and that often anchor the plant. Aboveground, other structures evolved into **shoot systems**. Among other things, shoot systems consist of stems and leaves, which function in the absorption of sunlight energy and carbon dioxide from the air.

The evolution of roots, stems, and leaves involved the development of pipelines through the plant. Such pipelines evolved as components of two vascular tissues—xylem and phloem. **Xylem** distributes water and dissolved ions through plant parts. **Phloem** distributes sugars and other photosynthetic products.

Extensive growth of stems and branches became possible when plants started producing **lignin**. This organic compound strengthens cell walls. Today, lignin-reinforced tissues structurally support plant parts that display the leaves and help increase the surface area that can intercept sunlight.

Life on land also depended on water conservation, which had not been a problem in most aquatic habitats. Stems and leaves became protected by a **cuticle**, a waxy surface covering that reduces water loss on hot, dry days. Numerous, tiny passageways called **stomata**

became the main route for absorbing carbon dioxide while controlling evaporative water loss. The next unit describes these tissue specializations.

From Haploid to Diploid Dominance

As early plants moved into higher, drier parts of the world, modifications accumulated in their life cycles. Think about the green algae. They spend most of their life producing and releasing gametes into the surrounding water. Actually, their gametes cannot get together *except* in liquid water. Figure 25.2*a* shows how you might diagram the life cycle of some green algae. The haploid (*n*) phase, which starts at meiosis and gamete formation, dominates the cycle. A diploid (2*n*) phase starts when gametes fuse at fertilization.

By contrast, the diploid phase dominates the life cycles of vascular plants (Figure 25.2*c*). Diploid dominance is an adaptation to land environments, most of which do not have unlimited supplies of water and nutrients. Long ago, through natural selection, complex sporophytes must have been favored. Recall that **sporophyte** means "spore-producing body." In plants, these form when a new zygote embarks on a course of mitotic cell divisions, which in time produce a large, multicelled body. A pine tree is an example. Such complex sporophytes have well-developed root systems. The young roots interact with mycorrhizal fungi, as described on page 386. This symbiotic relationship provides sporophytes with water and scarce nutrients, even in seasonally dry habitats.

For land plants, the haploid phase of the life cycle starts in reproductive parts of the sporophyte. There, haploid spores form by meiosis. Such spores divide by mitosis and give rise to gamete-producing structures called **gametophytes**. These multicelled, haploid game-

Figure 25.2 Comparison of life cycles for (**a**) some algae, (**b**) bryophytes, and (**c**) vascular plants. The diagrams highlight an evolutionary trend from haploid to diploid dominance during the colonization of land.

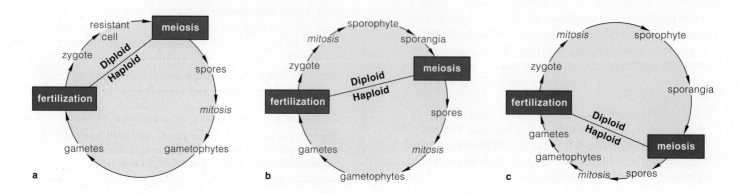

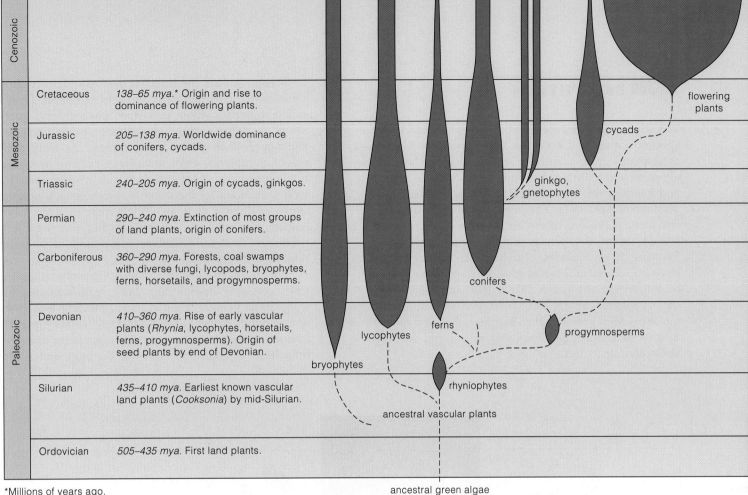

Era	Period	
Cenozoic		
Mesozoic	Cretaceous	*138–65 mya.* Origin and rise to dominance of flowering plants.
	Jurassic	*205–138 mya.* Worldwide dominance of conifers, cycads.
	Triassic	*240–205 mya.* Origin of cycads, ginkgos.
Paleozoic	Permian	*290–240 mya.* Extinction of most groups of land plants, origin of conifers.
	Carboniferous	*360–290 mya.* Forests, coal swamps with diverse fungi, lycopods, bryophytes, ferns, horsetails, and progymnosperms.
	Devonian	*410–360 mya.* Rise of early vascular plants (*Rhynia*, lycophytes, horsetails, ferns, progymnosperms). Origin of seed plants by end of Devonian.
	Silurian	*435–410 mya.* Earliest known vascular land plants (*Cooksonia*) by mid-Silurian.
	Ordovician	*505–435 mya.* First land plants.

*Millions of years ago.

Figure 25.3 Milestones in plant evolution.

tophytes nourish and protect the forthcoming generation. Thus, unlike algae, plants retain spores and gametophytes until environmental conditions favor dispersal and fertilization.

Evolution of Pollen and Seeds

Gymnosperms and angiosperms have radiated into nearly all high and dry habitats. The evolution of pollen grains and seeds were factors in their successful radiations. Both kinds of seed-bearing plants produce not one but two kinds of spores. This condition is called heterospory, as opposed to homospory. One kind of spore develops into **pollen grains**, which become mature, sperm-bearing male gametophytes. The other kind of spore develops into female gametophytes, the plant parts where eggs form and become fertilized. Pollen grains hitch rides on air currents or insects, birds, and so on. Unlike algae, they don't require liquid water to meet up with the eggs.

Also in seed-bearing plants, a parent sporophyte holds onto its female gametophytes. It nourishes and protects fertilized eggs as they develop into young embryos. The nutritive tissues, protective tissues, and the embryo itself constitute a **seed**. It's probably no coincidence that the dominant seed plants arose during Permian times, when climatic fluctuations were extreme (page 335). Seeds are packages that endure hostile conditions.

Putting the key points of this overview together,

1. The evolution of land plants involved structural adaptations to dry conditions. In different lineages, these adaptations included waxy cuticles, stomata, vascular tissues, and lignin-reinforced tissues.

2. Sporophytes with well-developed systems of roots and shoots (stems and leaves) came to dominate the life cycle of complex land plants. Such plants had the means to nourish and protect their spores and gametophytes through unfavorable conditions.

3. Some plants started producing two types of spores instead of one. This led to the evolution of male gametes specialized for dispersal without liquid water. It led to the evolution of seeds—embryos packaged with protective and nutritive tissues.

Before turning to the spectrum of diversity among these plants, take a look at Figure 25.3. You can use it as a map for the branching evolutionary roads.

The byrophyte lineage encompasses about 16,000 species of **mosses**, **liverworts**, and **hornworts**. Most of these plants grow in fully or seasonally moist habitats, although you will find some mosses growing in deserts and even on the excruciatingly cold, windswept high plateaus of Antarctica. The mosses especially are sensitive to air pollution. Where air is bad, mosses are often few or absent. Bryophytes are small plants, generally less than 20 centimeters (8 inches) tall. Although they have leaflike, stemlike, and rootlike parts, these do not contain xylem or phloem. Like lichens and some algae, the bryophytes can dry out, then revive after absorbing moisture. Most species have **rhizoids**, elongated cells or threads that attach gametophytes to soil and serve as absorptive structures.

Bryophytes are the simplest plants to display three features that emerged early in plant evolution. *First*, a cuticle prevents water loss from aboveground parts. *Second*, a cellular jacket around the sperm-producing and egg-producing parts holds in moisture. *Third*, of all plants, bryophytes alone have large gametophytes that hold onto sporophytes and do not depend on them for their nutrition. To the contrary, embryo sporophytes start developing in gametophyte tissues. Even when mature, a sporophyte remains attached to the gamete-producing body and derives some amount of nutritional support from it.

With 9,500 species, true mosses (Bryophyta) are the most common bryophytes. Their gametophytes are leafy. Some grow in tight clusters, forming low, cushiony mounds. Others show branched, feathery growth; in humid regions, masses of them often hang from tree branches. As Figure 25.4 shows, the eggs and sperm

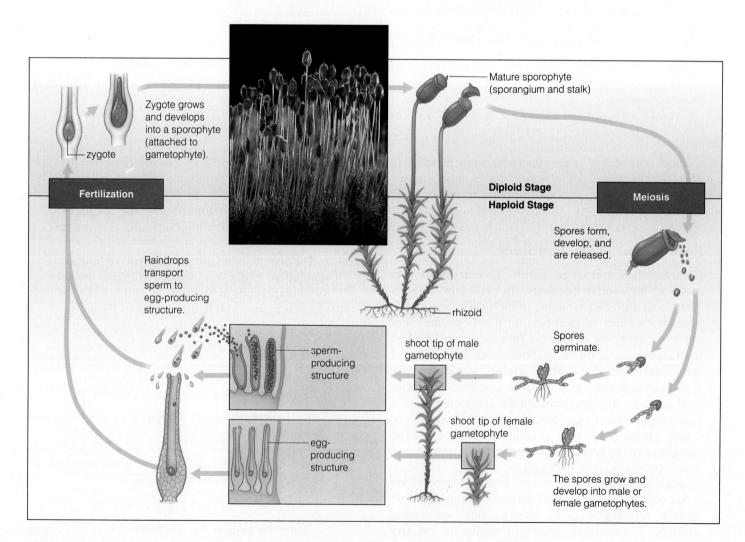

Figure 25.4 Life cycle of a moss (*Polytrichum*), a representative bryophyte. The sporophyte remains attached to and depends on the gametophyte for water and nutrients.

a

b

Figure 25.5 (**a**) From Alaska, a forest floor cloaked with a thick mat of peat—excessively moist, compressed organic matter that resists decomposition. (**b**) One of the peat mosses (*Sphagnum*). These gametophytes have several sporophytes attached to them. Along with grasses and other plants, such mosses have contributed to the formation of extensive peat bogs in cold and temperate regions. In Ireland and elsewhere, dried peat is used as a fuel source.

male gametophyte   female gametophyte   gemmae

a   b   c

Figure 25.6 *Marchantia*, a liverwort. Like other liverworts, it reproduces sexually. Unlike others, it produces (**a**) male and (**b**) female reproductive structures on separate plants. (**c**) *Marchantia* also can reproduce asexually by way of gemmae, multicelled vegetative bodies that develop in tiny cups on the plant body. Gemmae grow into new plants after splashing raindrops transport them to suitable sites.

develop at shoot tips, in **gametangia** (singular, gametangium, meaning "gamete vessel"). Sperm reach eggs by swimming through a film of water on the plant parts. After fertilization, zygotes give rise to sporophytes. Each sporophyte consists of a stalk and **sporangium** (plural, sporangia), a jacketed structure in which spores develop.

Figure 25.5 shows one of 350 species of peat mosses. These structurally simple mosses live in bogs, where turgor pressure keeps them erect. They soak up five times as much water as cotton does, owing to large, dead cells in their leaflike parts. Home gardeners use peat moss to increase the water-holding capacity as well as the acidity of soils.

So as not to slight the liverworts, Figure 25.6 shows *Marchantia*, a type that has some interesting ways of reproducing.

Bryophytes are rather unspecialized, nonvascular plants that require liquid water for fertilization. Their sporophytes develop in gametophyte tissues and remain attached to them, and so continue to receive nutrients.

25.4 SEEDLESS VASCULAR PLANTS

By the late Silurian, some 420 million years ago, seedless vascular plants were established. Diverse plants of this lineage flourished for the next 60 million years, then most kinds became extinct.

Cooksonia, one of the rhyniophytes, was among their earliest ancestors (Figure 25.7a). Although leafless, members of this genus resembled many existing vascular plants in internal structure. Some were pin-sized, and others were a few centimeters tall. *Psilophyton*, one of the trimerophytes, was taller and more complex. By Devonian times, the progymnosperms evolved; they may have been ancestral to seed-bearing plants. Figures 20.2c and 21.11 show a few fossils of these extinct plants.

Existing seedless vascular plants are commonly called **whisk ferns**, **lycophytes**, **horsetails**, and **ferns**. Like their ancestors, they differ from bryophytes in some important respects. First, the sporophyte doesn't remain attached to the gametophyte. Second, it has complex vascular tissues. Third, it is the larger, longer lived phase of the life cycle.

Most seedless vascular plants live in wet, humid regions. Their gametophytes have no vascular tissues for water transport, and their flagellated sperm require ample water to reach the eggs. The few species in deserts and other extreme habitats can reproduce sexually only during brief, seasonal pulses of heavy rains or summer thaws. Thus the whisk ferns, lycophytes, horsetails, and ferns are the "amphibians" of the plant kingdom. *They have not fully escaped the aquatic habitats of their ancestors.*

Whisk Ferns

Whisk ferns (Psilophyta), which are not true ferns, resemble whisk brooms. These common ornamental plants grow in moist tropical and subtropical regions, including Hawaii, Florida, Louisiana, Texas, and Puerto Rico. One genus, *Psilotum*, is unique among vascular plants. Its sporophytes have no roots and leaves. Its members have branching photosynthetic stems with scalelike projections, not leaves (Figure 25.7b). The stems contain xylem and phloem. Belowground are **rhizomes**. These short, branching, mostly horizontal stems function in absorption. Mycorrhizal fungi assist the rhizomes in this task.

Lycophytes

About 350 million years ago, lycophytes (Lycophyta) included tree-sized members of those swamp forests that eventually became transformed into the world's coal deposits (see the *Focus* essay). Today there are 1,000

Ancient Carbon Treasures

Between 360 and 280 million years ago, during the Carboniferous, swamp forests carpeted the wet lowlands of continents. Among the diverse species were the ancient ancestors of lycophytes, horsetails, ferns, and possibly the seed-bearing plants (Figure a). This was a period when sea levels rose and fell fifty times. When the seas moved out, swamp forests flourished. When the seas moved in, forest plants were submerged and became buried in sediments that protected them from decay. Gradually, the sediments compressed the saturated, undecayed remains into what we now call **peat**. As more sediments accumulated, increased heat and pressure made the peat even more compact. It became **coal** (Figure b).

Coal has a high percentage of carbon; it is energy-rich. It is one of our premier "fossil fuels." It took a fantastic amount of photosynthesis, burial, and compaction to form each major seam of coal in the earth. It has taken us only a few centuries to deplete much of the known coal deposits. Often you will hear about our annual "production rates" for coal or some other fossil fuel. But how much do we really produce each year? None. We simply *extract* it from the earth. Coal is a nonrenewable source of energy.

species. The most familiar are club mosses—tiny, inconspicuous members of communities in the Arctic, the tropics, and regions in between. Many types form mats on forest floors. Club mosses were once grouped in the same genus (*Lycopodium*), but as many as fifteen genera are now recognized.

Most club moss sporophytes have a branching rhizome that gives rise to vascularized roots and stems. All have tiny leaves. Sporophytes of some species also have nonphotosynthetic, cone-shaped clusters of leaves that contain spore sacs. Each cluster is a **strobilus** (plural, strobili). Figure 25.7c shows examples. Following dispersal, the spores germinate and develop into small, free-living gametophytes.

Members of one genus alone (*Selaginella*) are heterosporous. The two kinds of spores develop in the same cone-shaped cluster of leaves. In Texas, New Mexico, and Mexico, one species (*S. lepidophylla*) is commonly known as the resurrection plant.

a Reconstruction of a Carboniferous forest.

b Coal

a

Figure 25.7 (**a**) *Cooksonia*, the earliest known vascular plant, no more than a few centimeters tall. It probably grew in mud flats. Its upright, branching stems had a cuticle. Its spores were produced in sporangia at stem tips. Compare Figure 20.2*c*. (**b**) Sporophytes of a whisk fern (*Psilotum*), a seedless vascular plant. Pumpkin-shaped, spore-producing structures form at the ends of stubby branchlets. (**c**) Sporophytes of one of the lycophytes (*Lycopodium*).

b

c

a   b   c

Figure 25.8 (**a**) Vegetative shoots of *Equisetum*, the shape of which provides you with a clue to why someone thought "horsetails" would be a suitable name for these seedless vascular plants. (**b**) Nonphotosynthetic, fertile shoots of *Equisetum*. At the stem tips are strobili, the spore-bearing structures. (**c**) Closer look at a strobilus, showing the umbrella-shaped clusters of sporangia.

Horsetails

The seedless vascular plants called sphenophytes (Sphenophyta) barely squeaked through to the present. Gone are the tree-size calamites and other forms that flourished in ancient swamp forests. All that remain are fifteen species of a single genus (*Equisetum*). These are the horsetails. They may be the oldest of all existing plant genera; they have scarcely changed over the past 300 million years.

Horsetails grow in muds of streambanks and disturbed habitats, such as vacant lots, roadsides, and beds of railroad tracks. When their spores land on nutrient-rich mud, they grow into free-living, pinhead-sized gametophytes. Sporophytes usually have rhizomes, hollow photosynthetic stems, and scalelike leaves. In the stem tissue, clusters of xylem and phloem are arranged as a ring. Silica-reinforced ribs support stems and give them a gritty quality, like sandpaper. Pioneers of the American West used horsetails to scrub their cooking pots and pans.

Figure 25.8*a* shows the vegetative, photosynthetic stems of one horsetail species. Figure 25.8*b* shows the fertile stems, which have strobili at the tips. In the strobili, spores form along the edge of umbrella-shaped, spore-bearing branches (Figure 25.8*c*). Air currents disperse them.

Ferns

With 12,000 or so species, the ferns (Pterophyta) are the largest and most diverse group of plants. All but about 380 species are native to the tropics, but they are popular houseplants all over the world. Their size range is stunning. Some floating species are less than 1 centimeter across. Some tropical tree ferns are 25 meters (82 feet) tall. One climbing fern has a modified leaf stalk about 30 meters long.

Most species have underground, vascularized rhizomes that give rise to the leaves and roots. Exceptions include tropical tree ferns and epiphytes. ("Epiphyte" refers to any aerial plant that grows on tree trunks or branches.) Young fern leaves are coiled, in the shape of a "fiddlehead." The mature leaves (fronds) commonly are divided into leaflets.

You may have noticed rust-colored patches on the lower surface of many fern fronds. Each patch is a **sorus** (plural, sori), which is a cluster of sporangia. At dispersal time, the sporangia snap open, causing the spores to catapult through the air. Each germinating spore develops into a small gametophyte, such as the green, heart-shaped type shown in Figure 25.9.

We now leave the seedless vascular plants. Keep these summary points in mind:

Whisk ferns, lycophytes, horsetails, and ferns have sporophytes adapted to land, but they have not entirely escaped their aquatic ancestry.

The life cycle of these seedless vascular plants can be completed only when their flagellated sperm have ample water to reach the eggs.

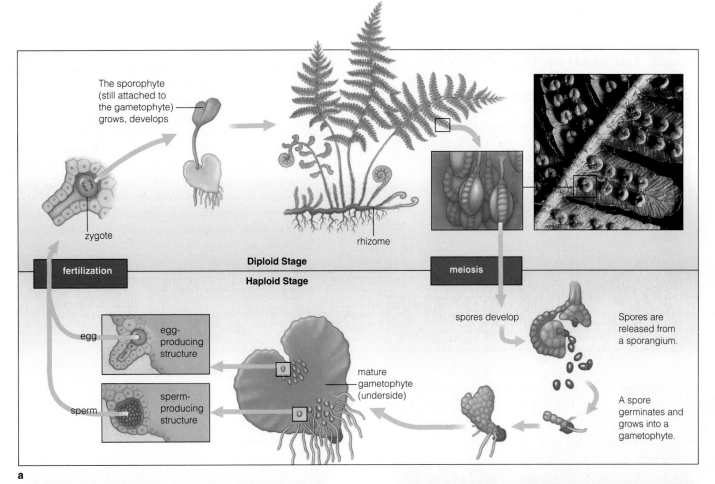

The sporophyte (still attached to the gametophyte) grows, develops

zygote

rhizome

Diploid Stage

fertilization

Haploid Stage

meiosis

spores develop

Spores are released from a sporangium.

egg

egg-producing structure

sperm

sperm-producing structure

mature gametophyte (underside)

A spore germinates and grows into a gametophyte.

a

b

c

Figure 25.9 (**a**) Life cycle of a fern. (**b**) Ferns in a moist habitat in Indiana. (**c**) Tree ferns in a temperate rain forest of Tasmania.

THE SEED-BEARING PLANTS

As indicated earlier, gymnosperms and angiosperms are seed-bearing plants that arose in Devonian times. In terms of diversity, numbers, and distribution, they have become the most successful plants, owing to an enormous advantage they have over seedless types. Air cur-rents or insects carry their sperm—packaged in pollen grains—to eggs. *Thus, seed-bearing plants have escaped dependency on free water for fertilization.* In addition, their sporophytes produce reproductive structures called ovules. An **ovule** contains the egg-producing female gametophyte, surrounded by nutritive tissue and a jacket of cell layers (Figure 25.10). As a fertilized egg develops, the outer layers form a coat. The mature, coat-enclosed ovule is the seed. *Thus, embryo sporophytes are protected by a seed coat during their dispersal, and they can tap stored nutrients at the critical time of germination, before their roots and shoots become fully functional.*

Figure 25.10 Life cycle of a gymnosperm (ponderosa pine).

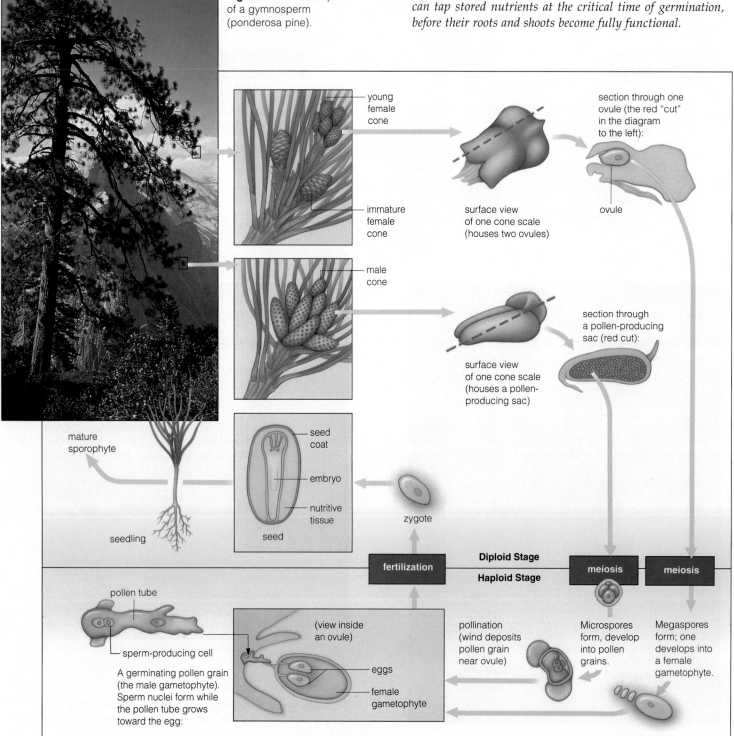

young female cone

immature female cone

male cone

section through one ovule (the red "cut" in the diagram to the left):

ovule

surface view of one cone scale (houses two ovules)

section through a pollen-producing sac (red cut):

surface view of one cone scale (houses a pollen-producing sac)

mature sporophyte

seedling

seed coat

embryo

nutritive tissue

seed

zygote

fertilization

Diploid Stage

Haploid Stage

meiosis

meiosis

pollen tube

sperm-producing cell

A germinating pollen grain (the male gametophyte). Sperm nuclei form while the pollen tube grows toward the egg:

(view inside an ovule)

eggs

female gametophyte

pollination (wind deposits pollen grain near ovule)

Microspores form, develop into pollen grains.

Megaspores form; one develops into a female gametophyte.

25.6 GYMNOSPERMS

"Gymnosperm" is a word derived from the Greek *gymnos* (meaning naked) and *sperma* (which is taken to mean seed). The name refers to the exposed location of their ovules and seeds. They are borne on surfaces of spore-producing reproductive structures. Conifers are familiar members of the lineage. Less well known are the cycads, ginkgos, and gnetophytes.

Conifers

Conifers (Coniferophyta) are "evergreen," woody trees and shrubs with needlelike or scalelike leaves. Most shed old leaves throughout the year but retain enough to distinguish them from "deciduous" species (which shed all their leaves in the fall). Conifers include diverse pines, spruces, firs, hemlocks, junipers, cypresses, and redwoods. Nearly all of their **cones** are clusters of modified leaves in which spore-producing structures develop. As the Figure 25.10 example suggests, conifers are heterosporous. Their **microspores** form in male cones and develop into pollen grains. Their **megaspores** form in the shelflike scales of female cones and develop into the female gametophytes.

Each spring, millions of pollen grains drift off male cones, and some land on ovules. The arrival of pollen on female reproductive parts is called **pollination**. Now pollen grains germinate. Each develops into a long, tubelike structure that grows down through the female tissues. This pollen tube transports sperm toward female gametophytes. In many species of conifers, fertilization does not occur until months to as much as a year after pollination.

Recall, from Chapter 21, that conifers radiated into many land environments during the Mesozoic. Their leisurely reproductive pace may have put them at a competitive disadvantage when flowering plants began their spectacular adaptive radiations.

Coniferous forests still cloak many regions, including the far north, high altitudes, and parts of the Southern Hemisphere. However, besides having to compete with flowering plants for resources, conifers are now vulnerable to global **deforestation**—the removal of all trees from vast tracts of land, as by clear cutting (Figure 25.11). Conifers are the premier source of lumber, furniture, paper, and many other products. We return to this topic in Chapter 50.

Conifers are woody, mostly evergreen shrubs and trees that produce pollen grains, bear seeds on cone scales, and are key sources of lumber, paper, and other manufactured products.

Figure 25.11 From North America, a small sampling of the rampant deforestation now under way throughout the world. (**a,b**) From the Nahmint Valley of British Columbia, an ancient forest before and after logging. From Alaska (**c**) and Washington (**d**), denuded mountains. (**e**) From America's heartland, logged-over acreage in Arkansas. From the eastern seaboard, a denuded piece of North Carolina (**f**). These are not isolated examples. In the early 1980s, 400 million board feet of timber were being cut annually in Washington's Olympic Peninsula. In Arkansas, approximately one-third of the Ouachita National Forest was clear-cut. Its once-diverse forest communities have been replaced by "tree farms" of a single species of pine. Throughout the world, large tracts that were deforested years ago still show no signs of recovery.

a

b

c

d

Lesser Known Gymnosperms

Figure 25.12 shows a few relatives of the conifers. The **cycads** (Cycadophyta) flourished with dinosaurs during the Mesozoic. About 100 species have survived to the present in tropical and subtropical regions. They superficially resemble palm trees (which are flowering plants). Cycads have massive cones that bear either pollen or ovules. Air currents or insects transfer pollen from "male" plants to developing seeds on "female" plants. In parts of Asia, people eat cycad seeds and a starchy flour made from cycad trunks, but only after they rinse out the plant's poisonous alkaloids.

Ginkgos (Ginkgophyta) were diverse during dinosaur times. The maidenhair tree, *Ginkgo biloba*, is the only surviving species. Several thousand years ago, ginkgo trees were planted in cultivated grounds around temples in China. Then the natural populations almost became extinct. This is puzzling, for ginkgos seem hardier than many other trees. Perhaps, as human populations grew, ginkgos were cut for firewood. Male ginkgo trees are now planted in cities. They have attractive, fan-shaped leaves and resistance to insects, disease, and air pollutants. Female trees are not favored; their fleshy-coated seeds produce an awful stench when stepped on.

The seventy species of **gnetophytes** (Gnetophyta) are divided into three genera: *Gnetum*, *Ephedra*, and *Welwitschia*. Thirty or so species of *Gnetum*, either trees or leathery leafed vines, grow in humid tropical regions. About thirty-five species of *Ephedra* grow in deserts and other arid regions around the world, including parts of California. Of all gymnosperms, *Welwitschia* is the most bizarre. This seed-producing plant grows in hot deserts of south and west Africa. The bulk of the plant is a deep-reaching taproot. The only exposed part, a woody disk-shaped stem, bears cone-shaped strobili and leaves. The plant never produces more than two strap-shaped leaves, which split lengthwise repeatedly as the plant ages.

Cycads, ginkgo, and gnetophytes bear seeds on exposed surfaces of spore-bearing structures, but they differ greatly from one another and from conifers in other features.

e

Figure 25.12 Lesser known gymnosperms. (**a**) Massive female cone of a cycad (*Encephalartos ferox*). (**b**) *Ephedra viridis*, growing in California. (**c**) Ginkgo trees. (**d**) Fleshy-coated ginkgo seeds. (**e**) A mature specimen of *Welwitschia mirabilis*. The inset shows the appearance of its female cones.

25.7 ANGIOSPERMS

Flowering Plant Diversity

Angiosperms, the flower-producing seed plants, have dominated the land for more than 100 million years. "Angiosperm" is derived from the Greek *angeion* (a vessel) and *sperma*. The "vessel" refers to tissues that surround and protect ovules and developing seeds. We know of about 235,000 existing species and are discovering new ones almost weekly in previously unexplored tropical forests. Untold numbers may soon be extinct, owing to deforestation.

Angiosperms are the most diverse of all plants. They live in mountains and deserts, wetlands, and seawater. They range in size from duckweeds (about a millimeter long) to *Eucalyptus* trees more than 100 meters tall. Most are free-living and photosynthetic, but there are exceptions (Figure 25.13). Some, including mistletoe, are parasites on other plants.

In terms of sheer diversity, numbers, and distribution, the angiosperms are the most successful plants. These seed-bearing plants alone produce flowers.

e

f

a

c

b

g

d

Figure 25.13 Sampling of flowering plant diversity. (**a**) Plants of alpine tundra. (**b**) Water lily (*Nymphea*). (**c**) Indian pipe (*Monotropa uniflora*), a nonphotosynthetic plant that obtains nutrients by way of mycorrhizal fungi that are partners with photosynthetic plants. (**d**) Dwarf mistletoe (*Arceuthobium*), a parasitic plant that limits the growth of forest trees in the western United States. Highly prized species: (**e**) *Theobroma cacao* fruits, source of cocoa beans; (**f**) an exotic orchid plant; and (**g**) sugarcane plants growing in Hawaii.

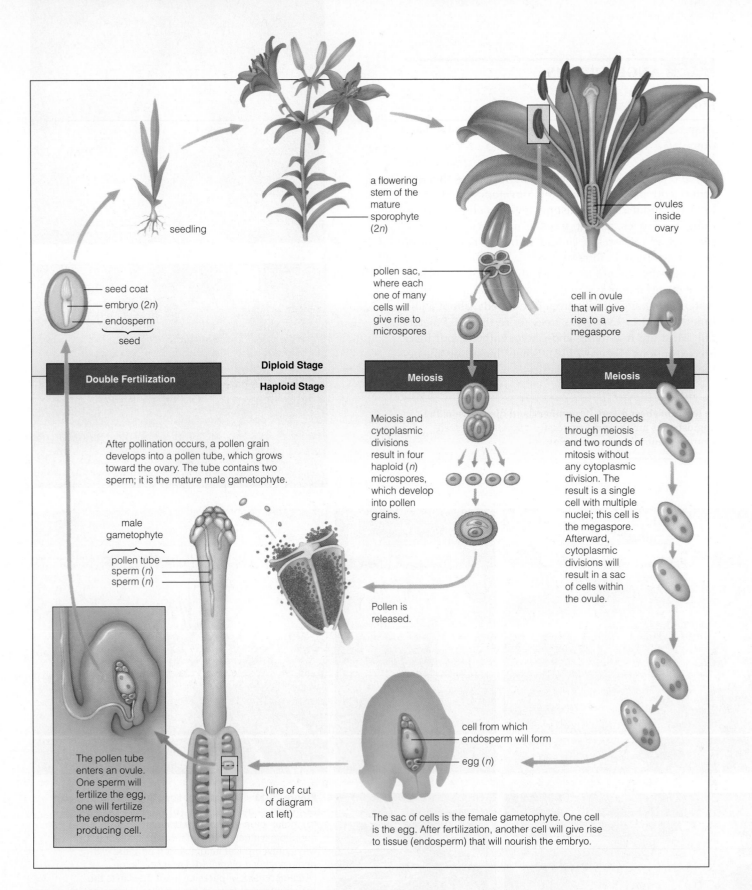

seedling

a flowering
stem of the
mature
sporophyte
(2n)

ovules
inside
ovary

seed coat
embryo (2n)
endosperm
seed

pollen sac,
where each
one of many
cells will
give rise to
microspores

cell in ovule
that will give
rise to a
megaspore

Diploid Stage

Double Fertilization

Haploid Stage

Meiosis

Meiosis

After pollination occurs, a pollen grain
develops into a pollen tube, which grows
toward the ovary. The tube contains two
sperm; it is the mature male gametophyte.

Meiosis and
cytoplasmic
divisions
result in four
haploid (n)
microspores,
which develop
into pollen
grains.

The cell proceeds
through meiosis
and two rounds of
mitosis without
any cytoplasmic
division. The
result is a single
cell with multiple
nuclei; this cell is
the megaspore.
Afterward,
cytoplasmic
divisions will
result in a sac
of cells within
the ovule.

male
gametophyte

pollen tube
sperm (n)
sperm (n)

Pollen is
released.

The pollen tube
enters an ovule.
One sperm will
fertilize the egg,
one will fertilize
the endosperm-
producing cell.

(line of cut
of diagram
at left)

cell from which
endosperm will form

egg (n)

The sac of cells is the female gametophyte. One cell
is the egg. After fertilization, another cell will give rise
to tissue (endosperm) that will nourish the embryo.

Figure 25.14 Life cycle of a monocot (*Lilium*). "Double" fertilization is a distinctive feature. The male
gametophyte delivers two sperm to an ovule. One sperm fertilizes the egg, and the other fertilizes a cell
that gives rise to a tissue (endosperm) that will nourish the forthcoming embryo. Figure 31.5 provides a
closer look at flowering plant life cycles, using a dicot as the example.

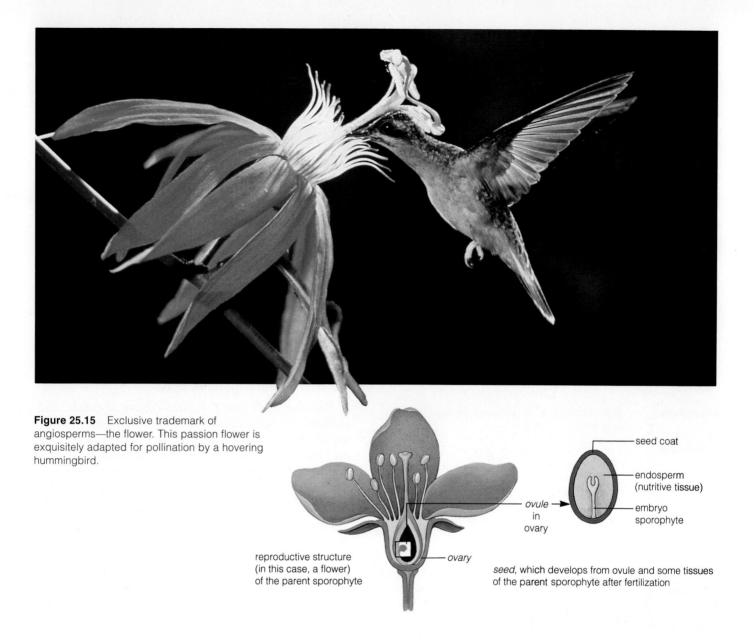

Figure 25.15 Exclusive trademark of angiosperms—the flower. This passion flower is exquisitely adapted for pollination by a hovering hummingbird.

reproductive structure
(in this case, a flower)
of the parent sporophyte

ovule in ovary

ovary

seed coat

endosperm
(nutritive tissue)

embryo
sporophyte

seed, which develops from ovule and some tissues of the parent sporophyte after fertilization

Classes of Flowering Plants

There are two classes of flowering plants—the Dicotyle-donae and Monocotyledonae (or, informally, dicots and monocots). Among the 200,000 species of dicots are most shrubs and trees, most nonwoody (herbaceous) plants, cacti, and water lilies. The 65,000 species of monocots include palms, lilies, and orchids. They also include grasses, such as wheat, corn, rice, rye, sugarcane, barley, and other major crop plants.

Key Aspects of the Life Cycles

Figure 25.14 shows a monocot life cycle. A typical dicot life cycle is described in detail in the next unit, which focuses on the structure and function of flowering plants. As with gymnosperms, a large diploid sporophyte dominates the cycle, it retains and nourishes gametophytes, and its sperm travel in pollen grains.

Unlike gymnosperms, flowering plants provide their seeds with a unique, nutrient-packed tissue (endosperm). They also package seeds in **fruits**, which aid in protection and dispersal. Figure 25.13e shows one of many diverse kinds of fruits. Generally, the flowers that develop during the life cycle have coevolved with insects, bats, birds, and other animals that assist in pollination (Figure 25.15). These **pollinators** go after nectar or pollen in flowers. As they do, they transfer pollen grains to female reproductive parts. The "recruitment" of animals probably was a key factor in the rise of angiosperms.

Flowering plants produce flowers, endosperm-rich seeds, and fruits. Most species have coevolved in mutually beneficial ways with animal pollinators.

SUMMARY

1. Nearly all members of the plant kingdom are multi-celled, photoautotrophic species that live on land. They probably evolved from multicelled green algae more than 400 million years ago. Table 25.1 summarizes the major phyla (divisions) of existing plants.

Table 25.1 Comparison of Major Plant Groups*	
Nonvascular land plants. Fertilization requires free water. Haploid dominance. Cuticle, stomata present in some.	
Bryophytes	16,000 species. Moist, humid habitats.
Seedless vascular plants. Fertilization requires free water. Diploid dominance. Cuticle, stomata present.	
Lycophytes	1,000 species with simple leaves. Mostly wet or shady habitats.
Horsetails	15 species of single genus. Swamps, disturbed habitats.
Ferns	12,000 species. Wet, humid habitats in mostly tropical, temperate regions.
Vascular plants with "naked seeds" (gymnosperms). Diploid dominance. Cuticle, stomata present.	
Conifers	550 species, mostly evergreen, woody trees and shrubs having pollen- and seed-bearing cones. Widespread distribution.
Cycads	100 slow-growing species. Tropics, subtropics.
Ginkgo	1 species, a tree with fleshy-coated seeds.
Gnetophytes	70 species, limited distribution in deserts and tropics.
Vascular plants with flowers and protected seeds (angiosperms). Diploid dominance. Cuticle, stomata present.	
Flowering plants:	
Monocots	65,000 species. Floral parts often arranged in threes or multiples of three; one seed leaf; parallel leaf veins common.
Dicots	Nearly 200,000 species. Floral parts often arranged in fours, fives, or multiples of these; two seed leaves; net-veined leaves common.

*More than 275,000 known species total.

2. As summarized in Table 25.2 and listed below, several trends in plant evolution can be identified by comparing the characteristics of different lineages:

a. Structural adaptations to dry conditions, especially the development of the vascular tissues called xylem and phloem.

b. A shift from haploid to diploid dominance. Complex sporophytes evolved that could hold onto, nourish, and protect spores and gametophytes.

c. A shift from the production of one kind to two kinds of spores (homospory to heterospory). Among gymnosperms and angiosperms (flowering plants), this led to the evolution of pollen grains and seeds.

3. Existing nonvascular land plants (those without well-developed xylem and phloem) include the bryophytes—the mosses, liverworts, and hornworts. Existing seedless vascular land plants include the whisk ferns, lycophytes, horsetails, and ferns. In all cases, their flagellated sperm require ample water, so that they can swim to and fertilize the eggs.

4. Vascular land plants typically have a cuticle and stomata that help control water loss. They have sporangia (protective tissue layers around their spores). They also have ovules or similar structures with protective and nutritive tissue layers around their gametophytes. The embryo sporophyte begins its development *within* gametophyte tissues.

5. Gymnosperms and angiosperms are seed-bearing vascular plants. The evolution of pollen grains freed them from dependence on water for fertilization. Their seeds are efficient means of dispersing the new generation and helping it through hostile conditions. Pollen grains and seeds were key adaptations underlying the move into high, dry habitats.

6. The female gametophytes of gymnosperms and angiosperms are attached to and protected by the

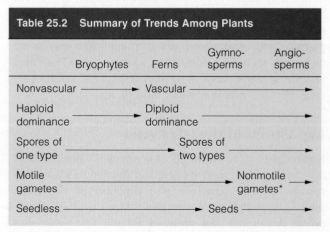

Table 25.2 Summary of Trends Among Plants			
	Bryophytes	Ferns	Gymno-sperms	Angio-sperms
Nonvascular ——→ Vascular ———————————————————————→			
Haploid dominance ——→ Diploid dominance ————————————→			
Spores of one type ———→ Spores of two types ————————————→			
Motile gametes ————————————————→ Nonmotile gametes* ————→			
Seedless ————————————————→ Seeds ————————————→			

*Require pollination by wind, insects, etc.

sporophyte. With its root and shoot systems, the sporophyte is well adapted to conditions on dry land.

7. Angiosperms alone produce flowers. Their seeds contain a distinctive nutritive tissue (endosperm) and are often surrounded by fruits, which aid in dispersal. Most flowering plants have coevolved with animal pollinators, which enhance the efficiency of transferring pollen grains to female reproductive parts.

Review Questions

1. Describe the evolutionary trends among plants that figured in the invasion of land. *392–393*

2. What are some differences between bryophytes and the vascular plants? *392, 394–395, 396*

3. How does the life cycle of a gymnosperm (such as pine) differ from that of an angiosperm (such as a lily)? *400, 404*

Self-Quiz *(Answers in Appendix IV)*

1. Which of the following was *not* a major trend in the evolution of complex land plants?
   a. from nonvascular to vascular plant bodies
   b. from sporophyte to gametophyte dominance
   c. from one spore type to two specialized types
   d. away from reliance on free-standing water for fertilization
   e. among some lineages, reliance on pollinators, seed formation, and seed dispersal mechanisms

2. Which of the following statements is *not* true?
   a. Monocots and dicots are two classes of angiosperms.
   b. Bryophytes are nonvascular plants.
   c. Lycophytes, horsetails, ferns, gymnosperms, and angiosperms are vascular plants.
   d. Horsetails and gymnosperms are the simplest vascular plants.

3. Of all land plants, bryophytes alone have independent _____ and attached, dependent _____ .
   a. sporophytes; gametophytes
   b. gametophytes; sporophytes
   c. rhizoids; zygotes
   d. rhizoids; sporangia with stalks

4. Psilophytes, lycophytes, horsetails, and ferns are _____ .
   a. multicelled aquatic plants
   b. nonvascular seed plants
   c. seedless vascular plants
   d. seed-bearing vascular plants

5. Gymnosperms and angiosperms are _____ .
   a. multicelled aquatic plants
   b. nonvascular seed plants
   c. seedless vascular plants
   d. seed-bearing vascular plants

6. Which does *not* apply to both gymnosperms and angiosperms?
   a. vascular tissues   c. single spore type
   b. diploid dominance   d. all of the above

7. A seed is _____ .
   a. a female gametophyte   c. a mature pollen tube
   b. a mature ovule   d. an immature embryo

8. Which does *not* apply to both gymnosperms and angiosperms?
   a. produce pollen grains and seeds
   b. sporophyte well adapted to conditions on dry land
   c. female gametophytes are attached to and protected by the sporophyte
   d. produce flowers

9. Which plant's life cycle shows haploid dominance?
   a. conifer   d. lycophyte
   b. bryophyte   e. monocot
   c. horsetail

10. Match the terms appropriately.
    _____ gymnosperms   a. produces haploid gametes
    _____ sporophyte   b. control water loss
    _____ seedless vascular c. "naked" seeds
          plants   d. protect, disperse embryo
    _____ ovary   sporophyte
    _____ bryophytes   e. produces haploid spores
    _____ gametophyte   f. nonvascular land plants
    _____ stomata   g. lycophytes
    _____ angiosperm seed   h. usually a fruit at maturity

Selected Key Terms

angiosperm *391*
bryophyte *391*
coal *396*
cone *401*
conifer *401*
cuticle *392*
cycad *402*
deforestation *401*
fern *396*
fruit *405*
gametangia *395*
gametophyte *392*
ginkgo *402*
gnetophyte *402*
gymnosperm *391*
hornwort *394*
horsetail *396*
lignin *392*
liverwort *394*
lycophyte *396*
megaspore *401*

microspore *401*
moss *394*
ovule *400*
peat *396*
phloem *392*
pollen grain *393*
pollination *401*
pollinator *405*
rhizoid *394*
rhizome *396*
root system *392*
seed *393*
shoot system *392*
sorus *398*
sporangium *395*
sporophyte *392*
stomata *392*
strobilus *396*
vascular plant *391*
whisk fern *396*
xylem *392*

Readings

Bold, H., and J. LaClaire. 1987. *The Plant Kingdom.* Fifth edition. Englewood Cliffs, New Jersey: Prentice-Hall. Paperback.

Gensel, P., and H. Andrews. 1987. "The Evolution of Early Land Plants." *American Scientist* 75: 478–489.

Raven, P., R. Evert, and S. Eichhorn. 1986. *Biology of Plants.* Fourth edition. New York: Worth. Lavishly illustrated.

26 ANIMALS: THE INVERTEBRATES

Animals of the Burgess Shale

It happened in Cambrian times, deep in a submerged basin between a massive reef and the coast of an early continent. Protected from ocean currents, great banks of sediments and mud had formed in the basin. About 500 feet below the water's surface, the waters were dimly lit, oxygenated, and clear. Tiny marine animals flourished in, on, and above muddy sediments piled against the steep flank of the reef.

Like castles built from wet sand along a seashore, their home was unstable. About 530 million years ago, part of the bank slumped suddenly (Figure 26.1). The underwater avalanche deposited sediments over an entire community, and these prevented scavengers from reaching and obliterating all traces of the dead animals. Muddy silt gradually rained down on the natural tomb. In time, through increased pressure and chemical changes, the sediments became compacted into finely stratified shale. And the soft parts of the flattened carcasses of animals became shimmering films of calcium aluminosilicate.

All the while, imperceptibly, the continents were on the move. By the dawn of the Cenozoic, a crustal plate under the Pacific Ocean was plowing under the western edge of the North American plate (page 341). The mountain ranges of western Canada were forming.

By the year 1909, the fossils were high up in the mountains that run along the eastern border of British Columbia. In that year a lucky fossil hunter, Charles Wolcott, tripped over a chunk of shale on a mountain path. The shale layers split apart easily, and Wolcott made the discovery of a lifetime. The Burgess Shale animals had come to light.

Wolcott and others soon recovered fossils of more than 120 species of animals. Some forms resembled existing species. Others were bizarre evolutionary experiments that apparently led nowhere. *Opabinia* was one of these. About as long as a tube of lipstick, *Opabinia* had five eyes perched on its head and a grasping organ that possibly was used in prey capture (Figure 26.1c). The smaller *Hallucigenia* sported seven pairs

Figure 26.1 (**a**) Reconstruction of a few marine organisms that ultimately became fossilized in the Burgess Shale. (**b**) Underwater slide off the coast of Baja California, reminiscent of the slumping that entombed these organisms. (**c**) Two Burgess Shale fossils: the five-eyed *Opabinia*, with a grasping organ on its head, and *Hallucigenia*, with multiple spines and soft protuberances.

a

b

of sharp spines on one side and seven soft organs on the other. For a long time, no one figured out which side was up.

The Burgess Shale animals are testimony to a great adaptive radiation, which may have been triggered by the breakup of a supercontinent just before the Cambrian era. Think of the adaptive zones that must have opened up in the waters along the vast new coastlines!

Yet where are signs of the *first* animals—the ones predating the Burgess Shale species? Here the fossil record is a closed book. One thing is clear, however. Animals representing every major animal phylum had evolved before the time when the Burgess Shale animals were buried. Some of these were on the evolutionary roads that led to modern sponges, cnidarians, flatworms, roundworms, mollusks, annelids, arthropods, echinoderms, and other invertebrates. Others were animals with muscles arrayed in a zigzag pattern and a stiffened rod running partway down their back. They were on or near the evolutionary road leading to vertebrates—to humans and all other animals with backbones. These are the roads traced in this chapter and the next.

KEY CONCEPTS

1. Animals are multicelled, motile heterotrophs that pass through embryonic stages during their life cycle.

2. There are probably more than 2 million existing species of animals. Fewer than 50,000 species are vertebrates (animals with a backbone). The rest are invertebrates (animals with no backbone).

3. The invertebrates described in this chapter range from *Trichoplax* and the sponges to cnidarians, comb jellies, flatworms, roundworms, ribbon worms, rotifers, mollusks, annelids, arthropods, and echinoderms.

4. Several trends occurred in animal evolution. These are evident through comparisons of the body plans of existing animals, in conjunction with the fossil record.

5. The most revealing aspects of an animal's body plan are its type of symmetry, gut, and cavity (if any) between the gut and body wall; whether it has a distinct head end; and whether it is divided into a series of segments.

6. Sponges have no body symmetry or tissue organization. Cnidarians are radial animals with some organized tissues. Nearly all other invertebrates are bilateral animals with well-developed tissues, organs, and organ systems.

7. In terms of numbers and distribution, insects are the most successful invertebrates. These arthropods have a hardened exoskeleton, specialized segments, and jointed appendages. Many have efficient respiratory and nervous systems and a division of labor in the life cycle.

c

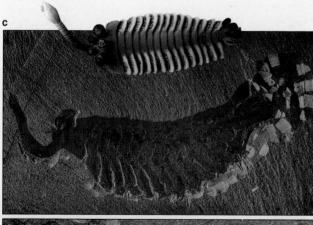

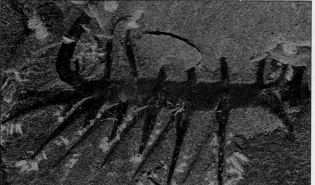

26.1 OVERVIEW OF THE ANIMAL KINGDOM

General Characteristics of Animals

Generally speaking, an **animal** is defined by the following characteristics:

1. Animals are multicelled. In most groups, cells form tissues that are arranged as organs and organ systems.

2. Animals are heterotrophs that must obtain carbon and energy by eating other organisms or absorbing nutrients from them.

3. Animals require oxygen (for aerobic respiration).

4. Animals reproduce sexually and, in many cases, asexually. Most are diploid organisms.

5. Animal life cycles include a period of embryonic development. In brief, cell divisions transform the zygote into a multicelled embryo. The embryo's cells become arranged as primary tissue layers: **ectoderm**, **endoderm**, and, in most species, **mesoderm**. These give rise to the adult's tissues and organs, as described on page 756.

6. Most animals are motile during at least part of the life cycle.

Diversity in Body Plans

The most familiar animals—mammals, birds, reptiles, amphibians, fishes—are **vertebrates**, the only animals with a "backbone." And yet, of probably more than 2 million species of animals, fewer than 50,000 are vertebrates! The **invertebrates** (animals whose ancestors evolved before backbones did) are lesser known, yet spectacularly diverse species.

Table 26.1 lists the groups of animals described in this chapter. Their shared characteristics arose early in time, before divergences from a common ancestor gave rise to separate lineages. Later, morphological differences accumulated among the lineages, and they were the foundation for today's bewildering diversity.

How can we get a conceptual handle on animals as different as flatworms, hummingbirds, platypuses, humans, and giraffes? We can compare similarities and differences with respect to five basic features. These are body symmetry, cephalization, type of gut, type of body cavity, and segmentation.

Body Symmetry and Cephalization Nearly all animals are radial *or* bilateral. Animals with **radial**

Table 26.1 Animal Phyla Described in This Chapter

Phylum	Some Representatives	Number of Known Species
Placozoa (*Trichoplax*)	Simplest animal; like a tiny plate, but with tissue layers	1
Porifera (poriferans)	Sponges	8,000
Cnidaria (cnidarians)	Hydrozoans, jellyfishes, corals, sea anemones	11,000
Ctenophora (comb jellies)	Radial animals with comblike structures of modified cilia	100
Platyhelminthes (flatworms)	Turbellarians, flukes, tapeworms	15,000
Nemertea (ribbon worms)	Proboscis-equipped worms closely related to flatworms	800
Nematoda (roundworms)	Pinworms, hookworms	20,000
Rotifera (rotifers)	Species with crown of cilia	1,800
Mollusca (mollusks)	Snails, slugs, clams, squids, octopuses	110,000
Annelida (segmented worms)	Leeches, earthworms, polychaetes	15,000
Arthropoda (arthropods)	Crustaceans, spiders, insects	1,000,000+
Echinodermata (echinoderms)	Sea stars, sea urchins	6,000
Chordata (chordates)	Invertebrate chordates: Tunicates, lancelets	2,100
	Vertebrates: Fishes	21,000
	Amphibians	3,900
	Reptiles	7,000
	Birds	8,600
	Mammals	4,500

symmetry have body parts arranged regularly around a central axis, like spokes of a bike wheel. Thus a cut down the center of a hydra (Figure 26.2*a*) divides it into equal halves; another cut at right angles to the first divides it into equal quarters. Radial animals are aquatic. With their body plan, they can respond to food suspended in the water all around them.

Animals with **bilateral symmetry** have right and left halves that are mirror images of each other. Most of these animals have an *anterior* end (head) and an opposite, *posterior* end. They have a *dorsal* surface (a back) and an opposite, *ventral* surface. Figure 26.2*b* shows this body plan, which evolved after forward-creeping animals emerged. Most often, their forward end was the first to encounter food and other stimuli. So we can

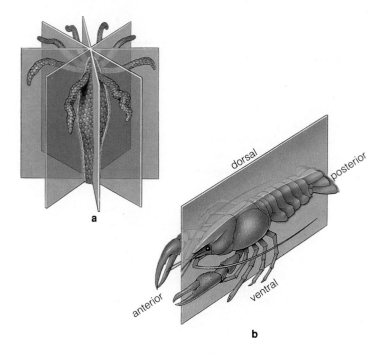

Figure 26.2 (**a**) Radial symmetry of a hydra. (**b**) Bilateral symmetry of a crayfish.

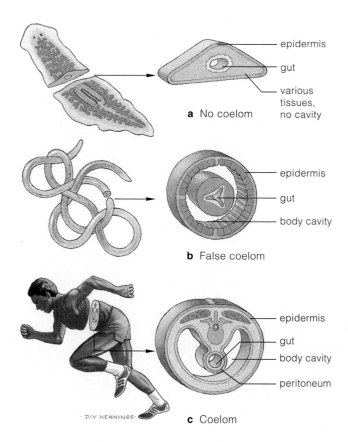

a No coelom

b False coelom

c Coelom

Figure 26.3 Type of body cavity (if any) in animals.

imagine there was strong selection for **cephalization**. By this evolutionary process, sensory structures and nerve cells became concentrated in a head. Bilateral body plans and cephalization evolved together. Their joint evolution resulted in *pairs* of muscles, sensory structures, nerves, and brain regions. We will return to this topic on page 576.

Type of Gut A **gut** is a region where food is digested and absorbed. Saclike guts have only one opening (a mouth) for taking in food and expelling residues. Other guts are part of tubelike systems with openings at both ends (mouth and anus). These "complete" digestive systems are more efficient. Their different regions have specialized functions, such as preparing, digesting, and storing food. The evolution of such systems helped pave the way for increases in body size and activity.

Body Cavities A body cavity separates the gut and body wall of most bilateral animals (Figure 26.3). One type of cavity, a **coelom**, has a distinctive lining called a peritoneum. The lining also covers organs in the coelom and helps hold them in place. You have a coelom. A sheetlike muscle divides it into an upper, *thoracic* cavity and a lower, *abdominal* cavity. Your thoracic cavity holds a heart and lungs; your abdominal cavity holds a stomach, intestines, and other organs.

Some types of worms do not have a body cavity. The region between their gut and body wall is packed with tissues. Other types have a "false coelom" (pseudocoel)—a body cavity but no peritoneum. A true coelom was a major step in the evolution of animals that were larger and more complex than worms. By cushioning and protecting organs, the coelom favored increases in their size and activity.

Segmentation Earthworms and other kinds of "segmented" animals consist of a series of body units that may or may not be similar to one another. Most earthworm segments have the same appearance on the outside. Insect segments are organized in three body regions (head, thorax, and abdomen), and they differ greatly from one another. Insects are fine examples of how diverse head parts, legs, wings, and other appendages evolved from less specialized segments.

The body plans of animals differ with respect to five features: body symmetry, cephalization, type of gut, type of body cavity, and segmentation.

26.2 AN EVOLUTIONARY ROAD MAP

Animals arose from protistanlike ancestors during the 200-million-year span before the Cambrian era. Most likely, the first ones were soft-bodied, so we may never find evidence of them. Yet by 560 million years ago, all major phyla were established.

There are more than thirty phyla of animals that range from placozoans and sponges to the chordates. Suppose we compare a number of these phyla in terms of body symmetry, cephalization, type of gut and body cavity, and segmentation. Doing so will reveal certain trends in animal evolution. As Figure 26.4 indicates, the trends did not show up in every line of descent. But their absence doesn't mean an animal is "primitive" or evolutionarily stunted. As you will see, even "simple" sponges are exquisitely adapted to their environment.

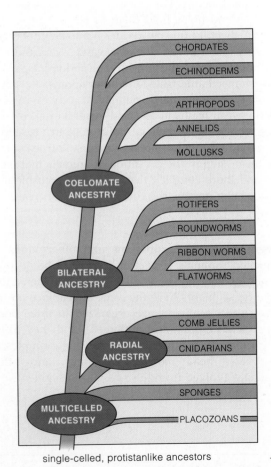

single-celled, protistanlike ancestors

Figure 26.4 Evolutionary relationships among major groups of animals. Take a moment to study this family tree. We will use it repeatedly as our road map through discussions of each group.

26.3 THE ONE PLACOZOAN

The first specimen of *Trichoplax adhaerens*, a soft-bodied marine animal no more than 3 millimeters across, was discovered gliding about in an aquarium tank. It is the only known **placozoan** (Placozoa, after the Greek *plax*, meaning plate; and *zoon*, meaning animal). *Trichoplax* is a plate-shaped, two-layered animal with no symmetry and no organs:

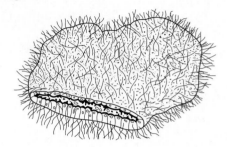

As this animal glides over food, glandular cells in its lower layer secrete digestive enzymes on it, then the cells absorb the breakdown products. Structurally and functionally, this is as simple as animals get.

26.4 SPONGES

Sponges (Porifera) are one of nature's success stories. Since Cambrian times, they have been abundant in the seas, especially in the waters off coasts and along tropical reefs. Of the 8,000 or so known species, about 100 live in freshwater habitats. Many animals, including a variety of worms, shrimps, small fishes, and barnacles, make their home in or on sponges.

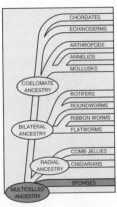

Some sponges are fingernail size; others are big enough to sit in. They are sprawling, flattened, compact, lobed, tubular, cuplike, or vaselike (Figure 26.5). Regardless of the shape, the sponge body has no symmetry or organs. Its outer surface and portions of cavities in the body wall have linings of flattened cells. But these are not much more complex than the cell layers of *Trichoplax*. And they certainly are not the same as the tissue linings of other animals.

Between the two linings is a semifluid matrix and glasslike elements (spicules), fibrous elements, and amoebalike cells. Spicules consist of silica or calcium carbonate. The pointed, sharp spicules function in support and protection. Fibers of a flexible protein give natural bath sponges a soft, squeezable texture.

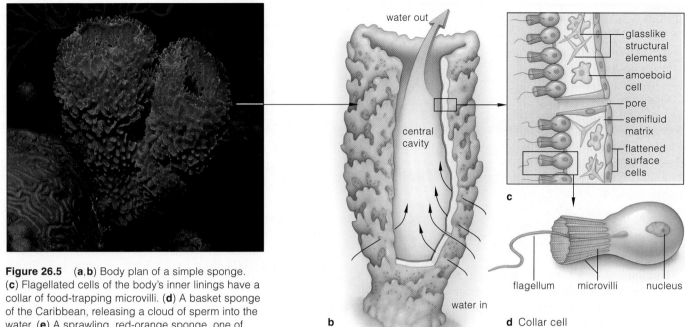

Figure 26.5 (**a**,**b**) Body plan of a simple sponge. (**c**) Flagellated cells of the body's inner linings have a collar of food-trapping microvilli. (**d**) A basket sponge of the Caribbean, releasing a cloud of sperm into the water. (**e**) A sprawling, red-orange sponge, one of many types that encrust underwater ledges in temperate seas.

Water flows into the sponge body through microscopic pores and chambers, then out through one or more openings called oscula (singular, osculum). The collective beating of thousands or millions of flagellated cells keeps water flowing this way. These **collar cells** help make up the sponge's inner linings. Besides flagella, they have "collars" of microvilli, which are absorptive structures (Figure 26.5c). Bacteria and other bits of food dissolved in the water become trapped in the collars. Some food gets transferred to the amoebalike cells for further breakdown, storage, and distribution.

Sponges reproduce sexually when collar cells give rise to eggs and to sperm, which are released into the water (Figure 26.5d). After being fertilized, the eggs are released. The young sponges pass through a microscopic, swimming larval stage. A **larva** (plural, larvae) is a sexually immature form that precedes the adult stage of many animal life cycles.

Some sponges reproduce asexually when small fragments break away from the parent body, then grow into new sponges. Most freshwater species also produce gemmules. These are clusters of sponge cells, some of which form a hard covering around others. The ones inside are protected from extreme cold or drying out. Under favorable conditions, gemmules germinate and establish a new colony of sponges.

Next to placozoans, sponges have the simplest body plan of all animals, yet they are among the most successful.

26.5 CNIDARIANS

All **cnidarians** (Cnidaria) are radial, tentacled animals, and most live in the seas. Of about 11,000 species, fewer than fifty live in freshwater habitats. Members of this phylum include scyphozoans (the jellyfishes), anthozoans (such as sea anemones and corals), and hydrozoans (such as *Hydra*). Figures 26.6 and 26.7 show examples. Of all the animals, cnidarians alone produce **nematocysts**. Their tentacles bristle with these thread-discharging capsules, which have roles in capturing prey and fending off predators. The toxin-laden threads of some species can give humans a painful sting (hence the phylum name Cnidaria, after the Greek word for "nettle").

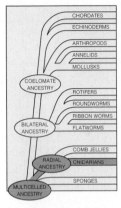

Cnidarian Body Plans

Two body forms are common among cnidarians. Both are organized around a saclike gut. As Figure 26.6*a* shows, one form is a **medusa** (plural, medusae). Some look like tentacle-fringed bells; others look like upside-down saucers. All medusas float in the water. Their mouth, centered under the bell, may have oral arms (extensions that assist in prey capture and in feeding). The other body form, a tubelike **polyp**, has a tentacle-fringed mouth at one end. The opposite end is usually attached to a substrate (Figure 26.6*b*).

Visualize the platelike *Trichoplax* draped over and digesting a mound of food, and you see that this simple animal is rather like a gut on the run. By comparison, the saclike cnidarian gut is a permanent food-processing chamber. It has a complex lining with glandular cells that secrete digestive enzymes. This sheetlike lining, a "gastrodermis," is one of the cnidarian's **epithelial tissues**. *All animals more complex than sponges have epithelial tissues.* Another lining, the "epidermis," covers the rest of the body's surfaces (Figure 26.7).

Threading through both epithelial tissues are interacting **nerve cells**. The ones in cnidarians form a "nerve net," a simple nervous system that coordinates responses to stimuli. Together with **sensory cells** and **contractile cells**, they control movements and shape changes. In some jellyfishes, structures composed of sensory cells detect changes in orientation.

Between the epidermis and the gastrodermis is a layer of secreted material, the mesoglea ("middle jelly"). Jellyfishes have enough mesoglea to impart buoyancy and serve as a firm yet deformable skeleton against which contractile cells act. Contractions narrow the bell, forcing a jet of water out from it, and the animal is propelled forward. Mesoglea is not abundant in most polyps, which typically use the water in their gut as a hydrostatic skeleton. A **hydrostatic skeleton** is a fluid-filled cavity or cell mass. When appropriate cells contract, its volume remains the same but is shunted about, and this changes the body's shape. Figure 26.7*d* shows how this feature can be put to use.

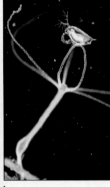
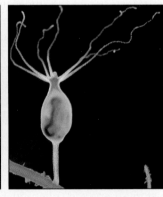
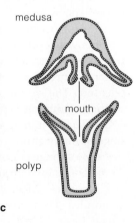

Figure 26.6 Representative cnidarians. (**a**) Medusa of the sea nettle (*Chrysophora*), one of the jellyfishes. Long, oral arms extend from the mouth, which is centered below the bell-shaped medusa. (**b**) A hydrozoan polyp (*Hydra*), firmly attached to a substrate. The polyp is shown capturing and then digesting a tiny aquatic animal. (**c**) This generalized diagram shows how a medusa and a polyp might look when sliced lengthwise, through their midsection.

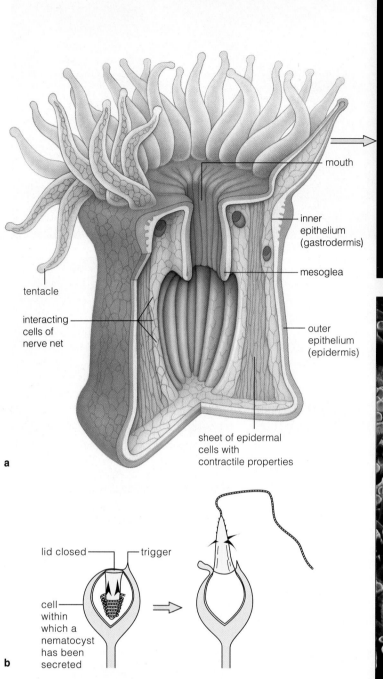

mouth

inner epithelium (gastrodermis)

mesoglea

tentacle

interacting cells of nerve net

outer epithelium (epidermis)

sheet of epidermal cells with contractile properties

a

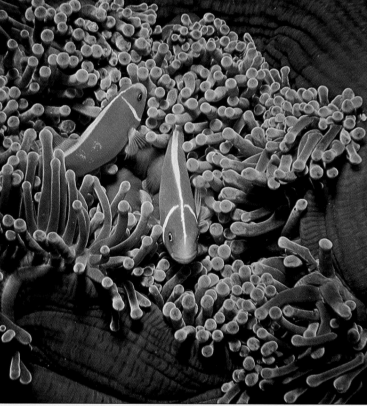

lid closed — trigger

cell within which a nematocyst has been secreted

b

c

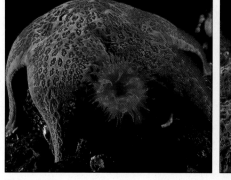

d

Figure 26.7 (**a**) Body plan of a sea anemone. The cutaway diagram shows aspects of its tissue organization.

(**b**) Nematocysts are embedded in the tentacles. These capsules contain an inverted tubular thread. The one shown has a bristlelike trigger. When prey touch the trigger, the capsule becomes more "leaky" to water. Water diffuses inward, pressure inside the capsule builds up, and the thread is forced to turn inside out. The thread's tip may penetrate prey and release a toxin.

(**c**) Tentacles fringing a sea anemone mouth. Sea anemones eat many fishes, but not clownfishes. Clownfishes swim away, capture food, and return to the tentacles, which protect them from predators. The sea anemone eats food scraps falling from fish mouths. This is a case of mutualism, a two-way flow of benefits between species.

(**d**) A sea anemone using its hydrostatic skeleton to escape from a sea star. The sea anemone closes its mouth, and the force generated by contractile cells in its epithelial tissues acts against water in the gut. The body changes shape, allowing the anemone to thrash about, generally in a direction away from the predator.

a

b

c   mouth

d

Figure 26.8 Corals. (**a**) A reef-building coral, with dinoflagellates in its tissues. In this mutually beneficial arrangement, the coral protects the dinoflagellates, which provide the coral with oxygen and recycle its mineral wastes. (**b**) A barrier reef, formed mainly of countless coral skeletons. Not all corals are reef builders. (**c**) A cup coral, the polyps of which look like tiny sea anemones. This one is *Tubestrea*. (**d**) *Telesto*, one of the tall, branching corals of deep or sheltered waters.

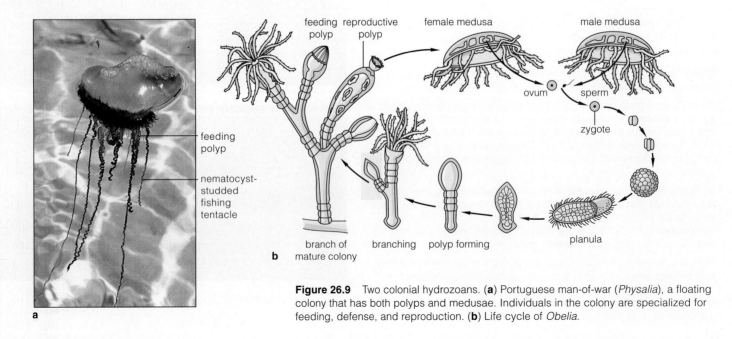

feeding polyp

nematocyst-studded fishing tentacle

feeding polyp   reproductive polyp   female medusa   male medusa

ovum   sperm

zygote

planula

branch of mature colony   branching   polyp forming

a

b

Figure 26.9 Two colonial hydrozoans. (**a**) Portuguese man-of-war (*Physalia*), a floating colony that has both polyps and medusae. Individuals in the colony are specialized for feeding, defense, and reproduction. (**b**) Life cycle of *Obelia*.

Some Colonial Cnidarians

Colonial anthozoans show interesting variations on the basic body plan. For example, most reef-building corals consist of interconnected polyps that secrete their own calcium-reinforced external skeletons. Uncounted numbers of coral skeletons serve as the main building material for reefs (Figure 26.8). Their most extravagant accomplishment, the Great Barrier Reef, parallels the eastern Australian coast for about 1,600 kilometers. The reef-forming corals have photosynthetic protistans in their tissues and must live in clear, warm water (at least 20°C).

Physalia, the hydrozoan shown in Figure 26.9*a*, has a remarkable colonial organization. You may have heard it called the Portuguese man-of-war. The colony consists of several kinds of polyps and medusae. They are specialized for different functions, such as feeding, defense, and reproduction.

Cnidarian Life Cycles

Many cnidarians have only a polyp stage or a medusa stage in the life cycle. *Physalia, Obelia,* and some other types have both, with the medusa being the sexual form (Figure 26.9*b*). Simple **gonads** (reproductive organs) are associated with the epidermis or gastrodermis. They just rupture and release the gametes. A zygote formed at fertilization nearly always develops into a swimming or creeping larva called a **planula**, which usually has ciliated epidermal cells. In time a mouth opens at one end, the larva is transformed into a polyp or medusa, and the cycle begins anew.

Cnidarians are tentacled, radial animals with a saclike gut, nerve net, and epithelial tissues. They are the only animals to produce nematocysts.

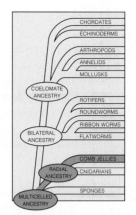

26.6 COMB JELLIES

The **comb jellies** belong to the phylum Ctenophora (a word that means "comb-bearing"). These are marine predators that show modified radial symmetry. You can slice them into two equal halves. Slice them into quarters, and two of the quarters will be mirror images of the other two.

Their name refers to eight rows of comblike structures that are made of thick, fused cilia (Figure 26.10). All the combs in each row beat in waves and propel the animal forward, usually mouth first. Some species have a feeding net (two long, muscular tentacles with many branches that are equipped with sticky cells). Comb jellies don't produce nematocysts, but sometimes they opportunistically save and use the ones from jellyfish they have eaten. Others use sticky lips to capture prey, including other comb jellies or jellyfishes.

Evolutionarily, comb jellies are interesting because they have cells with multiple cilia. This trait evolved in many of the complex animals. Besides this, they are the simplest animals with embryonic tissues that resemble mesoderm. In complex animals, mesoderm gives rise to muscles and to organs of the circulatory, excretory, and reproductive systems.

Comb jellies have a modified radial body and comblike structures of modified cilia. They have an embryonic tissue that resembles the mesoderm of more complex animals.

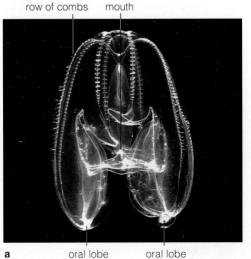

Figure 26.10 Two of the comb jellies. (**a**) *Mnemiopsis*, with its expanded, contractile mouth lobes. These waft prey into the mouth. (**b**) *Pleurobrachia*, with its two long, sticky tentacles. This one is about actual size.

row of combs   mouth

a   oral lobe   oral lobe   b

26.7 FLATWORMS

Among the 15,000 or so known species of **flatworms** (phylum Platyhelminthes) are the turbellarians, flukes, and tapeworms. Flatworms have a saclike gut, with food usually entering through a pharynx (a muscular tube). Most flatworms have a more or less flattened body, hence the name. Unlike sponges or cnidarians, the flatworms are bilateral and cephalized. Three primary tissue layers develop in flatworm embryos. The midlayer, mesoderm, is the source of many tissues. Flatworms in fact are the simplest animals with **organs**, which consist of different tissues that are organized in specific proportions and patterns. They have **organ systems**, which consist of two or more organs interacting in a common task.

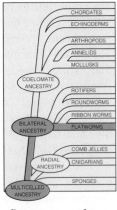

a

branching gut

pharynx (protruded)

protonephridia

b

brain

nerve cord

c

ovary   testis   oviduct   genital pore

penis

d

Turbellarians

Most turbellarians live in the seas; the planarians and a few others live in freshwater. The carnivores eat small animals or suck tissues from dead or wounded ones. Herbivorous types eat diatoms.

As is true of your urinary system, one planarian organ system regulates the composition and volume of body fluids. It has one or more protonephridia (singular, protonephridium). These branched tubes extend from pores at the body surface to many bulb-shaped flame cells in body tissues (Figure 26.11). Flame cells get their name from a tuft of cilia that "flickers" inside the bulbous, hollow interior. When excess water moves into a bulb, the flickering drives it through the tubes, to the outside. Some solutes may be reabsorbed from the tubes. Nitrogenous wastes leave with water and also are eliminated across the epidermis and gut lining.

Planarians commonly reproduce asexually by way of fission. A worm divides in half; then each half regenerates the missing part. However, most flatworms rely on sexual reproduction. Each worm is a **hermaphrodite**, with both female and male gonads, and two worms typically reproduce by the mutual transfer of sperm. The reproductive system includes a penis (a sperm-delivery structure) and glands that help produce a protective capsule around fertilized eggs.

Flukes

Flukes are parasites. A parasite, recall, lives on or in a living organism. It obtains nutrients from its host's tissues and may or may not end up killing it. Most adult flukes (trematodes) live in the vertebrate gut, liver, lungs, bladder, or blood vessels. Flukes have complex life cycles, with sexual and asexual phases and at least two kinds of hosts. They reach sexual maturity in a *primary* host. Their larval stages develop or become encysted in an *intermediate* host.

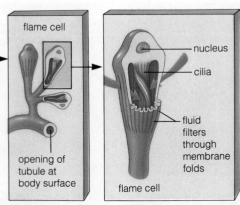

Figure 26.11 Organ systems in a planarian, a type of flatworm. (**a**) A pharynx opens to the gut. It projects to the outside, then retracts into a chamber between feedings. (**b**) Water-regulating system, (**c**) nervous system, and (**d**) reproductive system.

Tapeworms

Tapeworms (cestodes) are parasites of vertebrate intestines. Ancestral tapeworms probably had a gut, but they presumably lost it as they evolved in environments rich with predigested food. Existing tapeworms attach to the intestinal wall by a **scolex**, a structure with suckers, hooks, or both (Figure 26.12 and page 421). Just behind the scolex, **proglottids** (new tapeworm body units) form by budding. Proglottids are hermaphroditic; they mate and transfer sperm. Older proglottids (farthest from the scolex) store eggs; then they leave the body in feces and so carry the eggs to the outside. There, eggs may meet up with an intermediate host.

In some respects, the simplest turbellarians, larval flukes, and larval tapeworms resemble the planulas of cnidarians. The resemblance inspires speculation that bilateral animals evolved from planula-like ancestors, through increased cephalization and the development of tissues derived from mesoderm.

The **roundworms** (nematodes) live just about everywhere, from snowfields to deserts and hot springs. A cupful of rich soil has thousands of them; a dead earthworm or rotting fruit can hold many interesting scavenging types. Roundworms also parasitize living plants and animals. Thirty or so species, including pinworms and hookworms, parasitize humans.

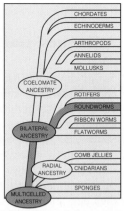

The roundworm body plan works well in many different habitats and has not changed much over time. All roundworms have a bilateral, cylindrical body, usually tapered at both ends. They have a **cuticle**, which in animals is a tough, flexible body covering. Roundworms are the simplest animals with a complete digestive system (Figure 26.13). Between the gut and body wall is a false coelom, often packed with reproductive organs. Fluid in the cavity circulates nutrients through the body.

We turn next to a *Focus* essay on some diseases caused by a few notorious flatworms and roundworms.

Flatworms and roundworms are among the simplest bilateral, cephalized animals with organ systems.

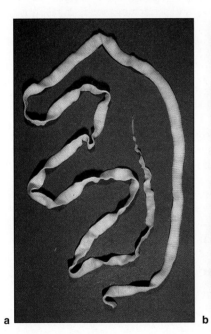
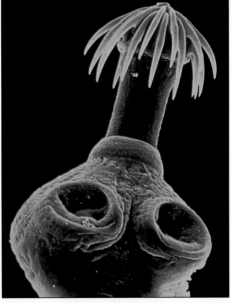

Figure 26.12 (**a**) A sheep tapeworm. (**b**) Hooks and suckers of a scolex, the part of a tapeworm that attaches to a primary host (in this case, a shorebird).

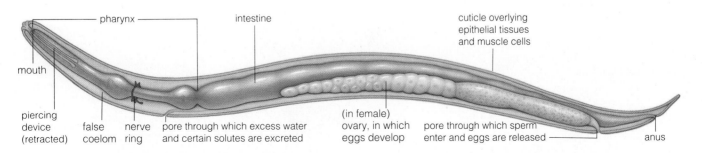

pharynx — intestine — cuticle overlying epithelial tissues and muscle cells

mouth

piercing device (retracted) — false coelom — nerve ring — pore through which excess water and certain solutes are excreted — (in female) ovary, in which eggs develop — pore through which sperm enter and eggs are released — anus

Figure 26.13 Body plan of a roundworm (*Paratylenchus*) that parasitizes the roots of certain plants.

A Rogues' Gallery of Parasitic Worms

To our enormous discomfort, many parasitic flatworms and roundworms call the human body home. In a given year, about 200 million people house blood flukes responsible for *schistosomiasis*. Figure *a* shows the life cycle of a Southeast Asian blood fluke (*Schistosoma japonicum*). The cycle requires a human primary host, standing water through which larvae can swim, and an aquatic snail that serves as intermediate host. Flukes reproduce sexually, and their eggs mature in the human body (1). Eggs leave the body in feces, then hatch into ciliated, swimming larvae (2). These burrow into a snail and multiply asexually (3). In time, many fork-tailed larvae develop (4). These leave the snail and swim until they encounter human skin (5). They bore inward and migrate to thin-walled intestinal veins, and the cycle begins anew. In infected humans, white blood cells that defend the body attack the masses of fluke eggs, and grainy masses form in tissues. In time, the liver, spleen, bladder, and kidneys deteriorate.

Tapeworms also parasitize humans. One species uses pigs as intermediate hosts; another uses cattle. Humans become infected when they eat insufficiently cooked pork or beef (Figure *b*). They become infected in other ways. For example, tiny crustaceans called copepods eat the larvae of one tapeworm species. The larvae escape digestion; then they mature in fishes that eat the copepods. Suppose humans dine on raw, improperly pickled, or insufficiently cooked fish that were infected by the larvae. They can become hosts for the adult tapeworms.

Or consider the roundworms. One species that infects humans produces thin, serpentlike ridges at the surface of skin. For thousands of years, healers removed the "serpents" by winding them out slowly, painfully, around a stick. The symbol of the medical profession continues to be a serpent wound around a staff.

Then there are the types of roundworms called pinworms and hookworms. *Enterobius vermicularis*, a type of pinworm, parasitizes humans in temperate regions. It lives in the large intestine. At night, the centimeter-long female pinworms migrate to the anal region and lay eggs. Their presence causes itching, and scratchings made in response will transfer some eggs to other objects. Newly laid eggs contain embryos, but within a few hours they are juveniles and ready to hatch if another human inadvertently ingests them.

Hookworms are a serious problem in impoverished parts of the tropics and subtropics. Adult hookworms live in the small intestine. They feed on blood and other tissues after teeth or sharp ridges bordering their mouth cut into the intestinal wall. Adult females, about a centimeter long, can release a thousand eggs daily. These leave the body in

a Life cycle of a dangerous blood fluke, *Schistosoma japonicum*. The micrograph shows an adult male fluke.

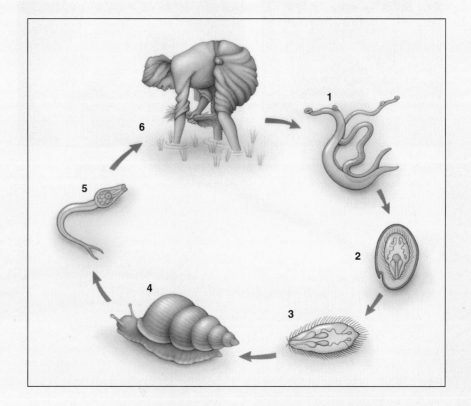

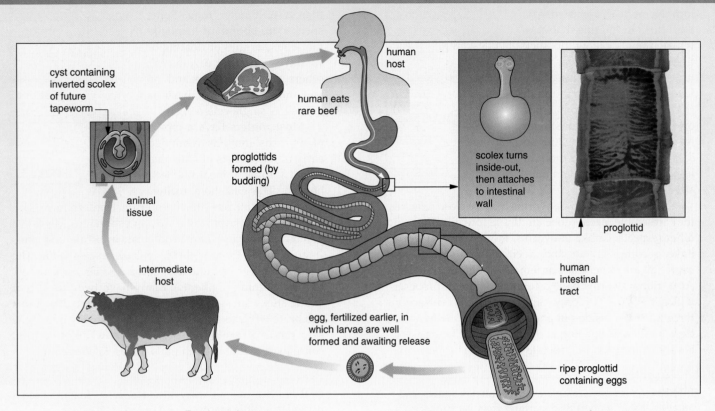

cyst containing inverted scolex of future tapeworm

animal tissue

intermediate host

human host

human eats rare beef

proglottids formed (by budding)

scolex turns inside-out, then attaches to intestinal wall

proglottid

human intestinal tract

egg, fertilized earlier, in which larvae are well formed and awaiting release

ripe proglottid containing eggs

b Life cycle of a beef tapeworm, *Taenia saginata*.

feces, then hatch into juveniles. A juvenile may penetrate the skin of a barefoot person. Inside a host, the parasite travels the bloodstream to the lungs, where it works its way into the air spaces. After moving up the windpipe, the parasite is swallowed. Soon it is in the small intestine, where it may mature and live for several years.

Another roundworm, *Trichinella spiralis*, causes painful, sometimes fatal symptoms. Adults live in the lining of the small intestine. Female worms release juveniles (Figure *c*), and these work their way into blood vessels and travel to muscles. There they become encysted (they produce a covering around themselves and enter a resting stage). Humans usually become infected by eating insufficiently cooked meat from pigs or certain game animals. The presence of encysted juveniles cannot easily be detected when fresh meat is examined, even in a slaughterhouse.

Figure *d* shows the results of prolonged, repeated infections by *Wuchereria bancrofti*, a roundworm. Adult worms live in the lymph nodes, where they can obstruct the flow of a fluid (lymph) that normally is returned to the bloodstream. The obstruction causes fluid to accumulate in legs and other body regions. These undergo grotesque enlargement, a condition called *elephantiasis*. A mosquito is the intermediate host. Female *Wuchereria* produce active young that travel the bloodstream at night. If a mosquito

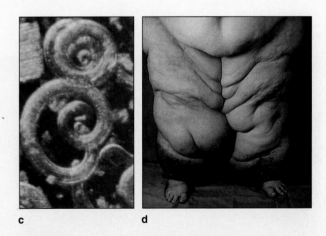

c   **d**

c   Juveniles of a roundworm, *Trichinella spiralis*, living inside the muscle tissue of a host animal. (**d**) Legs of a woman who was infected by *Wuchereria bancrofti*, a roundworm that causes the disease elephantiasis.

sucks blood from an infected human, the juveniles may enter the insect's tissues. In time they move near the insect's sucking device, where they are ready to enter another human host when the mosquito draws blood again.

26.9 RIBBON WORMS

Nearly all **ribbon worms** (phylum Nemertea) are predators that lurk in marine habitats, where they swallow or suck tissue fluids from worms, mollusks, and crustaceans. Very few species have been found in freshwater or in damp habitats on land. The specimen shown in Figure 26.14 gives you an idea of why someone thought to name these bilateral animals after ribbons.

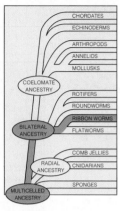

Ribbon worms and flatworms may be close relatives; they resemble each other in tissue organization and other respects. Unlike flatworms, though, ribbon worms have a complete gut and a circulatory system, and nearly all are either male or female, not hermaphrodites. Also, ribbon worms have a tubular device, a proboscis, which they use for prey capture. Muscle contractions force it to turn inside-out and project from the mouth or from a different opening at the head end. Glands associated with the proboscis secrete a paralytic venom.

Ribbon worms have a complete gut, circulatory system, and other traits that are notable departures from flatworms.

26.10 ROTIFERS

Most rotifers (Rotifera) are less than a millimeter long. Some species are abundant in plankton (communities of floating or weakly swimming organisms); others crawl on algae and on wet mosses. They eat bacteria as well as single-celled algae.

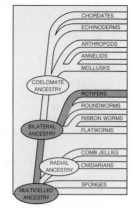

Most rotifers have a crown of cilia. Its motions reminded early microscopists of a turning wheel ("rotifer" means wheel-bearer). The cilia help rotifers swim and waft food to the mouth. Rotifers have salivary glands, a jawed pharynx, an esophagus, digestive glands, a stomach, and usually an intestine and anus (Figure 26.15). Protonephridia remove excess water. The head has nerve cell clusters. Some species have "eyes" (light-absorbing pigments). Many have two "toes" that exude sticky substances and allow rotifers to attach to a substrate while feeding. Rotifers are probably one of the side roads of evolution, but they show the structural complexity possible in animals with a false coelom.

Of the rotifers, we might say this: Seldom has so much been packed in so little space.

Figure 26.14 A ribbon worm, one of the lesser known invertebrates.

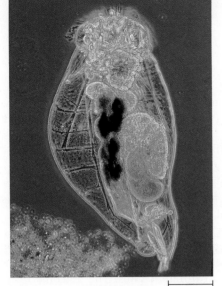

Figure 26.15 Lateral view of a rotifer, busily laying eggs. Males are unknown in many species; the females produce diploid eggs that develop into diploid females. Females of other species do the same, but they also produce haploid eggs that develop into haploid males. If a haploid egg happens to be fertilized by a male, it develops into a female. The male rotifers appear only occasionally, and are dwarfed and short-lived.

50 µm

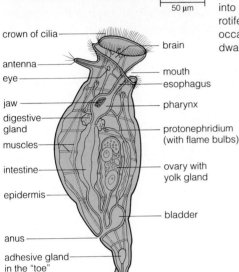

crown of cilia
brain
antenna
eye
mouth
esophagus
jaw
pharynx
digestive gland
muscles
protonephridium (with flame bulbs)
intestine
ovary with yolk gland
epidermis
bladder
anus
adhesive gland in the "toe"

26.11 A MAJOR DIVERGENCE

When we turn to the animals with coeloms, we find complexity on a spectacular scale. During Cambrian times, bilateral animals not much more complex than flatworms evolved. Not long afterward, they diverged into two distinct evolutionary lineages, the **protostomes** and **deuterostomes**. Mollusks, annelids, and arthropods represent the protostome lineage. Echinoderms and chordates represent the deuterostome lineage.

The two lineages differ partly in the way their embryos develop (Figure 26.16 and page 757). In deuterostomes, cell divisions divide a zygote as you might divide an apple, by cutting it into four wedges from top to bottom. Each wedge is cut in half at its midsection; then each piece is cut at *its* midsection. Cell divisions made parallel and perpendicular to an embryo's main body axis are called **radial cleavage**. In protostomes, the "wedges" are cut at oblique angles to the original axis, a pattern called **spiral cleavage**.

Besides this, the mouth and anus form differently, starting with openings at two indentations that appear in early embryos. In deuterostomes, the first opening becomes an anus, and the second becomes a mouth. In protostomes, the first opening becomes a mouth; an anus forms elsewhere. As a final example, a deuterostome coelom starts forming as outpouchings of the gut wall. A protostome coelom arises within solid masses of tissue at the sides of the gut.

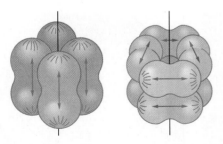

a Radial cleavage, a cell division pattern that transforms the zygotes of deuterostomes into the early embryo.

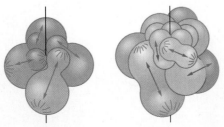

b Spiral cleavage, a cell division pattern that transforms the zygotes of protostomes into the early embryo.

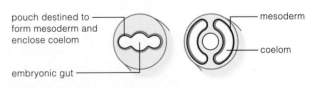

pouch destined to form mesoderm and enclose coelom
mesoderm
coelom
embryonic gut

c How the coelom forms in a deuterostome embryo. The diagrams show what you would see if you sliced an embryo through its midsection at two successive stages of development.

mesoderm
solid mass of mesoderm
coelom
embryonic gut

d How the coelom forms in a protostome embryo.

Figure 26.16 Two characteristics of the early embryonic development of species that belong to the deuterostome and protostome branches of the animal family tree.

26.12 MOLLUSKS

With well over 100,000 known species, the fleshy, soft-bodied **mollusks** are the second largest invertebrate group. (Mollusca means "soft-bodied.") In this phylum we find the greatest array of marine invertebrates as well as successful occupants of land habitats. Among its better-known members are snails, clams, oysters, octopuses, and squids. These mollusks and others share the same bilateral ancestry, although the details are highly modified from one group to the next. Most have a head, foot, and visceral mass (mainly a gut and a heart, kidneys, and reproductive organs). Species with a well-developed head also have eyes and tentacles.

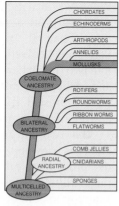

A distinctly molluscan feature is a **mantle**, a tissue fold that hangs, rather like a skirt, around some or all of the body (Figure 26.17). Another is a pair of gills, of a kind called ctenidia (singular, ctenidium). Most mollusks have these respiratory organs, each a series of thin-walled leaflets across which oxygen and carbon dioxide are exchanged. Most mollusks have a shell (or an evolutionary remnant of one) composed mainly of calcium carbonate. Many have a radula, a tonguelike organ with hard teeth that preshreds food.

Gastropods

Snails and slugs make up the largest molluscan group, the gastropods ("belly foots"). They are so named because their foot spreads out when the animal crawls about. Many of the snails have spirally coiled or cone-shaped shells (Figure 26.17a). Coiling is a way of compacting the organs into a mass that can be balanced above the rest of the body, much as you would balance a backpack full of books.

As most young gastropods grow, some body parts undergo a strange internal realignment. In a process called torsion, certain muscles contract and different body parts grow at different rates. The outcome? The posterior mantle cavity twists around to the right and then forward, until it is above the head:

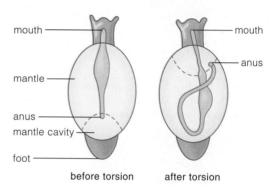

before torsion   after torsion

Figure 26.17 Features of the molluscan body plan, as represented by snails. (**a**) A land snail. (**b**) Body plan of an aquatic snail. (**c**) Snails and some other mollusks have a radula, which is protracted and retracted in a rhythmic way. Food is rasped off and drawn to the gut on the retraction stroke.

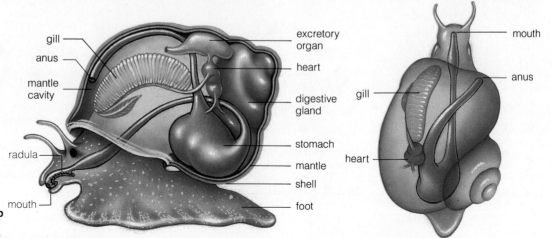

a

b

Figure 26.18 Two of the soft-bodied gastropods called nudibranchs.

(**a**) A sea hare (*Aplysia*), widely used in experimental studies of behavior and physiology. The two flaps over its dorsal surface are extensions of the foot. They undulate and help ventilate the underlying mantle cavity. Like many other gastropods, *Aplysia* is hermaphroditic. It may act as a male, a female, or both during encounters with others like itself. Neurobiologists are much taken with *Aplysia* and have tracked several of its nerve pathways. (**b**) These two sea slugs are "Mexican dancers," engaged in mating behavior.

The twisting produces a cavity into which the head can withdraw in times of danger. It also brings the gills, anus, and kidney openings above the head. This could create a sanitation problem, given that wastes are dumped near the respiratory structures and the mouth. In most species, the beating of cilia in this region creates currents that sweep wastes away.

Some of the gastropods, including the colorful nudibranchs (sea slugs), seem to have undergone "detorsion" during their evolution. To some extent, the larval body still twists. Then muscle contractions and asymmetrical growth make the body untwist. Also among nudibranchs, the mantle cavity largely disappeared. So did the ctenidia, although other outgrowths function as secondary gills in most species. Figure 26.18 shows the array of outgrowths on a few of the more striking nudibranchs.

Chitons

Chitons are slow-moving or sedentary marine mollusks. Most species use a radula to graze on algae, hydrozoans, and other low-growing organisms. A large, broad foot enables them to creep over and cling tenaciously to rocks and other hard surfaces. Chitons have a dorsal shell divided into eight plates (Figure 26.19). The divisions impart flexibility. When a chiton is disturbed or exposed by a receding tide, its foot muscles can pull down the body mass tightly. The mantle's edge, which more or less covers the plates, now functions like the rim of a suction cup, so the chiton is difficult to detach. If you manage to separate a pulled-down chiton from its substrate, it will roll up into a ball, protecting itself until it can safely unroll and become reattached elsewhere.

Figure 26.19 Chitons—one variation on the basic molluscan plan. Walk along a rocky shore and you will probably see chitons. The ones shown here live in the intertidal zone of Monterey Bay, California. Members of this class of mollusks show beautiful variations in the color and patterns of their elongated shells.

Bivalves

Clams, scallops, oysters, and mussels are well-known bivalves (animals with a "two-valved shell"). Humans have been eating one type or another since prehistoric times. The shell of many bivalves, including a few pearl-producing types, has an inner lining of iridescent mother-of-pearl. Some species are only 1 or 2 millimeters across. A few giant clams of the South Pacific are more than a meter across and weigh 225 kilograms (close to 500 pounds).

The bivalve head is not much to speak of, but the foot is usually large and specialized for burrowing. In nearly all bivalves, gills function in collecting food and in respiration. As cilia in the gills move water through the mantle cavity, mucus on the gills traps tiny bits of food (Figure 26.20). Tracts of cilia move the mucus and food to the palps, where final sorting takes place before acceptable bits are driven to the mouth.

Bivalves hunkered in mud or sand have a pair of siphons (extensions of the mantle edges, fused into tubes). Water is drawn into the mantle cavity through one siphon and leaves through the other, carrying wastes from the anus and kidneys. Siphons of the giant geoduck (pronounced "gooey duck") of the Pacific Northwest may be more than a meter long.

Scallops and a few other bivalves can swim by clapping the valves together and producing a localized jet of water that propels the animal for some distance (Figure 26.21).

Cephalopods

The smallest mollusks are snails and clams less than a millimeter long. Among the cephalopods we find species at the other extreme. The giant squid, which may measure 18 meters (about 60 feet) from tip to tip, is the largest invertebrate known. Also among the cephalopods are the swiftest invertebrates (jet-propelled squids) and the smartest ones (octopuses). Squids, octopuses, nautiluses, and cuttlefish—these are all fast-swimming predators of the oceans. Figures 26.22 and 26.23 show representative species.

Instead of a foot, cephalopods have tentacles, usually equipped with suction pads of the sort shown in Figure 26.23. They use the tentacles to capture prey, a radula to draw it into the mouth, and a beaklike pair of jaws to bite or crush it. Venomous secretions often speed the captive's death.

Cephalopods move rapidly by a kind of jet propulsion. They force a stream of water out of the mantle cavity through a funnel-shaped siphon. When muscles in the mantle relax, water is drawn into the mantle cavity. When they contract, a jet of water is squeezed out. If, at the same time, the free edge of the mantle is closed down on the animal's head and siphon, water shoots

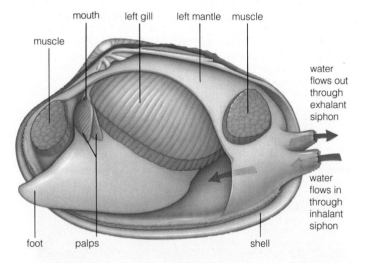

mouth · left gill · left mantle · muscle · muscle · water flows out through exhalant siphon · water flows in through inhalant siphon · foot · palps · shell

Figure 26.20 Body plan of a clam, with half of its shell (the left valve) removed for this diagram. The palps sort particles of food that become trapped in mucus on the gills. Cilia on the palps sweep suitable tidbits to the mouth.

Figure 26.21 A scallop moving away from a predator, a sea star. By clapping their valves together, scallops and a few other bivalves create a strong jet of water that can propel them for some distance.

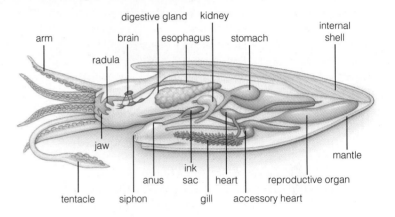

out in a jet through the siphon. By manipulating the siphon, the animal partly controls the direction of its own movement.

Being highly active, cephalopods have great demands for oxygen, and they alone among the mollusks have a *closed* circulatory system (page 654). In such systems, blood is confined within the walls of one or more hearts and blood vessels. In cephalopods, blood is pumped from the main heart to the two gills. Each gill has an accessory, "booster" heart at its base that speeds the flow and therefore the uptake of oxygen and elimination of carbon dioxide.

Even when a cephalopod rests, it rhythmically draws water into its mantle cavity and then expels it. The water circulation enhances gas exchange and also helps to eliminate wastes that excretory organs and the anus discharge into the mantle cavity.

Cephalopods have a well-developed nervous system. Compared to other mollusks, they have the largest brains relative to body size. Giant nerve fibers connect the brain with muscles used in jet propulsion, making it possible for a cephalopod to respond quickly to food or danger. Cephalopod eyes bear some resemblance to yours, although they are formed in a different way. Finally, cephalopods can learn. Give an octopus a mild electric shock after showing it an object with a distinctive shape, for example, and it will thereafter avoid the object. In terms of memory and learning, the cephalopods are the world's most complex invertebrates.

Cephalopods have separate sexes. Male squids even rush at each other in mock battles that may get serious enough for a loser to lose an arm.

Mollusks make up the second largest animal phylum, next to arthropods, and have the greatest array of marine species.

All mollusks are soft-bodied, bilateral animals with a shared ancestry, but they vary enormously in size, life-styles, and details of their body plan.

Figure 26.22 General body plan of a cuttlefish, one of the cephalopods. The tentacles are more slender than the arms and are specialized for prey capture. The photograph shows a female and a male cuttlefish (*Sepia*) head end to head end. They are engaged in mating behavior.

Figure 26.23 Another cephalopod—an octopus.

26.13 ANNELIDS

When heavy rains soak the soil, earthworms make their presence known by wriggling to the surface and so avoid drowning. At such times, you may suddenly notice these worms, which otherwise leave their burrows only at night. They are among the 15,000 or so species of **annelids** (Annelida). Also in this group are polychaetes (which are far more diverse but less familiar) and a few species of leeches (some of which are the most notorious). The name of their phylum means "ringed forms." However, what appear to be rings actually are a series of body segments.

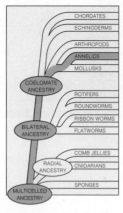

Annelids are bilateral worms with pronounced segmentation. Except for leeches, each side of their body has pairs or clusters of chitin-reinforced bristles on nearly every segment. Sometimes the bristles are called *setae* (a Latin term) or *chaetae* (a latinized Greek term), but these are merely formal names for bristles.

The bristles provide the traction required for crawling and burrowing through soil, and in swimming species they often are broadened into paddles. Although earthworms have only a few setae per body segment, the marine polychaete worms typically have many of them (*poly-* means many).

Earthworms: A Case Study of Annelid Adaptations

Earthworms, of the sort shown in Figure 26.24, are the usual textbook examples of annelids. They belong to a class of worms called oligochaetes, which has land-dwelling and aquatic members. The earthworms are scavengers on land. They burrow in moist soil and mud, where they feed on decomposing plant material and other organic matter. Every twenty-four hours, an earthworm can ingest its own weight in soil. Earthworms aerate soil, to the benefit of many plants. They also make nutrients available to other organisms by carrying subsoil to the surface.

As is true of most annelids, an earthworm's body segments have a cuticle, a layer of secreted material that wraps around the body's outer surface (Figure 26.25). The annelid cuticle is thin and flexible. It can bend easily, and it also permits the exchange of gases to and from the body surface. Inside the body, partitions separate the segments

from one another, so that there is a series of coelomic chambers. The gut extends continuously through all the chambers, from the earthworm's mouth to its anus.

As for other annelids, the coelomic chambers serve as a hydrostatic skeleton against which muscles can operate. As Figure 26.25a shows, circular muscles are located mostly within the body wall of each segment. Longitudinal muscles span several segments. When these contract and the circular muscles relax, segments shorten and fatten. The segments lengthen when the pattern reverses.

Muscle contractions are not enough to move an earthworm forward. Locomotion also involves the protraction and retraction of bristles on different segments. The body moves forward when its first few segments elongate. The segments just behind them plunge their bristles into the ground and hold their position. Then the first segments contract and plunge their bristles into the ground. Simultaneously, the segments behind them retract their bristles and are pulled forward. Alternating contractions and elongations proceed along the length of the body and move the worm forward.

A brain at the head end integrates sensory input for the whole worm (Figure 26.25c). Leading away from the brain is a double nerve cord that extends the length of the body. A **nerve cord** is a bundle of slender extensions of nerve cell bodies. In each segment, the cord broadens into a **ganglion** (plural, ganglia), a cluster of nerve cell bodies that controls local activity, as described in Chapters 34 and 35.

Figure 26.24 A well-known annelid, an earthworm.

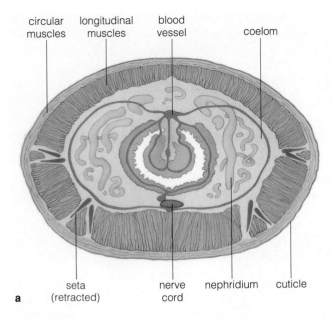

circular muscles · longitudinal muscles · blood vessel · coelom

seta (retracted) · nerve cord · nephridium · cuticle

a

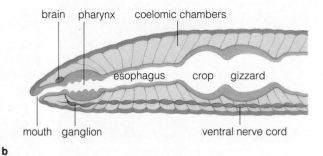

brain · pharynx · coelomic chambers

esophagus · crop · gizzard

mouth · ganglion · ventral nerve cord

b

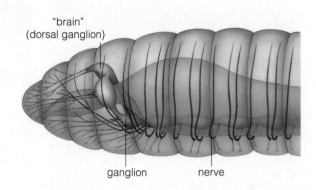

"brain" (dorsal ganglion)

ganglion · nerve

c

Figure 26.25 Body plan of an earthworm, a representative annelid. (**a**) Section through the midbody. (**b**) Head end of the digestive system. (**c**) A portion of the nervous system. (**d**) Portion of the closed circulatory system, which is functionally linked with a system of nephridia. The nephridium in (**e**) is one of many functional units that maintain the volume and composition of body fluids.

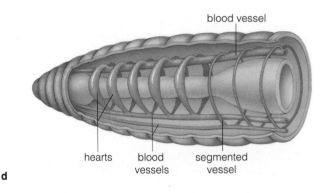

blood vessel

hearts · blood vessels · segmented vessel

d

Like most annelids, earthworms have a closed circulatory system (Figure 26.25*d*). Contractions of muscularized blood vessels keep blood circulating in one direction. Smaller blood vessels lead to and from the gut, nerve cord, and body wall.

A system of **nephridia** (singular, nephridium) regulates the volume and composition of body fluids. In many annelids, the units of this system have cells similar to flame cells, which may imply an evolutionary link between flatworms and annelids. More commonly, the beginning of each nephridium is a funnel-like structure that collects fluid from a coelomic chamber. The funnel leads into a tubular portion of the nephridium that carries the fluid to a pore at the body surface (Figure 26.25*e*). The funnel is located in one coelomic chamber, but its terminal pore is in the body wall of the next chamber in line.

From comparisons of the tissues and organ systems of annelids with those of other animals, it seems likely that annelids were an early offshoot of the protostome branch of the family tree. Their features remind us of developments that led to increased size and more complex internal organs.

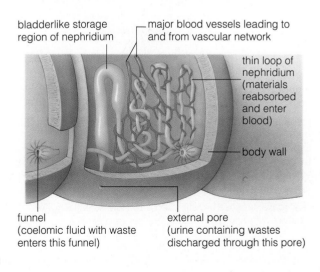

bladderlike storage region of nephridium

major blood vessels leading to and from vascular network

thin loop of nephridium (materials reabsorbed and enter blood)

body wall

funnel (coelomic fluid with waste enters this funnel)

external pore (urine containing wastes discharged through this pore)

e

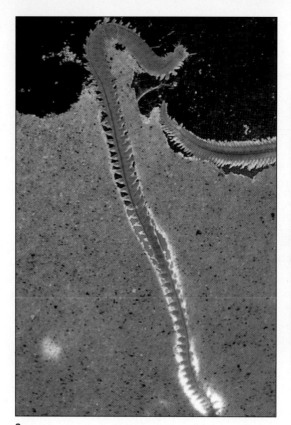

a

b

Figure 26.26 (**a**) A marine polychaete in a burrow of its own making. The profusion of bristles along the sides of its body grip the burrow wall and aid in the worm's movements. (**b**) A tube-dwelling polychaete with featherlike structures that are coated with mucus. After the mucus traps bacteria and other bits of food, the coordinated beating of cilia sweeps them to the nearby mouth.

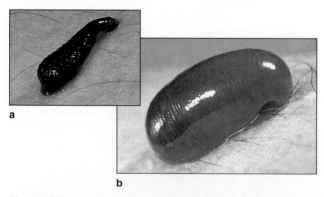

a

b

Figure 26.27 A leech before and after gorging itself with blood from a human host.

For at least 2,000 years, medical practitioners have used *Hirudo medicinalis*, a freshwater leech, as a blood-letting tool. Especially during the nineteenth century, doctors used leeches to "cure" problems ranging from nosebleeds to obesity. By one estimate, leeches were drawing more than 300,000 liters of blood annually from patients in France alone.

Nowadays, leeches are still used, but more selectively. For example, after surgeons reattach a severed ear, lip, or fingertip, leeches may be employed to draw off pooled blood. The body cannot do this on its own until the severed blood circulation routes are reestablished.

Polychaetes

The many-bristled polychaetes are the most diverse annelids. Most live in marine habitats. In fact, they are the most common animals along coasts, although you won't find them unless you turn over rocks or dig them up. Some excavate burrows in sand or soft mud; others build tubes from calcium carbonate secretions or from sand grains or bits of shells. Some are attached to substrates.

Sea nymphs and some other crawling polychaetes also can swim by undulating their long body. The swimmers typically have a specialized head end with notable sensory organs, sharp jaws and a muscular pharynx, and other structures. Their bristles project from well-developed parapods, a word taken to mean "closely resembling feet." Unlike your feet, parapods are fleshy-lobed appendages reinforced with chitin, not bone.

Many polychaetes dine on small invertebrates, some graze on algae, and others scavenge organic matter. The predatory types evert ("pop out") their jawed pharynx to capture prey. Herbivorous types grasp and tear algae. Most of the tube dwellers collect food suspended in the water with the help of ciliated, mucus-coated tentacles or featherlike structures (Figure 26.26).

Leeches

Leeches (Figure 26.27) are among the few annelids with no bristles; they inch about by using suckers at both ends of their muscular, segmented body. Different kinds

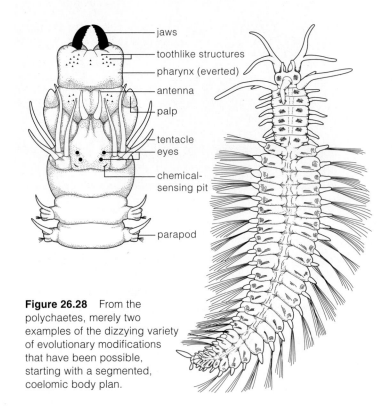

jaws
toothlike structures
pharynx (everted)
antenna
palp
tentacle
eyes
chemical-sensing pit
parapod

Figure 26.28 From the polychaetes, merely two examples of the dizzying variety of evolutionary modifications that have been possible, starting with a segmented, coelomic body plan.

Figure 26.29 A velvety "walking worm" of the small phylum Onychophora, the members of which resemble both the annelids and the arthropods. These worms live mostly in humid forests, especially of the tropics. They hide by day under rotting logs and in leaf litter and hunt for prey at night. The largest is about 15 centimeters (nearly 6 inches) long. Depending on the species, there are between fourteen and forty-three pairs of stumpy, unjointed legs with claws at the tip (onychophoran means "claw bearer").

Like annelids, these worms have ciliated nephridia. Like arthropods, they molt and they have an open circulatory system, with blood entering spaces in body tissues and reentering the heart through pores. Moreover, they have tracheal tubes similar to those of millipedes, centipedes, and insects.

Such similarities do not mean that onychophorans are "missing links" between annelids and arthropods. They are probably one of many invertebrates that evolved from annelid ancestors and have survived to the present.

live in freshwater, the seas, and, in tropical regions, moist habitats on land. Most swallow small animals or kill them and suck out their juices.

The leeches that most people have heard about feed on vertebrate blood (Figure 26.27). Blood-sucking leeches have sharp jaws. They plunge a sucking apparatus into incisions they make in prey, then secrete a substance that prevents blood from coagulating while they eat. The leech gut has many side branches for storing food taken in during a big meal, this being handy for an animal that may have long waits between meals.

Regarding the Segmented, Coelomic Body Plan

When you look carefully at Figure 26.28, you can sense that a segmented, coelomic body plan must be terrific working material for evolutionary forces. When the body is divided this way, different segments can become specialized in the performance of different tasks.

Most of the segments in an annelid are much the same, but among some species we have evidence of evolutionary forays into specialization. Leeches have suckers at both ends. Polychaetes have specialized parapods, and many have truly spectacular elaborations at their head end. In sea nymphs and their relatives, differentiated segments at the posterior end of the body perform reproductive functions.

Long ago, among the ancestors of annelids and other protostomes, evolutionary experimentation must have been rampant. We find further evidence of this in the onychophorans, a small, distinct phylum that seems to have evolutionary links to annelids and some adaptations like those of arthropods (Figure 26.29).

---

1. Annelids include the earthworms, polychaetes, and leeches. All are elongated, bilateral worms having segmentation and a true coelom.

2. Annelids resemble flatworms in some features, including their water-regulating system. They resemble arthropods in certain aspects of their segmented, coelomic body plans.

3. Given their array of traits, the annelids were probably one of the first groups to evolve on the protostome branch of the animal family tree.

26.14 ARTHROPODS

Of all invertebrate phyla, the arthropods (Arthropoda) show the most diversity. We have discovered more than a million species, most of which are insects. New ones are being discovered weekly, especially in the tropics. It seems likely that ancient annelids (or animals very much like them) gave rise to arthropod lineages. The trilobites, one of the four major lineages, are now extinct (Figure 21.10*d*).

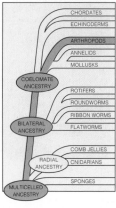

The other three lineages are the chelicerates (such as spiders), crustaceans (such as crabs), and uniramians (the centipedes, millipedes, and insects).

Adaptations of Insects and Other Arthropods

Arthropods are truly abundant, they engage in enormously different life-styles, and they have been evolutionary success stories in nearly all habitats. Six adaptations contributed to the success of the group in general and insects in particular:

1. A hardened exoskeleton
2. Specialized segments
3. Jointed appendages
4. Specialized respiratory structures
5. Efficient nervous system and sensory organs
6. Division of labor in the life cycle

Hardened Exoskeletons The arthropod cuticle is actually an external skeleton, or **exoskeleton**. Although it may be hardened, its protein and chitin components make it light, flexible—and protective. Think of the calcium-stiffened exoskeleton of a lobster or crab. It is like armor plating. Probably exoskeletons evolved as defenses against predators. They took on added functions when certain arthropods first invaded the land. An exoskeleton restricts evaporative water loss, and it can support a body deprived of water's buoyancy. It does restrict increases in size. But arthropods grow in spurts, by **molting**. At different stages in the life cycle, they grow a new, soft exoskeleton under the old one, which they shed (Figure 26.30). Before this new one hardens, aquatic arthropods enlarge by swelling with water; land-dwelling types swell with air.

Specialized Segments As arthropods evolved, body segments became fewer in number, grouped or

Figure 26.30 Molting of a centipede exoskeleton.

fused in various ways—and more specialized in function. For example, in insects, different segments combined to form a head, thorax, and abdomen. In spiders, segments fused as a forebody and hindbody.

Jointed Appendages Arthropods have highly specialized, jointed appendages. (Arthropod means "jointed foot.") For example, some paired appendages are used in feeding and detecting stimuli. Others are used in locomotion, feeding, sensing stimuli, transferring sperm to females, or spinning silk.

Respiratory Structures Aquatic arthropods rely on gills for gas exchange. The evolution of tracheas helped bring about diversity among the land-dwelling arthropods, insects especially. Insect **tracheas** begin as pores on the body surface and branch into narrow tubes that deliver oxygen directly to body tissues (page 698). They help support such energy-consuming activities as insect flight.

Specialized Sensory Structures Intricate eyes and other sensory organs contributed to arthropod success. Many species have a wide angle of vision and can process visual information from many directions.

Division of Labor Many insects have a division of labor among different stages of their life cycle. They first develop as sexually immature larvae that molt and change as they grow. Then larvae enter a pupal stage, which involves massive reorganization and remodeling of body tissues. Growth and major transformation of a larva into the adult form is called **metamorphosis**.

Moths, butterflies, beetles, and flies are examples of metamorphosing insects (Figure 1.4). Their larval stages specialize in *feeding and growth*, whereas the adult is concerned mostly with *dispersal and reproduction*. This division of labor is adaptive to seasonal changes in the environment, such as variations in food sources.

With this overview in mind, let's now sample the diversity among the major groups of arthropods.

a

Figure 26.31 Chelicerates. (**a**) Horseshoe crab, one of the few living aquatic chelicerates. Beneath its hard, shieldlike cover are five pairs of legs, a defining feature of the group. (**b**) Wolf spider. Like most spiders, it plays a role in keeping insect populations in check and is harmless to humans. (**c**) Brown recluse. Its bite can be severe to fatal for humans. This North American spider lives under bark and rocks, and in and around buildings. A violin-shaped mark is present on its forebody. (**d**) A female black widow. Its bite can be painful and sometimes dangerous. (**e**) A book lung, a type of respiratory organ of many spiders and scorpions.

Chelicerates

Chelicerates originated in shallow seas in the early Paleozoic. Except for a few mites, the only marine survivors are the horseshoe crabs (Figure 26.31*a*) and sea spiders. Of the familiar existing chelicerates—spiders, scorpions, ticks, and mites—we might say this: never have so many been loved by so few.

Like scorpions, spiders are predators, and many help control populations of insect pests. Ever so often, venomous types bite humans, and they have given the whole group a bad reputation. Ticks are blood-sucking parasites of vertebrates. Some transfer the bacteria that cause Rocky Mountain spotted fever, Lyme disease, and other diseases to humans (page 354). Most mites are free-living scavengers.

Spiders and scorpions pump digestive enzymes from their gut into prey, then suck up the liquefied remains. Their forebody has several eyes, four pairs of legs, a pair of chelicerae (which inflict wounds and discharge venom), and a pair of pedipalps (for grasping). Appendages on the spider hindbody spin out threads of silk for webs and egg cases. Most webs are netlike, but one spider spins a vertical thread with a ball of sticky material at the end. It uses one of its legs to swing the ball at passing insects!

Spiders and scorpions have pocketlike respiratory organs called book lungs (Figure 26.31*e*). The folded walls of the pockets resemble book pages, and they greatly increase the surface area available for gas exchange. Blood circulating between the moist folds picks up oxygen diffusing in from the surrounding air.

b

c

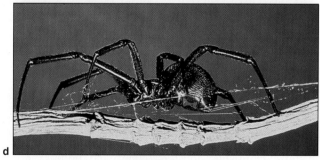

d

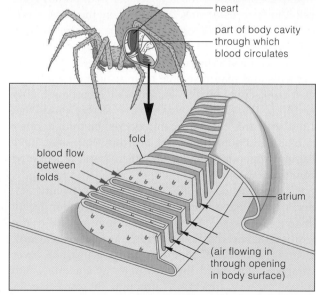

heart

part of body cavity through which blood circulates

fold

blood flow between folds

atrium

(air flowing in through opening in body surface)

e

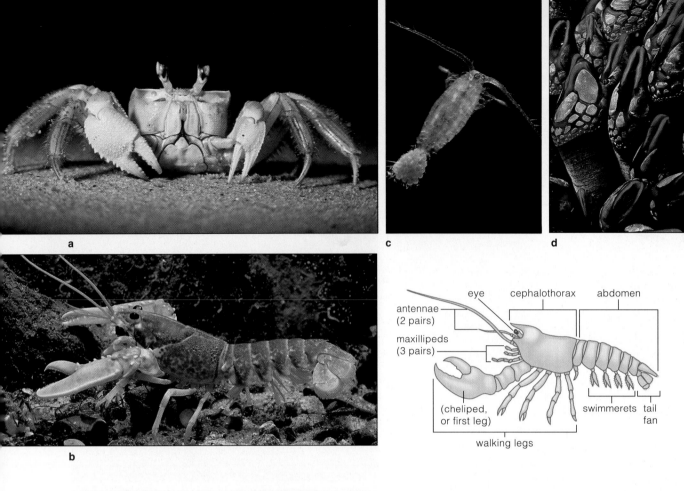

Figure 26.32 Representative crustaceans. (**a**) A crab and (**b**) a lobster, displaying their splendid appendages. (**c**) A marine copepod. (**d**) Stalked barnacles that grow in intertidal zones of the Pacific Ocean.

Crustaceans

Shrimps, crayfishes, lobsters, crabs, and pillbugs are well-known crustaceans. Most live in the seas; some live in freshwater or on land. All have roles in food webs, and humans harvest many edible types.

The simplest crustaceans have unspecialized body segments and may resemble their ancient ancestors. In most crustaceans, many segments have become highly modified. For example, crabs have strong claws that shred seaweed, collect organic debris, intimidate other animals, and sometimes dig burrows. Barnacles have feathery appendages that comb microscopic bits of food from the water.

Crustaceans commonly have sixteen to twenty segments; some have over sixty. The segmentation is not obvious in crabs, lobsters, and some other types because the head's dorsal cuticle extends backward, covering some or all segments with a shieldlike cover (carapace). The head has two pairs of antennae with mostly sensory functions, although in some species they are used in swimming. It has a pair of jawlike

appendages (mandibles) and a pair of food-handling appendages (maxillae). Figure 26.32*a* and *b* shows how these parts are arranged in crabs and lobsters.

With their ten pairs of legs projecting from the thorax, the crabs, lobsters, shrimps, crayfish, and their relatives are decapods (*deca-* means ten). Compared to other decapods, crabs are rather broad and flat, but their weight is distributed so evenly over their legs that they are agile walkers.

Figure 26.32*c* shows one of the 7,500 species of copepods, which are less than 2 millimeters long. A single eye in the middle of their head is a distinctive feature. These crustaceans are the most abundant aquatic animals. About 1,500 species are parasites of fishes and various invertebrates. The rest are weakly swimming members of plankton. Huge numbers of them feed on tiny photosynthetic protistans. By converting this food into their own tissues, they become food themselves for other organisms, and so on up through vital aquatic food webs.

Of all the arthropods, only barnacles have a strong, calcified "shell." It is actually a modified exoskeleton, and it protects them from predators, drying out, and the force of waves or currents. Some barnacles attach to rocks and other surfaces by an elongated stalk (Figure 26.32*d*). Others are stalkless forms that attach directly to rocks, wharf pilings, ship hulls, and similar surfaces. A few species prefer to attach themselves only to the skin of whales.

Insects and Their Relatives

Millipedes and Centipedes Closely related to insects are the millipedes and centipedes, which typically live under damp rocks, logs, or forest litter. As Figure 26.33 shows, both kinds of arthropods have a long, segmented body with many legs. Millipedes have a cylindrical or slightly flattened body. Different species range from about 2 millimeters to nearly 30 centimeters in length. All are slow-moving, nonaggressive scavengers of decaying plant material.

Centipedes have a streamlined, flattened body. All are fast-moving, aggressive carnivores, complete with fangs and venom glands. They prey on insects, earthworms, and snails, and some tropical species can subdue small lizards and toads. A few centipedes inflict painful bites.

Insects As a group, insects share the adaptations listed on page 432. Here we add a few details to the list. Insects have a head, a thorax, and an abdomen. The head has a pair of sensory antennae and paired mouthparts, specialized for biting, chewing, sucking, or puncturing (Figure 26.34). The thorax has three pairs of legs and usually two pairs of wings. Usually the only abdominal appendages are reproductive structures, such as egg-laying devices.

The insect gut is divided into foregut, midgut, and hindgut. Most digestion proceeds in the midgut; water is reabsorbed in the hindgut. Insects get rid of waste material through **Malpighian tubules**, small tubes that connect with the midgut. Nitrogen-containing wastes from protein breakdown diffuse from the blood into the tubules, where they are converted into harmless crystals of uric acid. The crystals enter the midgut and are eliminated with feces. This system allows land-dwelling insects to get rid of potentially toxic wastes without losing precious water.

a

b

Figure 26.33 (**a**) A mild-mannered millipede that scavenges on decaying plant parts. (**b**) An aggressive centipede of Southeast Asia that preys on small frogs and lizards.

Figure 26.34 From the insects, examples of specialized arthropod appendages. These headparts are used in feeding.

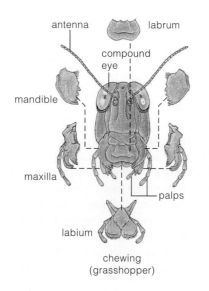

chewing
(grasshopper)

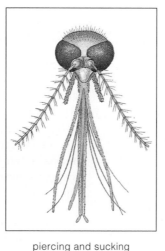

piercing and sucking
(mosquito)

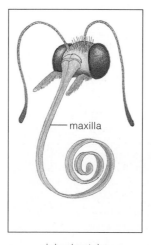

siphoning tube
(butterfly)

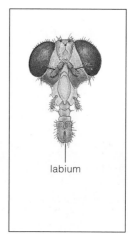

sponging
(housefly)

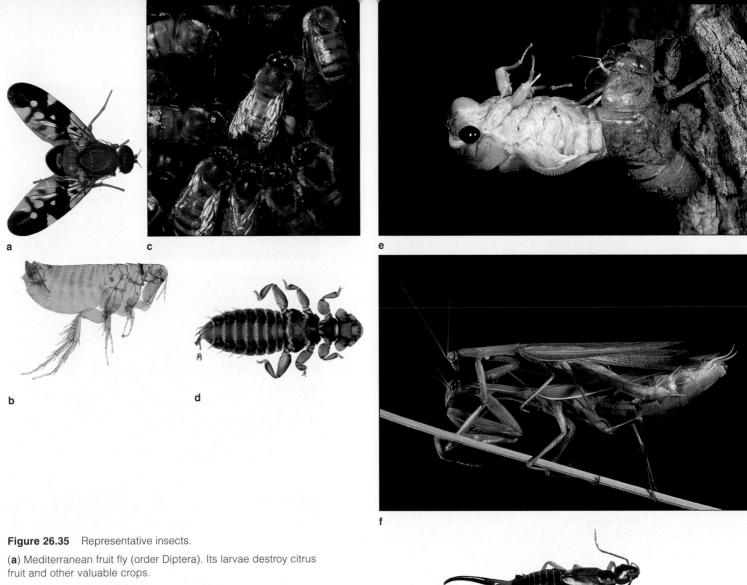

Figure 26.35 Representative insects.

(**a**) Mediterranean fruit fly (order Diptera). Its larvae destroy citrus fruit and other valuable crops.

(**b**) Flea (order Siphonaptera), with big strong legs for jumping onto and off animal hosts.

(**c**) A honeybee (order Hymenoptera) attracting its hive mates with a dance, as described in Chapter 52.

(**d**) Duck louse (order Mallophaga). It eats feather particles and bits of skin.

(**e**) Off with the old, on with the new. On the side of a tree trunk, a green cicada (order Homoptera) wriggling out of its outgrown cuticle during a molting cycle.

(**f**) A male praying mantid (order Mantodea) mating with a larger female. He may be eaten during or immediately after mating with her.

(**g**) European earwig (order Dermaptera), a common household pest.

(**h**) Stinkbugs (order Hemiptera), newly hatched.

(**i**) Ladybird beetles (order Coleoptera) swarming. These beetles are raised commercially and released in great numbers as biological controls of aphids and other pests. Also in this order, the scarab beetle (**j**). With more than 300,000 named species, Coleoptera is the largest order in the animal kingdom.

(**k**) Luna moth (order Lepidoptera), a flying insect of North America. Like most other moths and butterflies, its wings and body are covered with microscopic scales.

(**l**) Dragonfly (order Odonata). This swift aerialist captures and eats other insects in midflight.

Insect Diversity

More than 800,000 species of insects have already been catalogued. Figure 26.35 shows just a few representatives from the major orders of insects.

If we use sheer numbers and distribution as the criteria, which are the most successful insects? The most successful types are small in size and have a staggering reproductive capacity. Many might grow and reproduce in great numbers on a single plant that might be only an appetizer for another animal. By one estimate, if all the progeny of a single female fly were to survive and reproduce through six more generations, that fly would have more than 5 trillion descendants! Metamorphosis also is common in insect life cycles. This developmental

route allows the young and the adults to use different resources at different times.

Besides this, the most successful insect species are winged. In fact, they are the *only* invertebrates that have wings. They are able to move among food sources that are too widely scattered to be exploited by other animals. Their capacity for flight contributed greatly to their success on land.

The factors that contribute to their success also make insects our most aggressive competitors. They destroy vegetable crops, stored food, wool, paper, and timber. They draw blood from us and from our pets, and transmit pathogenic microorganisms as they do so. On the bright side, many insects pollinate flowering plants in general and crop plants in particular. And many "good" insects attack or parasitize the ones we would rather do without.

This completes our survey of the insects and other arthropods. Although our sampling necessarily has been limited, it has been enough to reinforce the following key point:

Arthropods, especially insects, owe their evolutionary success to hardened exoskeletons, specialized segments with jointed appendages, efficient respiratory and sensory structures, and often a division of labor in the life cycle.

26.15 ECHINODERMS

We turn now to a different lineage of coelomate animals, the deuterostomes. The major invertebrate members of this lineage include **echinoderms** (Echinodermata, meaning spiny skinned). Sea urchins, sea cucumbers, feather stars, brittle stars, and sea stars belong to this phylum (Figure 26.36). All contain calcium-stiffened spines, spicules, or plates in their body wall.

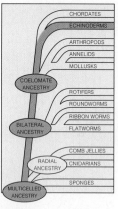

Notice the radially arranged arms of the sea star in Figure 26.36e. Oddly, adult echinoderms are radial with some bilateral features. Some even produce bilateral larvae during their life cycle. (Did bilateral invertebrates give rise to the ancestors of echinoderms? Maybe.)

Adult echinoderms have no brain. However, their decentralized nervous system does allow them to respond to stimuli coming from different directions. For instance, any arm of a sea star that senses food or danger can become the leader, directing the rest of the body to move in a suitable direction.

If you turn a sea star over, you will see **tube feet**. These are fluid-filled, muscular structures with sucker-like adhesive disks. Tube feet are used for walking, burrowing, clinging to a rock, or gripping a clam or snail about to become a meal. They are part of a **water vascular system** unique to echinoderms (Figure 26.37). In sea stars, the system includes a main canal in each arm. Short side canals extend from them and deliver water to the tube feet. Each tube foot has an ampulla, a fluid-filled, muscular structure something like the rubber bulb on a medicine dropper. When the ampulla contracts, it forces fluid into the foot, which thereby lengthens.

Tube feet change shape constantly as muscle action redistributes fluid through the water vascular system. Hundreds of tube feet may move at a time. After being released, each one swings forward, reattaches to the substrate, then swings backward and is released before swinging forward again.

Some sea stars swallow their prey whole. Many can push part of their stomach outside the mouth and around their prey, then start digesting the prey even before swallowing it. Sea stars get rid of coarse, undigested residues through the mouth. They do have a small anus, but this is of no help in getting rid of empty clam or snail shells.

With their curious assortment of traits, the echinoderms are a suitable point of departure for this chapter.

tube feet   spine

a

b

Figure 26.36 Echinoderms. (**a**) Sea urchin, which moves about on spines and tube feet. The spines also protect it from predators. If you step on a sea urchin and a spine breaks off under your skin, it can trigger inflammation. (**b**) Sea cucumber, with rows of tube feet along its body. (**c**) Feather star, with its finely branched food-gathering appendages. (**d**) Brittle stars. Their arms (rays) make rapid, snakelike movements. (**e**) Five-armed sea star, and a close view of its tube feet (**f**) and of the feeding apparatus on its ventral surface (**g**).

Even though we can identify broad trends in animal evolution, we should keep in mind that there are always confounding exceptions to the general rule.

Echinoderms, one of evolution's puzzling experiments, are mostly radial (but partly bilateral) animals with a coelom and with spines, spicules, or plates in the body wall.

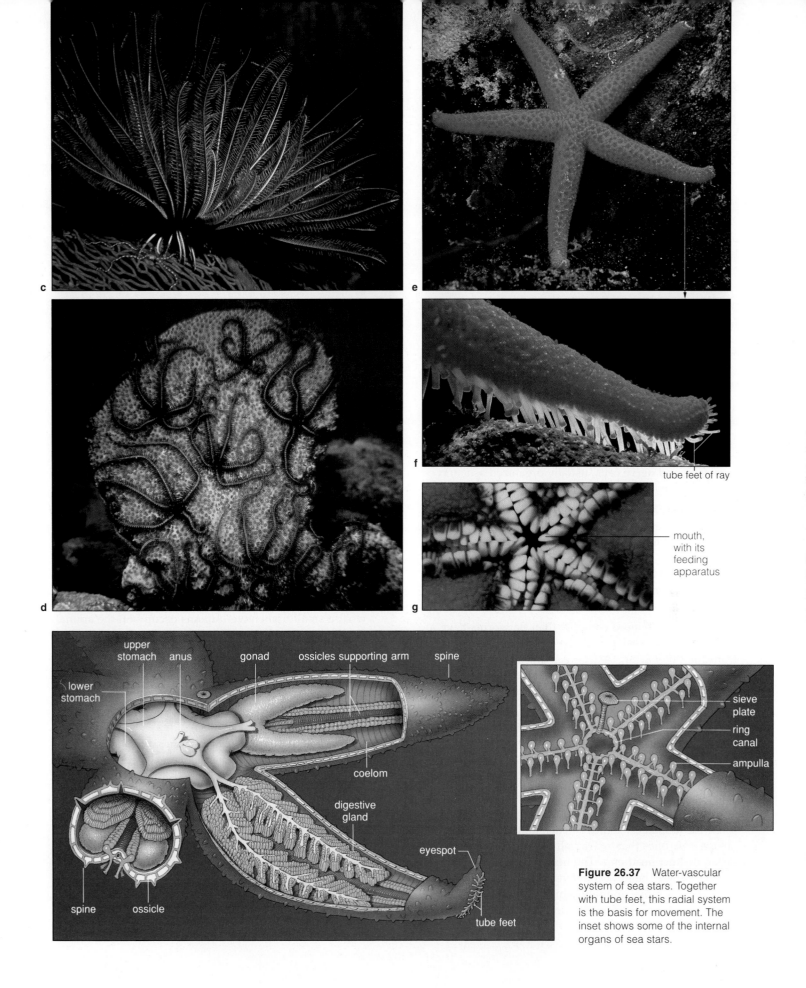

c

d

e

f

tube feet of ray

g

mouth, with its feeding apparatus

upper stomach   anus   gonad   ossicles supporting arm   spine

lower stomach

coelom

digestive gland

eyespot

spine   ossicle

tube feet

sieve plate

ring canal

ampulla

Figure 26.37 Water-vascular system of sea stars. Together with tube feet, this radial system is the basis for movement. The inset shows some of the internal organs of sea stars.

SUMMARY

1. Multicellular animals range from sponges to vertebrates. The members of each animal phylum share certain characteristics. By comparing phyla and integrating the information with fossil evidence, we can identify major trends that unfolded during the evolution of different groups.

2. The most revealing aspects of the animal body plan are its type of symmetry, gut, and cavity (if any) between the gut and body wall; whether it has a distinct head end; and whether it is divided into a series of segments (Figure 26.38).

3. Sponges, the simplest multicellular animals, do not have a gut or nervous system. Although they have several kinds of cells, these are not organized into tissues. Some cells trap food particles from the surrounding water. The beating of their flagella moves water through a system of pores, canals, and chambers.

4. Jellyfishes, sea anemones, and their relatives have radial symmetry, and their cells form tissues. The outer and inner tissue layers are separated by mesoglea, a secreted material. Cnidarians have contractile cells, sensory cells, and nerve cells, usually organized as a nerve net. They alone produce stinging capsules (nematocysts).

5. Most animals more complex than cnidarians are bilaterally symmetrical, and a type of tissue called mesoderm forms in their embryos. Mesoderm gives rise to muscle and other tissues that lie between the lining of the gut and the epidermis. The gut may be saclike, as it is in flatworms, but usually it is complete, with an anus as well as a mouth.

6. Most animals more complex than flatworms have a coelom or false coelom (cavities that separate the gut from the body wall). A true coelom originates within pouches or masses of mesoderm, and is lined by a peritoneum.

7. Most notably among the annelids and arthropods, the body is segmented (partitioned into a series of units). Arthropod segments tend to be highly specialized, and groups of successive segments form distinct body regions, such as the head, thorax, and abdomen of an insect. Arthropods also have a well-developed exoskeleton and jointed appendages.

8. Of the animals with a true coelom, the echinoderms are the only ones that are radially symmetrical and that lack a distinct head end. The nervous system is decentralized, there being no brain. This arrangement is favorable for radial animals, which can respond to food and danger coming from any direction.

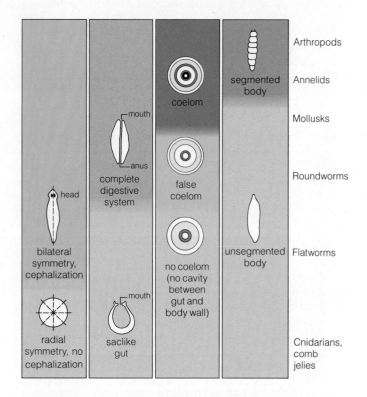

Figure 26.38 Summary of key trends in animal evolution, identified by comparing body plans of major phyla.

9. Table 26.2, on pages 442 and 443, summarizes the characteristics of the major groups of invertebrates described in this chapter. To provide perspective on what a small branch we humans occupy on the animal family tree, the phylum Chordata is included as the last entry for this table.

Review Questions

1. In what ways do invertebrates and vertebrates differ? *410*

2. Describe the features of one group of invertebrates. *428–431*

3. Name some of your paired body parts that evolved in your bilateral, cephalized ancestors. *410–411*

4. What is a coelom? Why was it important in the evolution of certain animal lineages? *411*

5. Name some animals with a saclike gut. Evolutionarily, what advantages does a complete gut afford? *414–418*

6. Choose an insect that thrives in your neighborhood and describe some of the adaptations that underlie its success. *432, 435–437*

1. Five body features that help us identify trends in animal evolution are _____ , _____ , _____ , _____ , and _____ .

2. Animals are _____ .
   a. autotrophs
   c. chemotrophs
   b. heterotrophs
   d. saprobes

3. Which is *not* characteristic of the animal kingdom?
   a. multicellularity; cells form tissues, organs
   b. exclusive reliance on sexual reproduction
   c. motility at some stage of the life cycle
   d. embryonic development during the life cycle

4. Jellyfishes, sea anemones, and their relatives have _____ symmetry, and their cells form _____ .
   a. radial; mesoderm
   b. bilateral; tissues
   c. radial; tissues
   d. bilateral; mesoderm

5. In sheer numbers and distribution, _____ are the most successful animals.
   a. arthropods
   d. sea stars
   b. sponges
   e. vertebrates
   c. snails and clams

6. Bilateral, segmented bodies and hardened exoskeletons occur among the _____ .
   a. arthropods
   d. sea stars
   b. sponges
   e. vertebrates
   c. snails and clams

7. Which phylum contains members that are notorious for causing serious diseases in humans?
   a. cnidarians
   c. segmented worms
   b. flatworms
   d. chordates

8. More complex animals have a _____ between the gut and body wall.
   a. pharynx
   c. coelom
   b. peritoneum
   d. archenteron

9. Most animals that are more complex than cnidarians have _____ symmetry, and _____ forms in their embryos.
   a. radial; mesoderm
   b. bilateral; endoderm
   c. bilateral; mesoderm
   d. radial; endoderm

10. Match the terms with the appropriate groups.
    ____ sponges   a. spiny-skinned
    ____ cnidarians   b. vertebrates and kin
    ____ flatworms   c. flukes and tapeworms
    ____ roundworms   d. no tissue organization
    ____ rotifers   e. no males for some
    ____ mollusks   f. nematocysts, radial symmetry
    ____ annelids   g. hookworms, elephantiasis
    ____ arthropods   h. jointed appendages
    ____ echinoderms   i. "belly-foots" and kin
    ____ chordates   j. segmented worms

Selected Key Terms

animal *410*
annelid *428*
bilateral symmetry *410*
cephalization *411*
cnidarian *414*
coelom *411*
collar cell *413*
comb jelly *417*
contractile cell *414*
cuticle *419*
deuterostome *423*
echinoderm *438*
ectoderm *410*
endoderm *410*
epithelial tissue *414*
exoskeleton *432*
flatworm *418*
ganglion *428*
gonad *417*
gut *411*
hermaphrodite *418*
hydrostatic skeleton *414*
invertebrate *410*
larva *413*
Malpighian tubule *435*
mantle *424*
medusa *414*

mesoderm *410*
metamorphosis *432*
mollusk *424*
molting *432*
nematocyst *414*
nephridium *429*
nerve cell *414*
nerve cord *428*
organ *418*
organ system *418*
placozoan *412*
planula *417*
polyp *414*
proglottid *419*
protostome *423*
radial cleavage *423*
radial symmetry *410*
ribbon worm *422*
roundworm *419*
scolex *419*
sensory cell *414*
spiral cleavage *423*
sponge *412*
trachea *432*
tube foot *438*
vertebrate *410*
water-vascular system *438*

Readings

Hickman, C. P., and L. S. Roberts. 1993. *Integrated Principles of Zoology.* Ninth edition. St. Louis: Mosby.

Kozloff, E. 1990. *Invertebrates.* Philadelphia: Saunders.

Pearse, V. and J., and Buchsbaum, M. and R. 1987. *Living Invertebrates.* Palo Alto, California: Blackwell.

Pough, F. H., J. Heiser, and W. McFarland. 1989. *Vertebrate Life.* Third edition. New York: Macmillan.

Table 26.2 Summary of Characteristics of the Major Animal Phyla

Phylum (and some representatives)	Typical Environment	Typical Life-Style of Adult Form	Nervous System	Support and Movement
Porifera (8,000)* sponges	Most marine, some freshwater	Attached filter feeders	No nervous system (only cell-to-cell communication)	Support by spicules, protein fibers, or both; contractile cells change openings at body surfaces
Cnidaria (11,000) hydras, sea anemones, jellyfishes, corals	Most marine, some freshwater	Attached, creeping, swimming, or floating carnivores	Nerve net, cell-to-cell transmission	Hydrostatic support (by fluid in gut), by secreted jellylike mesoglea or by skeletal elements; contractile fibers in epithelial cells
Ctenophora (100) comb jellies	Marine	Mostly planktonic; few attached or creeping	Nerve net	Support by mesoglea; muscles separate from epithelia; locomotion by cilia in eight rows of comblike structures; sometimes by muscular activity
Platyhelminthes (15,000) flatworms, flukes, tapeworms	Marine, freshwater, some terrestrial in moist places; many parasitic in or on other animals	Herbivores, carnivores, scavengers, parasites	Brain, nerve cords	Hydrostatic support (no secreted skeleton); well-developed muscle tissue
Nemertea (800) ribbon worms	Most marine; few freshwater, terrestrial	Mostly carnivores	Brain, nerve cords	Hydrostatic support; well-developed muscle tissue
Nematoda (20,000) roundworms	Marine, freshwater, terrestrial; many parasitic	Scavengers, carnivores parasites	Nerve ring, nerve cords	Hydrostatic support; (by false coelom); tough cuticle; longitudinal muscle in body wall
Rotifera (1,800) rotifers	Marine, freshwater, moisture on mosses	Mostly filter feeders, capturing bacteria, unicellular algae	Brain, nerve cords	Locomotion by cilia; muscles for shape changes
Mollusca (110,000) snails, slugs, clams, squids, octopuses	Marine, freshwater, terrestrial	Herbivores, carnivores, scavengers, detritus or filter feeders; mostly free-moving, some attached	Brain, nerve cords, major ganglia other than brain	Hydrostatic skeleton in most; well-developed musculature in foot, mantle, other structures
Annelida (15,000) earthworms, leeches, polychaetes	Marine, freshwater, terrestrial in moist places	Herbivores, carnivores, scavengers, detritus or filter feeders; mostly free-moving	Brain, double ventral nerve cord	Hydrostatic skeleton (using coelom); well-developed musculature in body wall
Arthropoda (1,000,000+) crustaceans, spiders, insects	Marine, freshwater, terrestrial	Herbivores, carnivores, scavengers, detritus or filter feeders, parasites; mostly free-moving	Brain, double ventral nerve cord	Exoskeleton (of cuticle); jointed appendages; muscles mostly in bundles
Echinodermata (6,000) sea stars, brittle stars, sea urchins, sea lilies, sea cucumbers	Strictly marine	Mostly carnivores, detritus feeders, few herbivores; most free-moving, some attached	Radially arranged nervous system	Endoskeleton (of spines, etc.); muscles for body movement, tube feet often used in locomotion
Chordata (47,100) tunicates, lancelets, jawless fishes, jawed fishes, amphibians, reptiles, birds, mammals	Marine, freshwater, terrestrial	Herbivores, carnivores, scavengers, filter feeders; generally free-moving (most tunicates attached as adults)	Well-developed brain, dorsal and tubular nerve cord in most	Notochord or a bony or cartilaginous endoskeleton; well-developed musculature in most

*Number in parentheses indicates approximate number of known species.

Digestive System	Respiratory System	Circulatory System	Mode of Reproduction
No gut; microscopic food particles secured by individual cells	None; respiration by individual cells	None	Sexual (certain cells become or produce gametes); some asexual budding; production of resistant bodies (gemmules)
Saclike gut (may be branched)	None; respiration by individual cells; gut may distribute oxygen	None, other than via gut	Sexual (usually separate sexes; gonads discharge gametes into gut or to exterior); asexual
Saclike, but branched	None; respiration by individual cells; gut may distribute oxygen	None, other than via gut	Sexual (usually hermaphroditic); gonads closely associated with gut
Saclike gut (may be branched)	None; gas exchange across body surface	None	Sexual (usually hermaphroditic, with complex reproductive system); some asexual
Complete gut	None; gas exchange across body surface	Closed system	Sexual (sexes usually separate; reproductive system simple)
Complete gut	None; gas exchange across body surface	False coelom	Sexual (sexes separate; reproductive system fairly complex)
Usually complete gut (sometimes saclike)	None; gas exchange across body surface	False coelom	Mostly parthenogenetic (eggs develop without fertilization); males appear only occasionally
Complete gut	Ctenidia, other gills; mantle can be modified as lung; gas exchange across body surface	Usually open (closed system in cephalopods)	Sexual (hermaphroditic or separate sexes; reproductive system usually complex)
Complete gut	Gas exchange across body surface; varied outgrowths of surface in many	Usually closed; coelom also may function in distribution	Sexual (hermaphroditic or sexes separate; reproductive system simple or complex); asexual also in many
Complete gut	Gills; tracheal tubes; book lungs; general body surface in some	Open system	Sexual (usually separate sexes; reproductive system fairly complex)
Usually complete gut (sometimes saclike)	Gas exchange across general body surface or surface of outgrowths of it (such as tube feet)	Coelom around viscera, also water-vascular coelom	Sexual (reproductive system simple; gonads usually discharge gametes directly to exterior); some asexual
Complete gut	Lungs in most vertebrates other than fishes; perforated pharynx; gills; gas exchange across body surface	Closed system in most (open system in most tunicates); lymphatic system in many vertebrates	Sexual (sexes usually separate, except in most tunicates); asexual in some tunicates

ANIMALS: THE VERTEBRATES

Making Do (Rather Well) With What You've Got

It has taken the platypus nearly two centuries to earn a little respect. In 1798, skeptical naturalists at the British Museum in London poked and probed a specimen, looking for signs that a prankster had stitched the bill of an oversized duck onto the pelt of a small furry mammal. No wonder. At first blush the platypus looks like one of nature's practical jokes (Figure 27.1*a*).

The platypus is a web-footed mammal, about half the size of a housecat. Like other mammals, it has mammary glands and hair. Yet, like birds and reptiles, it has a cloaca, a single external opening that functions in both reproduction and excretion. Like birds and most reptiles, it lays shelled eggs. Its young hatch from eggs early in their development, pink and helpless (Figure 27.1*b*). And about that fleshy platypus bill! It looks like it belongs to a duck. The broad, flat, furry tail seems borrowed from a beaver.

With its unusual array of traits, the platypus is a fine example of an animal that has been judged according to *our* ideas of what "an animal" is supposed to be. Those who assume that the platypus doesn't quite measure up are not using the real yardstick. *Its collection of traits happens to be exquisitely adapted for survival and reproduction under a particular set of environmental conditions.*

The platypus is a predator of streams and lagoons in remote parts of Australia and Tasmania. It dives into the water at night, forages skillfully for small edible mollusks and other prey, then hides during the day in underground burrows.

Its dense, blackish-brown fur functions in insulating the body. When a platypus is submerged all night long in cold water or sequestered all day in a cool burrow, the fur holds in heat and so helps maintain body temperature within tolerable limits. The broad, thick tail functions as a rudder, enhancing maneuverability through the water. It also functions as a storehouse for

Figure 27.1 One of evolution's success stories—the platypus, underwater (**a**) and in its burrow (**b**).

a

b

1. The chordate branch of the animal family tree includes invertebrate and vertebrate species. All are bilateral animals.

2. All chordates have a notochord (a supporting rod for the body), dorsal nerve cord, pharynx, and gill slits in the pharynx wall. These features appear in all chordate embryos, and some or all persist in the adult forms of different species. Vertebrate chordates additionally have a cartilaginous or bony backbone and protective chamber for the brain.

3. The existing invertebrate chordates include tunicates and lancelets.

4. The existing vertebrates include jawless fishes, cartilaginous fishes, bony fishes, amphibians, reptiles, birds, and mammals. Another group, the placoderms, became extinct early in vertebrate history.

5. Five major trends occurred during the evolution of vertebrates, although not in every group. Structural support and locomotor functions came to depend less on the notochord and more on a vertebral column. Jaws evolved. The nerve cord expanded into a spinal cord and brain. Among aquatic ancestors of land-dwelling vertebrates, gas exchange came to depend less on gills and more on lungs and an efficient circulatory system. Also, fleshy fins with skeletal supports evolved into legs, which became modified further in specialized ways in amphibians, reptiles, birds, and mammals.

energy-rich fats. The webbed feet can function as paddles when they flare out. The strong, clawed hind feet are great for digging.

When submerged in water, the platypus nostrils close, a fleshy groove around the eyes and ears snaps shut, and the bill takes over sensory functions. The platypus has a sensory system for land and a separate sensory system for underwater! More than 800,000 sensory receptors are built into the bill. Many detect mechanical pressure—waves produced by swimming prey. Others detect weak electric currents; even the tiny electric field created by the flick of a shrimp tail will do the trick. A platypus can zero in on a meal with awesome precision. It uses the flattened bill to scoop up freshwater snails, mussels, insect larvae, worms, and shrimps, then grinds them up on the horny pads of its jaws.

This remarkable body plan started evolving about 100 million years ago, when ancestors of the platypus coexisted with dinosaurs on Gondwana (page 336). That supercontinent broke apart into huge fragments, and the ancestral forms happened to be on the land mass that eventually became Australia. The platypus lineage has endured ever since. With that track record, who are we to chuckle over the collection of platypus traits? As is the case for all other animals, the traits meet nature's most important test. They work.

27.1 THE CHORDATE HERITAGE

Characteristics of Chordates

The preceding chapter concluded with a look at the echinoderms, some of our distant relatives on the deuterostome branch of the animal family tree. We turn now to the vertebrates and their close relatives. All are bilateral animals, grouped together as **chordates** (phylum Chordata).

There are more than 47,000 species of "vertebrate chordates." These have a backbone of cartilage or bone, and they have skull bones that protect the brain. The 2,100 or so species of "invertebrate chordates" share certain features with the dominant members of this phylum, although a backbone isn't one of them.

Four features are evident in chordate embryos, and in many species these persist into adulthood. First, a **notochord**, a long rod of stiffened tissue (not cartilage or bone), helps support the body. Second, the nervous system is based on a tubular, dorsal **nerve cord** that forms parallel to the notochord and gut. The anterior end of the cord expands during development, forming the brain. Third, a muscular tube called a **pharynx** functions in feeding, respiration, or both. The chordate pharynx has slits in the wall. Fourth, a tail forms and extends past the anus.

Chordate Classification

Nearly all chordates belong to three subphyla: the Urochordata (such as tunicates), Cephalochordata (lancelets), and Vertebrata (vertebrates). As Figure 27.2 shows, there are eight classes of vertebrates:

Agnatha	*Jawless fishes*
Placodermi	*Jawed, armored fishes (extinct)*
Chondrichthyes	*Cartilaginous fishes*
Osteichthyes	*Bony fishes*
Amphibia	*Amphibians*
Reptilia	*Reptiles*
Aves	*Birds*
Mammalia	*Mammals*

Appendix I has an expanded classification scheme for vertebrates. Unit VI provides details of their body plans and functions. Here we become acquainted with major trends in their evolution. And we find clues to those trends among invertebrate chordates.

The embryos of chordates alone have this combination of features: a notochord, a tubular dorsal nerve cord, a pharynx with slits in the wall, and a tail extending past the anus.

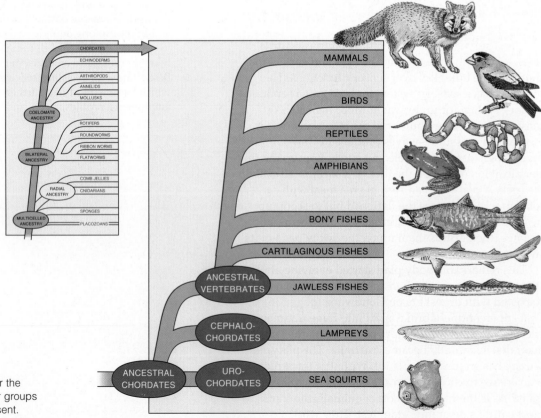

Figure 27.2 A family tree for the chordates, showing the major groups that have survived to the present.

INVERTEBRATE CHORDATES

Tunicates

The 2,000 species of existing urochordates are baglike animals, at least 2 centimeters long. They are often called **tunicates**, a name referring to the gelatinous or leathery "tunic" that adults secrete around themselves. The most common species are called sea squirts, because the adults tend to squirt out water through a siphon when something irritates them.

All tunicates live in marine habitats, ranging from the intertidal zone to surprising depths. Most adults remain attached to rocks, ship hulls, and other suitably hard substrates. Some live solitary lives; others are colonial. Figure 27.3 shows the body plan of one of the solitary types. It starts out as a bilateral, free-swimming larva that looks rather like a tadpole. (A larva, recall, is an immature stage between embryonic and adult stages of the life cycle.) The firm yet flexible notochord develops as a series of fluid-filled cells in the tail. It functions like a torsion bar. When muscles on one side or the other of the tail contract, the notochord bends. When the muscles relax, it springs back. The strong, side-to-side motion propels the animal forward. Most fishes use this kind of propulsive motion.

A sea squirt larva swims briefly, then attaches to a substrate and undergoes metamorphosis. The major tissue reorganization and remodeling transforms the larva into an adult. Its tail and notochord disappear; a tunic is secreted. The pharynx enlarges, and perforations in the pharynx wall become subdivided into many small slits. The tubular nerve cord regresses, leaving the adult with a much simplified nervous system. Figure 27.3 shows these events.

Water flows into and out of the adult body through two siphons. Incoming water enters a ciliated pharynx that is finely divided into openings called **gill slits**. As water flows through the sievelike pharynx, diatoms and other bits of food are strained from it. A heart and blood vessels service the thin-walled pharynx, which also serves as a respiratory organ.

Figure 27.3 (**a**) An adult tunicate. (**b**) The tadpolelike larva. It swims for a few minutes or days until it locates a suitable living site. (**c–e**) It attaches its head to a substrate. And now metamorphosis begins. The tail, notochord, and most of the nervous system are resorbed (recycled to form new tissues). The slits in the pharynx multiply. Organs become rotated until the openings through which water enters and leaves are directed away from the substrate.

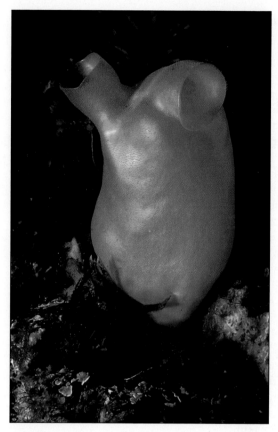

a Adult tunicate

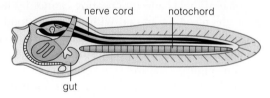

nerve cord   notochord

gut

b A tunicate larva

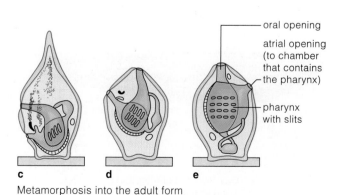

oral opening

atrial opening
(to chamber
that contains
the pharynx)

pharynx
with slits

c   d   e

Metamorphosis into the adult form

Lancelets

Just offshore, on seafloors around the world, we find twenty-five or so species of cephalochordates. These translucent, fish-shaped animals are seldom more than 5 centimeters (about 2 inches) long. They spend much of the time buried almost up to their mouth in sand and sediments. Their common name, **lancelets**, refers to the sharp tapering of their body at both ends. As is true of tunicate larvae, the notochord and muscles of a lancelet can work together to produce swimming motions.

Figure 27.4 includes a diagram of the lancelet body plan. The four distinctly chordate features are present. Besides this, its muscles are arranged in a segmented pattern on both sides of the notochord. In addition, lancelets have a closed circulatory system (but no red blood cells). Slow waves of contraction drive blood forward through some blood vessels and backward through others. Blood from all parts of the body collects in a sac that resembles the vertebrate heart, but this does not function as a muscular pump. Although a complex brain is nowhere in sight, the head end of the dorsal nerve cord is expanded and pairs of nerves extend into each muscle segment. Lancelets do not have an internal respiratory system as you do. Carbon dioxide and oxygen simply diffuse across their comparatively thin skin.

Lancelets are more like tunicates in their feeding mechanisms. They are filter feeders. Cilia lining the mouth cavity create a current that draws water through the mouth. The water passes through a pharynx, where food becomes trapped in mucus. Trapped food is delivered to the rest of the gut, where digestion proceeds, nutrients are absorbed, and waste material is compacted for elimination.

Cilia are microscopically small structures, and the driving force of their collective beating cannot by itself provide a filter-feeding animal with sufficient food. In such animals, great numbers of cilia must work in conjunction with a large food-trapping surface area. As you can see from Figure 27.4, the lancelet pharynx is quite large, relative to the overall body length, and it is perforated by up to 200 ciliated, food-trapping gill slits.

27.3 ORIGIN OF VERTEBRATES

Are lancelets "living fossils," the earliest members of the vertebrate lineage? You might think so, given their fishlike shapes, segmented muscles, circulatory system, pairs of nerves, and other features. However, even though lancelets might share some similarities with the early vertebrates, they are on a different evolutionary road. Several lines of evidence indicate that the similarities are an outcome of convergent evolution. By this process, recall, different body parts in separate lineages evolve in similar directions, because they are put to similar uses in similar kinds of environments.

Where, then, did the vertebrate road begin? We find tantalizing clues among the members of a rather obscure invertebrate phylum, the **hemichordates** (*hemi*-meaning half, as in "halfway to chordates"). Most people have never seen a hemichordate. The 100 or so species of these soft-bodied marine animals live on the seafloor. The acorn worm shown in Figure 27.5 is an example.

Evolutionarily, the hemichordates seem to be midway between the echinoderms and chordates. Like tunicates, they rely on a ciliated filter-feeding apparatus (Figure 27.6). They don't have a notochord, but they are decidedly like chordates in having a gill-slitted pharynx and a dorsal tubular nerve cord. Also, their larval stages resemble those of echinoderms and tunicates. Do the resemblances suggest a pathway of evolution? Possibly.

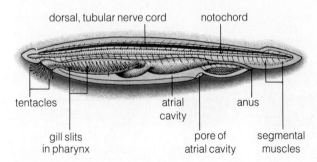

Figure 27.4 Cutaway view and photograph of a lancelet, showing nerve cord and flexible notochord.

Suppose, in ancient echinoderms, mutations in regulatory genes brought about an increase in the rate at which sex organs developed and matured. If sex organs became functional early on, in the *larval* body, there would no longer be much advantage to metamorphosis, the wholesale reconstruction leading to the adult. Over evolutionary time, the original adult form simply could be dispensed with. This scenario is not so far-fetched. Among some existing tunicates—even among some amphibians—we find larvae with functional sex organs. These sexually precocious larvae can reproduce, generation after generation!

Ancestral tunicates, lancelets, and hemichordates were not on the evolutionary road leading to vertebrates, but they provide tantalizing clues to the animals that were.

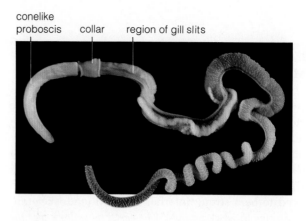

conelike proboscis   collar   region of gill slits

Figure 27.5 An acorn worm, of the phylum Hemichordata. Acorn worms often are found in mud and tidal flats, where they live in U-shaped burrows. One species was discovered near a hydrothermal vent, deep on the ocean floor. Many acorn worms are filter feeders. Food suspended in the water becomes trapped by sticky mucus on their proboscis and is carried to the mouth. The gill slits of acorn worms open from the pharynx and resemble the gill slits of chordates.

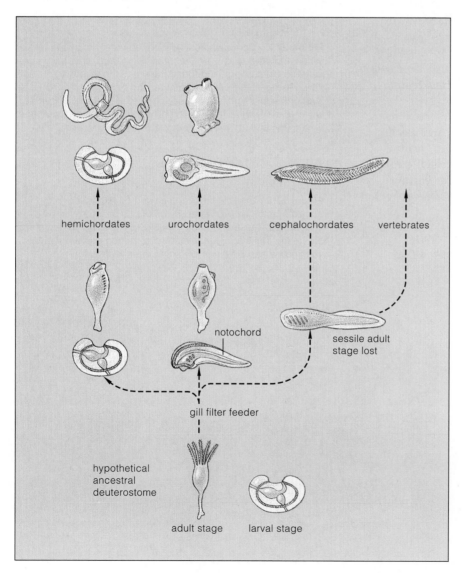

hemichordates   urochordates   cephalochordates   vertebrates

notochord

sessile adult stage lost

gill filter feeder

hypothetical ancestral deuterostome

adult stage   larval stage

Figure 27.6 One hypothesis regarding the evolutionary origins of hemichordates and chordates.

27.4 EVOLUTIONARY TRENDS AMONG THE VERTEBRATES

How do we get from tadpole-shaped chordates to the vertebrates? Let's start with a broad evolutionary picture. One evolutionary trend involved a shift away from the notochord to reliance on a skeletal column of separate, hard segments. We call the skeletal elements **vertebrae** (singular, vertebra). The vertebral column proved to be a strong internal skeleton (endoskeleton) for muscles to work against. *The vertebral column was the foundation for fast-moving predators—some of which were ancestral to all other vertebrate animals.*

In a related trend, part of the nerve cord expanded and developed into a complex brain. The expansion began after **jaws** evolved. In early fishes, the first jaws arose through modification of the first in a series of structural elements that helped support the gill slits (Figure 27.7). Jaws led to new feeding possibilities—and to intensified competition among predators. Fishes able to recognize food or predators from a distance were favored. Over time, they developed better senses of smell and vision. The brain became better at processing information (Figure 27.8). *The trend toward complex sensory organs and nervous systems began in fishes and continued among land vertebrates.*

Another trend began when paired fins evolved. **Fins** are appendages that help propel, stabilize, and guide the body through water. Among some fishes, ventral fins became fleshy and equipped with skeletal supports—the forerunners of limbs. *Paired, fleshy fins were the starting point for the legs, arms, and wings seen among amphibians, reptiles, birds, and mammals.*

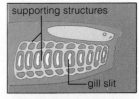

a Early jawless fish (an agnathan)

supporting structures

gill slit

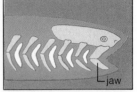

b Early jawed fish (a placoderm)

jaw

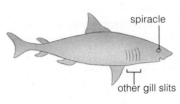
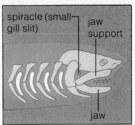

spiracle

other gill slits

c Modern jawed fish (a shark)

spiracle (small gill slit)

jaw support

jaw

Figure 27.7 Comparison of gill-supporting structures in jawless fishes and jawed fishes. In the placoderms and other early jawed vertebrates, cartilage supported the rim of the mouth. In modern jawed fishes, the gill slit between the jaws and an adjacent supporting element serves as a spiracle, an opening through which water is drawn. In (**a**), the gill supports are just under the skin. In (**b**) and (**c**), they are internal to the gill surface.

Figure 27.8 Evolutionary trend toward an expanded, more complex brain, as suggested by comparisons of existing vertebrates. These are dorsal views (looking down on the top of the head). Think about the head size of a frog and a horse, and you know these drawings are not to the same scale.

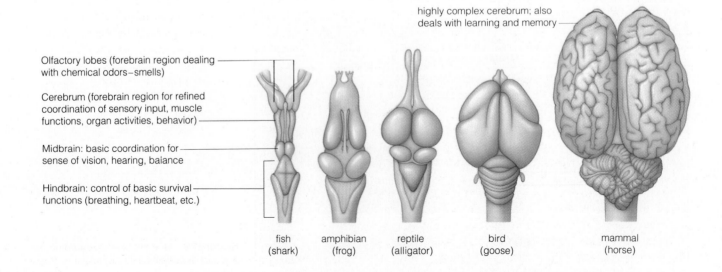

highly complex cerebrum; also deals with learning and memory

Olfactory lobes (forebrain region dealing with chemical odors–smells)

Cerebrum (forebrain region for refined coordination of sensory input, muscle functions, organ activities, behavior)

Midbrain: basic coordination for sense of vision, hearing, balance

Hindbrain: control of basic survival functions (breathing, heartbeat, etc.)

fish (shark)   amphibian (frog)   reptile (alligator)   bird (goose)   mammal (horse)

Figure 27.9 Circulatory systems of vertebrates. All vertebrates have a "closed" circulatory system, in which blood is confined within blood vessels and a heart. Some of these systems are more efficient than others at providing body tissues with oxygen.

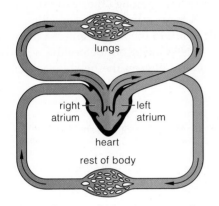

capillaries (network of fine blood vessels where gas exchange occurs)

a In fishes, a heart with two chambers (atrium and ventricle) pumps blood in one circuit. Blood picks up oxygen in gills, then delivers it to rest of body. Oxygen-poor blood flows back to the heart.

b In amphibians, the heart pumps blood through two partially separated circuits. Some blood is pumped to lungs. It picks up oxygen, returns to heart, and mixes with oxygen-poor blood still in heart. The heart pumps this partly oxygenated blood to rest of body.

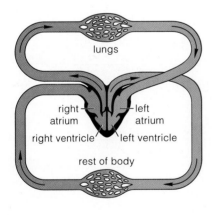

c In some reptiles and all birds and mammals, a four-chambered heart pumps blood in two completely separate circuits. In one circuit, the heart's right ventricle pumps oxygen-poor blood from the body to lungs. Oxygen-rich blood travels from lungs back to heart. In the other circuit, the left ventricle pumps this blood to rest of body.

richly endowed with blood vessels, and they provide a large surface area for gas exchange.

For example, five to seven pairs of a shark's gill slits contain gills, which extend from the pharynx to the outer body surface. When a shark opens its mouth and closes the external gill openings, its pharynx expands and oxygen-rich water flows in through the mouth. Now oxygen diffuses into the gills, and carbon dioxide diffuses out. When muscle action constricts the pharynx, water depleted of oxygen but loaded with carbon dioxide is forced out through the mouth.

As fishes became larger and more active, oxygen uptake and distribution improved. Gills became more efficient in aquatic lineages. But gills cannot work out of water. They stick together unless water flows through them and keeps them moist. In the fishes that were ancestral to land vertebrates, pouches developed on the gut wall. These pouches evolved into **lungs**—internally moistened sacs for gas exchange. In a related trend, modifications to the heart enhanced the pumping of oxygen and carbon dioxide through the body (Figure 27.9). *Ancestors of land vertebrates relied less on gills and more on lungs. More efficient circulatory systems accompanied the evolution of lungs.*

Another trend involved modifications to respiratory structures. Think of the gill slits of a lancelet. Except when this chordate buries itself in sand, oxygen dissolved in the surrounding water and carbon dioxide from metabolically active tissues simply diffuse across the body's surface, down their concentration gradients. In most vertebrate lineages, however, **gills** of one sort or another evolved. All of these respiratory structures have a moist, thin, intricately folded surface, they are

The emergence of a vertebral column, jaws, paired fins, and lungs were pivotal in the evolution of certain vertebrate groups.

Submerged as they are in the waters of the earth, the fishes might not seem to be the dominant vertebrates. And yet, in sheer numbers, they surpass all other vertebrate groups. And fishes show far greater diversity; there are more than 21,000 existing species of bony fishes alone.

The body form and behavior of a fish tell us about the kinds of challenges it faces in the water. Being about 800 times denser than air, water resists rapid movements. Predatory fishes of the oceans are streamlined for pursuit, with a long, trim body that reduces friction. Their tail muscles are organized for propulsive force and forward motion. Bottom-dwelling fishes spend most of their lives hiding from predators or prey. We can deduce that their flattened body is easy to conceal, and that these fishes are sluggish, not the Ferraris of the deep.

Or consider a motionless trout, suspended in shallow water. It is a fine example of an adaptation to water's density. Like many fishes, a trout can maintain neutral buoyancy with a **swim bladder**. This is an adjustable flotation device that exchanges gases with the blood. When a trout gulps air at the water's surface, it is busily adjusting the gas volume in its swim bladder.

The First Vertebrates

As Figure 27.10 indicates, free-swimming vertebrates originated during the Cambrian period. They soon gave rise to two kinds of fishes—those without and those with jaws. One or the other kind probably gave rise to all vertebrate lineages that followed.

Ostracoderms were among the earliest jawless fishes (Agnatha). They were bottom-dwelling filter feeders. Water drawn into the mouth entered the pharynx, where food was strained out before the water left through gill openings between the head plates (Figure 27.11).

Figure 27.10 Evolutionary history of the fishes.

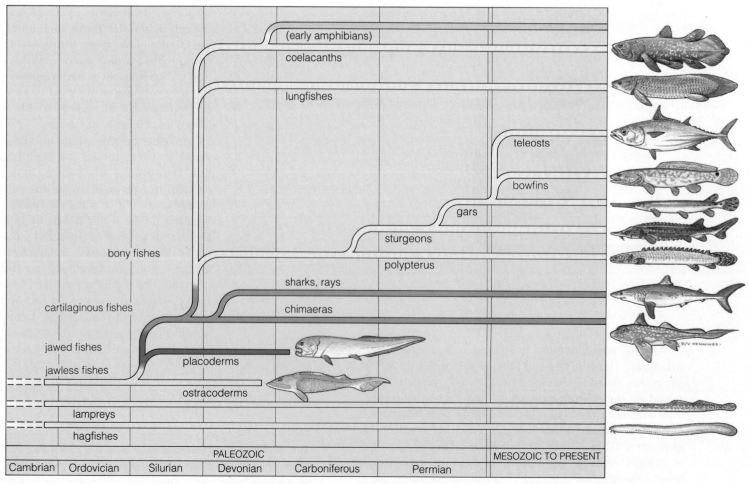

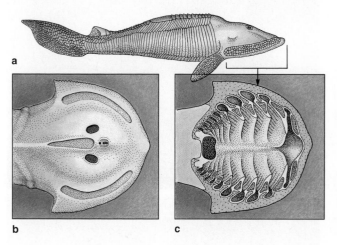

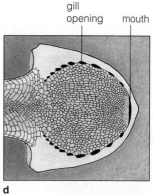

Figure 27.11 (**a**) Reconstruction of an ostracoderm and its skull. This one lived about 100 million years after the earliest known ostracoderms. The dorsal view of the skull in (**b**) shows the openings for three eyes (*blue*) and a single nostril (*red*). The ventral view (**c**) shows the small mouth and relatively large, food-straining pharynx; the flattened bones covering all but the gill openings and a slitlike mouth have been removed, but are shown in place in (**d**).

The ostracoderms did not have much of an internal skeleton; it probably consisted of a notochord and a protective covering for the newly enlarged brain and sensory organs. Armorlike plates covered the body surface. The plates were composed of bony tissue and dentin, a hard tissue that is still present in the teeth of living vertebrates. The armor afforded some protection against the giant pincers of sea scorpions, but it apparently was not much good against jaws. Ostracoderms disappeared when jawed fishes began their adaptive radiations.

Among the first fishes with jaws and paired fins were the **placoderms**. These fishes were bottom-dwelling scavengers and predators. Bony elements reinforced their notochord. The first in a series of gill-supporting structures had become enlarged and equipped with bony projections, something like teeth (Figure 27.7*b*). They functioned as jaws.

Before placoderms evolved, feeding strategies had been limited to filtering, sucking, or rasping bits of food material. Now, placoderms could bite and tear up large chunks of prey—and so get a splendid return on the energy they invested into securing food.

Placoderms with scalelike or bony-plated armor diversified during Silurian and Devonian times. Then, during the Carboniferous, they all became extinct. New kinds of predators—the cartilaginous and bony fishes—replaced them in the seas.

Existing Jawless Fishes

Descendants of some of the early jawless fishes made it to the present. We call them the **lampreys** and **hagfishes**. All of the seventy-five or so species have a cylindrical, eel-like body with no paired fins. They all have a skeleton—an internal structural framework—of cartilage.

Lampreys are specialized predators, almost parasites. Their suckerlike oral disk has horny, toothlike parts that rasp flesh from prey (Figure 27.12). Some

Figure 27.12 A lamprey, pressing its toothed oral disk to the glass wall of an aquarium.

types latch onto salmon, trout, and other commercially valuable fishes, then suck out juices and tissues. Just before the turn of the century, lampreys began invading the Great Lakes of North America. Wherever they have been introduced, populations of lake trout and other large fishes have collapsed. Today, in fisheries of the Great Lakes and other regions, the water is treated with a chemical that poisons lamprey larvae. But the battle goes on.

Visualize a large worm with feelers around the mouth and you have an idea of what the scavenging hagfishes look like. They are not the favorite of fishermen. They burrow into fish trapped by setlines or nets. The ones brought on deck secrete copious amounts of sticky, slimy mucus.

The first vertebrates were jawless and then jawed fishes. Among them were the ancestors of all existing vertebrate species.

Cartilaginous Fishes

Cartilaginous fishes (Chondrichthyes) include about 850 species of skates, sharks, and chimaeras. Most are specialized marine predators with a streamlined body. Figure 27.13 shows examples. All these fishes have pronounced fins, a skeleton of cartilage, and five to seven gill slits on both sides of the pharynx. Most species have a few to many rows of **scales**, small, bony plates at the body surface that often provide protection without weighing the fish down.

The skates and rays are mostly bottom dwellers with flattened teeth, suitable for crushing hard-shelled invertebrates. Both types have distinctive, enlarged fins that extend onto the side of the head. The largest species, the manta ray, measures up to 6 meters from fin tip to fin tip. Other rays have electric organs in the tail or fins that can deliver up to 200 volts of electricity—quite enough to stun prey. The stingray tail has a spine (a modified scale) with a venom gland at its base. Stingrays eat invertebrates, mostly, so the spine is probably used in defense against predators.

At 15 meters from head to tail, some sharks are among the largest living vertebrates. That is longer than two pickup trucks parked end to end. Sharks have formidable jaws, they can detect even traces of blood in water, and the relatively few shark attacks on humans have given the whole group a bad reputation. Yet most of the larger types feed on small invertebrates. The "man-eaters" obviously have not survived by eating humans alone. For many millions of years the large sharks have been eating large fishes and marine mammals, including seals, but not surfboards with legs dangling over the side. They use their sharp, triangular teeth to capture prey and rip off chunks of flesh. They continually shed and replace these teeth, which are modified scales.

The thirty or so species of chimaeras inhabit moderately deep waters, and most feed on hard-shelled mollusks. With their bulky body and long, slender tail, they do resemble a rat (hence their common name, ratfishes). A venom gland associated with a spine is present in front of the dorsal fin.

Bony Fishes

In terms of diversity and sheer numbers, the **bony fishes** (Osteichthyes) are the most successful of all vertebrates. Their ancestors arose during the Silurian, and before that period drew to a close, they had diverged into three lineages: ray-finned fishes, lobe-finned fishes, and lungfishes. Their descendants radiated into nearly every aquatic habitat. With at least 21,000 existing species, the bony fishes represent all but a paltry 4 percent of the modern fishes.

Figure 27.13 Cartilaginous fishes: (**a**) shark, (**b**) blue-spotted reef ray, and (**c**) chimaera (ratfish).

Figure 27.14 is a sampling of the stunning morphological diversity among the ray-finned fishes. The members of this group have paired fins supported by rays that originate from the dermis (one of the skin's layers). Most have notably maneuverable fins and light, flexible scales. Both adaptations contribute to the ability to make complex movements. A well-developed respiratory system rapidly delivers oxygen to metabolically active tissues.

Enormous variations exist on the basic body plan. Predators of the open ocean typically have torpedo shapes; strong, flexible bodies; and powerful tail fins that function in swift pursuit. Many reef dwellers have boxy shapes and fins adapted for maneuvering among narrow passageways of their habitats. The long, flexible bodies of eels are suitable for wriggling through mud

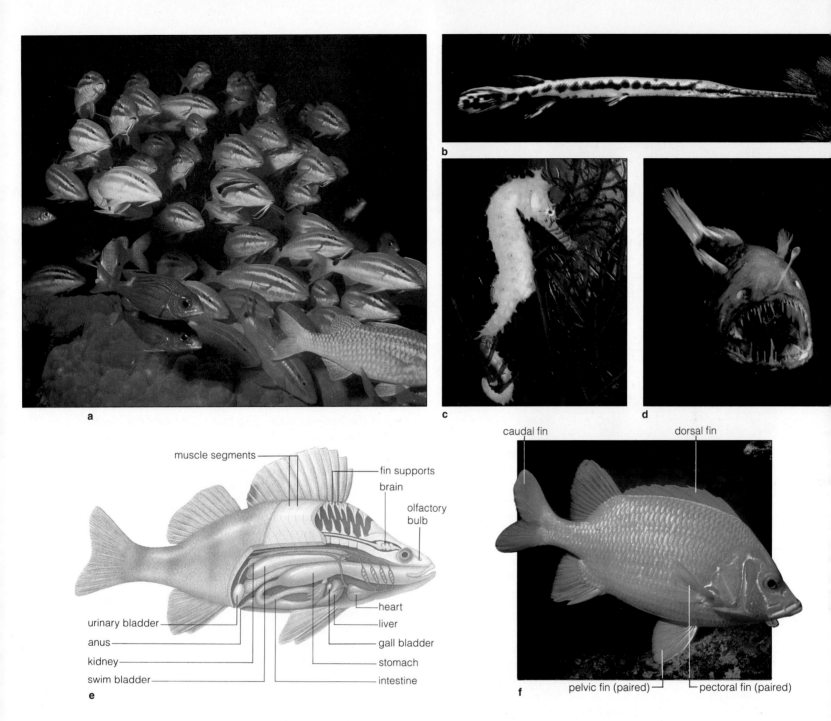

Figure 27.14 Bony fishes. (**a**) Goatfish showing schooling behavior, (**b**) a long-nose gar, (**c**) a seahorse, and (**d**) a deep-sea angler fish. Most ray-finned fishes have the same general body plan, represented here by a perch (**e**) and soldierfish (**f**).

and into nooks and crannies. Bottom dwellers have flattened bodies; sea horses and other types have bizarre body forms. All of these are suitable for concealment from prey and predators.

One group of ray-finned fishes still resembles the ancestral stock. It includes sturgeons and the paddlefishes of the Mississippi River basin. Another group, the thin-scaled or scaleless teleosts, are the most abundant. Existing members of the thirty-five orders include such diverse types as salmon, tuna, rockfish, catfish, minnows, moray eels, perch, flying fish, sculpins, scorpionfish, blennies, and pikes.

By contrast, there is only one existing species of lobe-finned fishes and three genera of lungfishes. As their name suggests, a **lobe-finned fish** is unique in having paired fins that incorporate fleshy extensions from the body. As you will see next, neither it nor the lungfishes have changed much from ancient forms.

Of all existing vertebrates, the bony fishes (ray-finned fishes especially) are the most spectacularly diverse in their morphology and behavior.

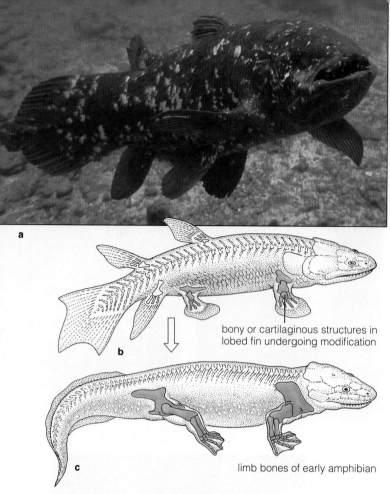

bony or cartilaginous structures in lobed fin undergoing modification

limb bones of early amphibian

Figure 27.15 (**a**) Coelacanth (*Latimeria*), a "living fossil" that resembles early lobe-finned fishes. (**b,c**) Proposed evolution of the skeletal elements inside the lobed fins of certain fishes into the limb bones of early amphibians.

About Those Lobed-Finned, Air-Gulping Fishes . . .

Figure 27.15 shows a coelacanth, the only existing species of lobe-finned fishes. Along with lungfishes, it is a relic of a pivotal time in evolution, when certain vertebrates first ventured onto land.

The ancestors of lobe-finned fishes arose during the Devonian, when sea levels rose and fell repeatedly and swamps that fringed the land were alternately flooded and drained. They must have used their lobed fins to pull themselves from disappearing pond to pond (Figure 21.12). But it was not only lobed fins that made pond-to-pond lurchings possible. These animals also had sac-shaped surfaces inside their body that supplemented respiration. The lobe-finned fishes had simple lungs.

An Australian lungfish provides additional clues to how the ancestral forms might have made it through stressful times. It lives in stagnant water but swims to the surface to gulp air. During the dry season, when streams shrink to mud, the lungfish encases itself in a mixture of mud and slime that protects it from drying out until the next rainy season.

Challenges of Life on Land

The lobe-finned fishes of the Devonian traveled across land simply as a way to reach more water. Yet their crucial travels favored the evolution of more efficient lungs and stronger fins. Among the evolving forms were the ancestors of amphibians. An **amphibian** is a vertebrate that is somewhere between fishes and reptiles in its body plan and reproductive mode. An amphibian has a mostly bony endoskeleton and four legs (or a four-legged ancestor).

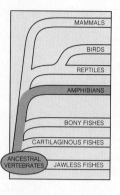

The ancestral amphibians began their forays onto land at the dawn of the Carboniferous. For them, the land was dangerous—and promising. Temperatures fluctuated more on land than in water, air didn't support the body as water did, and water was not always available. But air has far more oxygen. Lungs continued to evolve in ways that enhanced uptake of oxygen. Circulatory systems evolved and became better at distributing oxygen. Both modifications increased the energy base for more active life-styles.

The first amphibians also encountered novel forms of sensory information. Swampy forests abounded with tasty insects and other invertebrate prey. Animals with good vision, hearing, and balance—senses that are advantageous in land environments—were favored. Brain regions concerned with those senses expanded dramatically.

There are now three groups of amphibians—the salamanders, frogs and toads, and caecilians (Figure 27.16). None has escaped water entirely. Their thin skin dries out easily. To develop properly, the eggs of most species must be shed into water or laid in moist places. The larvae are still adapted to aquatic habitats. The species that spend most of their life in shallow water have gills and lungs, and they also use the skin as a respiratory surface. Among them we find somewhat flattened bodies, strong tails, and hindlimbs with webbed feet. The ones that spend most of their life on land use lungs, skin, and the lining of the pharynx to breathe.

Salamanders

Like fishes and the first amphibians, salamanders bend from side to side when they walk (Figure 27.16*a*). The first four-legged vertebrates probably walked this way, also. The adults of some species retain many larval features. Also, the larvae of some groups are sexually pre-

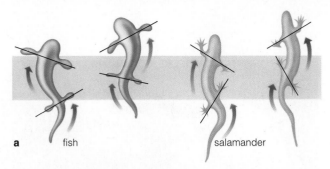

a   fish   salamander

Figure 27.16 Amphibians. (**a**) Forward movement of a salamander compared to that of a coelacanth. (**b**) Terrestrial stage in the life cycle of the red-spotted salamander. (**c**) A frog, splendidly jumping. (**d**) American toad. (**e**) A caecilian.

c

d

e

cocious; they can breed. This is true of the Mexican axolotl, for example. It retains the larval tail and external gills, and the development of its teeth and bones is arrested at an early stage.

Frogs and Toads

With more than 3,000 species, frogs and toads are the most successful amphibians. Their long hindlimbs and powerful muscles allow them to catapult into the air or move forcefully through water. Most often, their sticky-tipped, prey-capturing tongue flips out from the front of the mouth. An adult eats just about any animal it can catch; only its head size dictates the upper limit of prey size. One frog has such a large head, it is called a walking mouth. Skin glands of some species produce toxins, and poisonous types often have bright coloration that advertises their inedibility. The skin of the South African clawed frog (*Xenopus laevis*) is known to contain antibiotics that afford protection against the stew of microbes in its swampy habitat.

Frogs have a closed circulatory system with a three-chambered heart, in which oxygen-rich and oxygen-poor blood mix in the third chamber (Figure 27.9*b*). This may seem inefficient, compared with your four-chambered heart, but it keeps some oxygen circulating when these animals are underwater and cannot breathe. At that time, some gas exchange also takes place at the skin and pharynx.

Caecilians

The ancestors of caecilians lost their limbs and most of their scales, and they gave rise to decidedly worm-shaped amphibians (Figure 27.16*e*). Nearly all of the 150 or so species burrow through soft, moist soil in pursuit of insects and earthworms. A few live in shallow fresh-water habitats.

Regardless of how far they ventured onto land, amphibians have not fully escaped dependency on water.

27.7 REPTILES

The Rise of Reptiles

During the Late Carboniferous, insects began an adaptive radiation into lush habitats on land. At about the same time, the **reptiles** (Reptilia) evolved from amphibians. Like modern species, the amphibious forms probably were carnivores. The huge quantities and selections of edible insects represented a major, untapped food source.

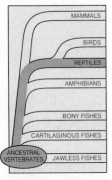

Compared to amphibians, early reptiles pursued prey with greater cunning and speed. Their jaw bones and muscles were better at applying sustained, crushing force. They had well-developed teeth, suitable for securing insects and fellow vertebrates. The limbs of nearly all types were better adapted to support the body on land. The circulatory system of those with a four-chambered heart was more efficient (Figure 27.9). Early reptiles relied fully on more efficient lungs, and they were the first vertebrates to suck in air rather than force it in by mouth muscles. Their cerebrum's thin surface layer, the **cerebral cortex**, was more highly developed (Figure 27.8). This brain region is responsible for the most complex integration of sensory information.

Reptiles were the first vertebrates to escape dependency on standing water. They did so through four adaptations that still distinguish reptiles from fishes and amphibians. *First*, they have tough, scaly skin that limits moisture loss. *Second*, they have some type of copulatory organ that permits internal fertilization. That is, sperm are deposited into a female's body; they do not have to swim through water to reach eggs. *Third*, reptilian kidneys are good at conserving water (the urine of many species is a concentrated paste, not liquid). *Fourth*, reptiles produce **amniote eggs**, in which the embryo develops to an advanced stage before being hatched or born into dry habitats. Such eggs have specialized membranes, and most (but not all) have a leathery or calcified shell (Figure 27.17). The membranes retain water and protect or metabolically support the embryo.

As Figure 27.18 indicates, the reptiles underwent a major adaptive radiation in the Mesozoic era. The types called dinosaurs as well as related forms emerged and ruled the land for the next 125 million years. Their domination finally ended when the Cretaceous drew to a close. The *Focus* essay on page 337 describes their dramatic story. Reptiles that survived to the present day include the turtles, lizards and snakes, tuataras, and crocodilians.

a

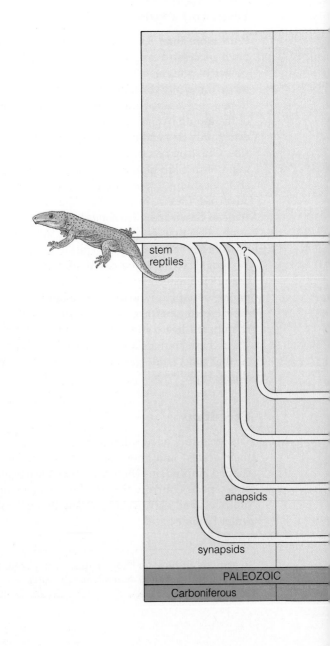

liveborn snake in egg sac

b

Figure 27.17 (**a**) Eastern hognose snakes, emerging from leathery-shelled amniote eggs. Shelled or not, this type of egg contributed to the successful colonization of land by reptiles and, later, birds and mammals. (**b**) Not all snakes lay eggs. For example, offspring of this female copperhead were nourished by yolk reserves in an egg sac, then were born live.

Figure 27.18 (*Below*) Evolutionary history of the reptiles.

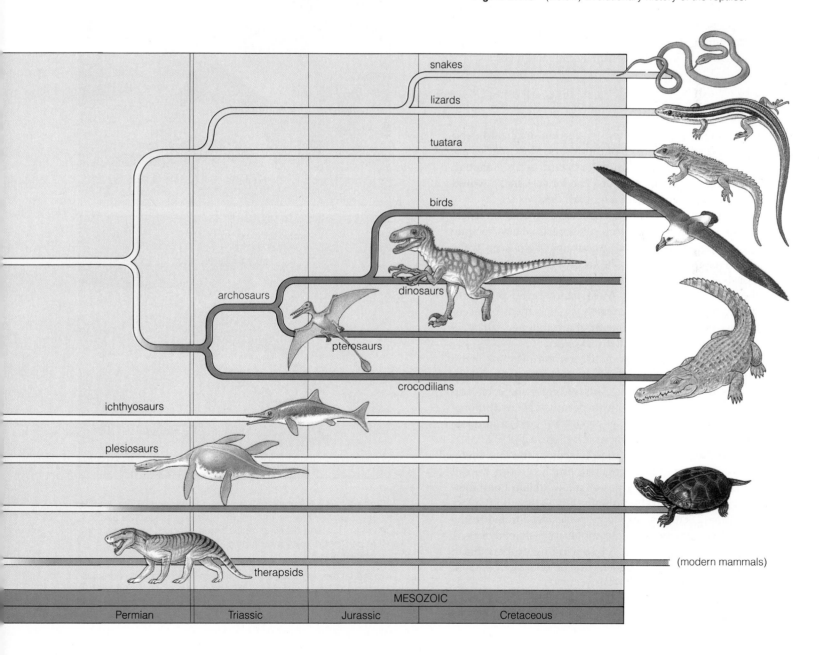

Turtles

The 250 existing species of turtles have inherited a body plan that works well; it has been around since Triassic times. Turtles live in a mobile home, a shell that actually is attached to the skeleton (Figure 27.19a and b). The shell consists of an inner, bony layer and an outer, horny layer of keratin. When threatened, most species can pull their head and limbs into their shell. Although turtles are toothless, they have tough, horny plates suitable for gripping and chewing food. They also have powerful jaws and sometimes a fierce disposition that helps keep predators at bay. For land-dwelling turtles, the shell also holds in moisture and body heat. Only among sea turtles and other notably mobile types is the shell reduced in size and strength.

All turtles lay eggs on land, then leave them. Predators eat most of the eggs, so few new turtles hatch. Prospects are especially bad for sea turtles. They are endangered species, on the brink of extinction (page 710). Humans kill them for their shell and meat.

Lizards and Snakes

Of all living reptiles, about 95 percent are lizards and snakes—distant relatives of dinosaurs. Most species are small, but the Komodo monitor lizard is large enough to hunt young water buffalo. The longest snake would stretch across 10 yards of a football field.

Most of the 3,750 known species of lizards are insect eaters of deserts and tropical forests. They include aggressive, adhesive-toed geckos that help keep walls and ceilings of homes in the tropics free of spiders and insects. They include iguanas; page 545 shows one of the more colorful types. Most lizards grab prey with small, peglike teeth. Chameleons catch them by accurately flicking at them with their tongue, which is longer than their body (Figure 47.12a).

Being small themselves, lizards are prey for many other animals. When grabbed by a predator, many lizards give up their tail. This wriggles for a bit and may be distracting enough to permit a getaway. Some lizards attempt to startle predators—and intimidate rivals—by flaring their throat fan (Figure 27.19c).

Short-legged, long-bodied lizards of the late Cretaceous gave rise to the elongated, limbless forms we call snakes. Some modern types of snakes retain bony remnants of the ancestral hindlimbs. Given their body plan, all of the 2,300 or so existing species are limited in the way they can move. Most move in S-shaped waves, much like salamanders. "Side-winding" vipers make surprisingly rapid J-shaped movements across loose sand and sediments.

Besides rattlesnakes, coral snakes are among the species that use venom to subdue prey (Figure 27.19d

a

b

c

Figure 27.19 A sampling of reptiles. (**a**) A marine turtle, with its shell streamlined for swimming. (**b**) A heavily shelled Galápagos tortoise. This land-dwelling reptile made a lasting impression on Charles Darwin. (**c**) A frilled lizard, flaring a ruff of neck skin in a defensive display.

On the facing page: (**d**) Coral snake, one of the most venomous types. (**e**) Tuatara, a "living fossil" that hasn't changed much since the age of dinosaurs. (**f**) Rattlesnake of the American Southwest. (**g**) An American alligator with offspring.

venom
gland

hollow
fang

d

e

f

g

and *f*). Pythons and boas coil tightly around a prey animal and suffocate it. Snake jaws are highly movable—some species swallow animals wider than they are. Generally, snakes are not aggressive toward humans, but each year as many as 40,000 people die from bites by the venomous ones.

Tuataras

The two existing species of tuataras (*Sphenodon*) are restricted to small, windswept islands near New Zealand. Figure 27.19*e* shows their body plan, which hasn't changed much since the Mesozoic. Although they resemble lizards, their lineage is more ancient. Tuataras don't engage in sex until they are twenty years old. This works well because tuataras, like turtles, may live for sixty years or more.

Crocodilians

Among the modern crocodiles and alligators are the largest living reptiles. All live in or near water. The feared "man-eater" of southern Asia and the Nile crocodile weigh as much as 1,000 kilograms. They drag a mammal or bird into the water, tear it apart by violently

turning over and over, then gulp down the torn chunks. Alligators of the southern United States were once hunted for their belly skin. They were placed under protection after they neared extinction. Their now-increasing numbers put them at odds with some rural community developments.

All crocodilians have a slender snout, powerful jaws, and sharp teeth (Figure 27.19*g*). Even though they look like big lizards, they are evolutionarily closer to birds. For example, unlike most reptiles, which have conserved the amphibian circulatory system, crocodilians have a four-chambered heart.

Like other reptiles, crocodilians adjust their body temperature through behavioral and physiological mechanisms (page 746). They show complex social behavior, as when parents guard nests and assist hatchlings in their move out of the eggshell and into the water. An unguarded nest is vulnerable to egg-eating mammals, and hatchlings are appetizing to large fishes.

With their scaly skin, reliance on internal fertilization, water-conserving kidneys, and amniote eggs, the reptiles were the first vertebrates to escape dependency on standing water.

27.8 BIRDS

By definition, **birds** alone are animals with feathers (Figure 27.20). **Feathers** are lightweight structures used in flight, insulation, or both. Judging from fossils of *Archaeopteryx*, birds descended from two-legged reptiles that lived about 160 million years ago (page 268). They still resemble reptiles in many internal structures, their horny beaks and scaly legs, and their habit of laying eggs. One of Darwin's champions, Thomas Huxley, argued that birds are glorified reptiles. Today, many biologists do indeed classify birds as a branch of the reptilian lineage.

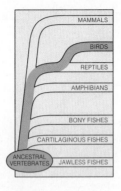

There are nearly 9,000 known species of birds. They show stunning variation in size, proportions, coloration, and capacity for flight (Figure 27.21). A small hummingbird barely tips the scales at 2.25 grams (0.08 ounce). The largest bird, the ostrich, weighs about 150 kilograms (330 pounds). As Figure 17.2 illustrates, ostriches cannot fly, but they are impressively long-legged sprinters. Warblers and other perching birds

Figure 27.20 Feathers—the defining characteristic of birds. This male pheasant has flamboyant plumage, an outcome of sexual selection. It is a native of the Himalaya Mountains of India, and an endangered species. As is the case for many birds, its jewel-colored feathers end up adorning humans—in this case, on the caps of native tribespeople.

Figure 27.21 A few more characteristics of birds. (**a**) Of all living vertebrates, only birds and bats fly by flapping wings. (**b**) Many birds, including these Canada geese, show migratory behavior. These birds are at their wintering grounds in New Mexico. (**c**) Speckled eggs of a magpie. All birds lay hard-shelled eggs of the sort shown in the generalized diagram.

a

b

c

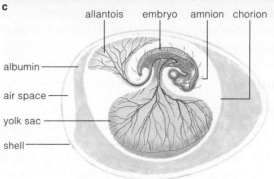

albumin

air space

yolk sac

shell

allantois embryo amnion chorion

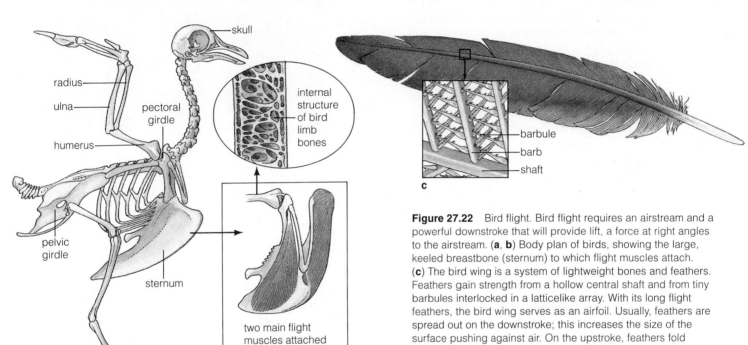

Figure 27.22 Bird flight. Bird flight requires an airstream and a powerful downstroke that will provide lift, a force at right angles to the airstream. (**a**, **b**) Body plan of birds, showing the large, keeled breastbone (sternum) to which flight muscles attach. (**c**) The bird wing is a system of lightweight bones and feathers. Feathers gain strength from a hollow central shaft and from tiny barbules interlocked in a latticelike array. With its long flight feathers, the bird wing serves as an airfoil. Usually, feathers are spread out on the downstroke; this increases the size of the surface pushing against air. On the upstroke, feathers fold somewhat, so the wing presents the least possible profile against the air.

especially differ markedly in feather coloration and in their territorial songs. Bird songs and other splendidly complex social behaviors are topics of later chapters.

Flight demands high metabolic rates, which require a good deal of oxygen. All birds have a large, strong, four-chambered heart that pumps oxygen-enriched blood to the lungs. They also have a unique respiratory system that greatly enhances oxygen uptake. This system is described on page 699. Flight also demands low weight and high power. The bird wing (a forelimb) consists of feathers, powerful muscles, and lightweight bones. Flight muscles attach to an enlarged breastbone

and the upper limb bones adjacent to it (Figure 27.22). When they contract, they produce the powerful downstroke for flight. Bird bones are strong and yet weigh very little because of air cavities in the bone tissue. The skeleton of a frigate bird, which has a 7-foot wingspan, weighs only 4 ounces. That's less than the feathers weigh!

Of all animals, birds alone have feathers, which they use in flight, in heat conservation, and in socially significant visual displays.

27.9 MAMMALS

We turn now to the **mammals**, the most recent of the existing classes of vertebrates. Ancestors of mammals evolved during the Carboniferous, from synapsid reptiles. As Figure 27.18 shows, those ancestral forms diverged from other lineages that led, ultimately, to the dinosaurs and to the modern reptiles and birds. Mammal-like reptiles called **therapsids** appeared on their evolutionary road. They gave rise to many small mammals that survived in the shrubbery during the late Mesozoic, when dinosaurs dominated the land. Mammal-eating dinosaurs probably kept them from moving into more diverse habitats.

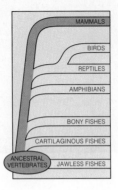

When the Cretaceous period abruptly crashed to a close, the last of the dinosaurs became extinct. Pages 336 through 341 cover this part of the evolutionary story. Afterward, a great mammalian radiation began that continued on into the modern era (Figure 19.11). It was then that the mammalian brain began to reveal its potential. Especially among primates, interconnected masses of information-encoding and information-processing cells expanded dramatically. The expansion was the foundation for our own capacity for memory, learning, and conscious thought.

With well over 4,500 existing species, mammals show great morphological diversity. They range in size from Kitti's hognosed bat, which weighs in at a mere 1.5 grams, to whales that exceed 100 tons. Like other vertebrates, all of these mammals have an internal skeleton. The skeleton's main axis is a bony column (backbone) that has a cordlike bundle of nerves threading through it. Perched above the backbone is a skull. Inside the skull is a complex brain that is functionally divided into three regions (hindbrain, midbrain, forebrain). Also inside are sensory organs concerned with sight, sound, balance, and smell.

Of all vertebrates, only female mammals feed the young with milk, a nutritious fluid produced by their mammary glands (Figure 27.23a). Hence the name of this vertebrate class, from the Latin *mamma*, meaning breast. The females and males care for the young for an extended period and serve as models for their behavior. Young mammals have an inborn capacity to learn and to repeat a set of behaviors that have survival value. Mammals in general show behavioral flexibility. That is, they can expand on the basics with novel forms of behavior.

Mammals typically have hair or thick skin that conserves heat (the trait was lost in most whales). And unlike reptiles, which generally swallow prey whole,

a

most mammals secure, cut, and sometimes chew food before they swallow it. As Figure 27.23c shows, reptiles and mammals differ in their **dentition** (the type, number, and size of teeth). Mammals have four types of upper and lower teeth that match up and work together to crush, grind, or cut food. Their incisors are like flat chisels or cones; they nip or cut food. Horses and other mammals that graze in open grasslands have pronounced incisors. Canines have piercing points. Meat-eating mammals use long, sharp canines to pierce prey. Premolars and molars (cheek teeth) are a platform for food. Their surface bumps (cusps) help crush, grind, and shear food. If a mammal has large, flat-surfaced cheek teeth, you can assume its ancestors evolved in places where fibrous plants were abundant food sources.

Teeth fossilize very well. As you will see in the chapter that follows, fragments of jaws and teeth from human ancestors give clues to their life-styles.

Mammals alone feed their young with milk from mammary glands. They have distinctive dentition and, typically, hair as well as an internal skeleton, a nerve cord, a three-part brain, and sensory organs inside the skull. Their young require an extended period of dependency and learning.

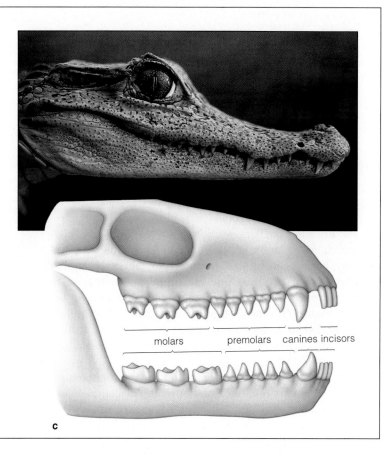

Figure 27.23 Distinctly mammalian traits. (**a**) A human baby, busily demonstrating the key defining feature. It derives nourishment from mammary glands. (**b**) An arctic fox, which survives murderously low winter temperatures with the help of hair. Its thick, insulative coat, which remains white during winter, also helps the fox "hide in the open" (camouflage itself) from prey by blending in with its snow-covered habitat. (**c**) Mammals descended from reptiles, which have peglike upper teeth that don't match up with peglike lower ones. Mammalian teeth do, as shown by the generalized diagram.

molars   premolars   canines   incisors

We turn now to the three major groups of this vertebrate lineage: the egg-laying, pouched, and placental mammals.

Egg-Laying Mammals

Two major types of egg-laying mammals (subclass Prototheria) have survived to the present day. They are the platypus, described at the start of this chapter, and several species of spiny anteaters, which are burrowing animals of Australia and New Guinea (Figure 27.24). Both types lost most or all of their teeth during the course of evolution. The loss is correlated with their specialized diets. Whereas the platypus feeds mostly on small aquatic invertebrates, the spiny anteaters feed on termites and ants, which they capture with their long, sticky tongue.

Several rather archaic traits have persisted in both types of egg-laying mammals. Besides being egg layers, these animals do not show the same kind of skeletal developments of other mammals, and their metabolic rates are lower. Yet their young are suckled after hatching. Like other mammals, the platypus is covered with hair, and the spiny anteaters are covered with spines derived from hair. Despite their low metabolic rates, both mammals maintain a relatively constant body temperature that is well above that of their surroundings.

Figure 27.24 An egg-laying mammal of arid habitats of Australia: the short-nosed anteater (*Tachyglossus aculeatus*). The platypus in Figure 27.1 is another egg-laying mammal.

Pouched Mammals

There are about 260 species of pouched mammals, or marsupials (subclass Metatheria). Nearly all are native to Australia and nearby islands; a few live in North and South America. Marsupial young are born live but not quite "finished"—they are tiny, blind, and hairless. However, the newborns have an excellent sense of smell and strong forelimbs, which they use to locate and reach the mother's pouch, which is some distance away from the birth canal. They are suckled in that pouch, and there they complete their early development.

The kangaroos, probably the most familiar marsupials, are also the largest. Some species weigh 90 kilograms. Even larger species once lived in Australia and may have been hunted to extinction by early human populations. At least one marsupial has managed to coexist successfully with humans, however. During the past century, the Virginia opossum has greatly expanded its range in the United States. It now thrives in urban areas as well as in its native forest habitats (Figure 27.25).

For at least 50 million years, the Australian marsupials evolved in relative isolation from placental mammals. Their ancestors had crossed a narrow sea that opened up between two land masses, but placental mammals stayed behind. In the absence of competition, the marsupials radiated freely into adaptive zones in their own land mass, which became Australia.

Kangaroos, wallabies, and other plant-eating (herbivorous) mammals now occupy adaptive zones comparable to the ones filled by deer, antelope, and other herbivores on other continents. Siberian and North American forests have wolves; Australia has (or had)

Figure 27.26 Representative placental mammals, which occupy diverse aquatic and terrestrial habitats. (**a**) The manatee lives in the ocean and eats submerged seaweed. (**b**) Bats, the only flying mammals, dominate the night sky vacated by birds. (**c**) Camels traversing an extremely hot desert with ease. (**d**) Zebras, sometimes called walking hamburgers for predators of the African savanna. (**e**) Walruses swim in frigid waters and sunbathe on ice. (**f**) Raccoons climb trees.

Figure 27.25 A pouched mammal—a female opossum with her young.

the Tasmanian "wolf," a marsupial that is now probably extinct. Other marsupials glide like flying squirrels, and some climb like monkeys. In Australia, humans have introduced cattle, sheep, horses, and other placental mammals from other places. Many native marsupials are being threatened with displacement.

Placental Mammals

More than 4,500 placental mammals (subclass Eutheria) are known. One or more members of this group live in virtually every kind of aquatic and terrestrial environment (Figure 27.26).

Chapter 45 will describe the reproduction of humans, a placental mammal, in detail. For now, it is enough to know that a **placenta** is a spongy tissue in a pregnant female's uterus that develops functional connections with the embryo. This tissue, composed of maternal tissue and embryonic membranes, is the

c

d

e

means by which the embryo receives nutrients and oxygen and gets rid of metabolic wastes. Placental mammals grow faster in the uterus than marsupials do in a pouch. At birth, many are fully developed and can move about almost immediately. Even then, they all remain with the mother for some time.

Take a quick look at Appendix I, and you will see that placental mammals include a great variety of species—many familiar to us, and others obscure. In terms of diversity, the rodents are most successful. These include rats, mice, squirrels, and prairie dogs. Bats are next in terms of diversity. Other familiar placental mammals are the carnivorous dogs, bears, cats, walruses, and dolphins, as well as the herbivorous horses, camels, deer, and elephants. Exotic types, including manatees and anteaters, also hold membership in this group.

The physiology, behavior, and ecology of many of these mammals will occupy our attention in chapters to follow.

f

SUMMARY

1. Chordates are animals with four distinguishing characteristics. Nearly all chordate embryos, and commonly the adults, have a notochord, a dorsal hollow nerve cord, a pharynx with gill slits (or hints of these), and a tail that extends past the anus.

2. Invertebrate chordates include tunicates (such as sea squirts) and lancelets.

3. There are eight classes of vertebrates:
    a. Jawless fishes, such as lampreys (Agnatha).
    b. Jawed armored fishes (Placodermi); extinct.
    c. Sharks, rays, and other cartilaginous fishes (Chondrichthyes).
    d. Salmon, tuna, coelacanths, and other bony fishes (Osteichthyes).
    e. Frogs, toads, and other amphibians (Amphibia).
    f. Turtles, snakes, crocodiles, and other existing and extinct reptiles (Reptilia).
    g. Robins, eagles, ostriches, penguins, and other flying and flightless birds (Aves).
    h. Platypuses, kangaroos, opossums, camels, bats, humans, and other mammals (Mammalia).

4. We find clues to vertebrate origins in the body plan and embryonic development of tunicates as well as the wormlike hemichordates. In several respects, the earliest vertebrates may have resembled some of the larval forms and adults of these animals.

5. The first vertebrates were jawless fishes that arose during the Cambrian era, as represented by the ostracoderms. Their only living descendants are lampreys and hagfishes.

6. Jawed fishes also arose during the Cambrian. A dominant lineage, the placoderms, is now extinct. Other lineages gave rise to the cartilaginous and bony fishes.

7. The following trends occurred during the evolution of certain vertebrate lineages:
    a. A vertebral column supplanted the notochord as the structural element against which muscles act. The vertebral column foreshadowed the evolution of fast-moving, predatory animals.
    b. Jaws evolved from gill-supporting elements. They led to increased predator-prey competition. This in turn favored the evolution of more efficient nervous systems and sensory organs.
    c. In one lineage of bony fishes, paired fins evolved into fleshy lobes with internal structural elements—the forerunners of paired limbs.
    d. In certain fishes, lungs evolved and supplemented respiration by way of gills. Lungs proved adaptive in the invasion of land. In a related development, the circulatory system became more efficient at distributing oxygen.

8. With staggering numbers of individuals distributed among 21,000 or so known species, the bony fishes (especially the ray-finned species) are the most successful vertebrates. They exhibit spectacular variation in morphology and behavior.

9. Amphibians were the first vertebrates to invade the land, but they never fully escaped the water. Their skin dries out, and aquatic stages persist in the life cycle of all species.

10. Reptiles evolved from amphibious forms. They were the first vertebrates to escape dependency on standing water, owing to these adaptations:
    a. Tough, scaly skin that conserves body moisture.
    b. Copulatory organs that permit internal fertilization.
    c. Extremely efficient, water-conserving kidneys.
    d. Amniote eggs, often leathery or shelled, that protect and metabolically support the embryos.

11. Reptiles, and the birds and mammals that descended from certain reptilian lineages, have efficient circulatory and respiratory systems and a well-developed nervous system and sensory organs.

12. Of all vertebrates, birds alone have feathers, which they use in flight, heat conservation, and social displays.

13. Mammals alone have milk-producing mammary glands, and they have hair or thick skin that functions in insulation. They have distinctive dentition and a highly developed brain. Adults nurture their young through an extended period of dependency and learning.

Review Questions

1. List and describe the features that distinguish chordates from other animals. *446*

2. Name and describe the characteristics of an organism from each of the two groups of invertebrate chordates. *447–448*

3. List four evolutionary trends that occurred in certain vertebrate lineages. *450–451*

4. Which evolutionary modifications in fishes set the stage for the emergence of amphibians? *452–456*

5. List some of the identifying characteristics of reptiles, birds, and mammals. *458–467*

6. List the eight classes of vertebrates and cite examples of each. *446*

7. Which class of vertebrates was the first to invade the land? *456*

8. What specific evolutionary adaptations allowed reptiles to evolve from amphibians? *458*

1. The embryos and often the adults of _____ have a notochord, a tubular dorsal nerve cord, a pharynx with slits in the wall, and a tail extending past the anus.
   - a. echinoderms
   - b. tunicates and lampreys
   - c. vertebrates
   - d. both b and c
   - e. all are correct

2. Gill slits function in _____ .
   - a. respiration
   - b. circulation
   - c. food trapping
   - d. water regulation
   - e. both a and c

3. Existing aquatic vertebrates include _____ .
   - a. jawed, armored fishes
   - b. jawless fishes
   - c. bony fishes
   - d. both a and b
   - e. both a and c
   - f. both b and c

4. _____ appeared early on the evolutionary road leading to vertebrates.
   - a. Tunicates and lancelets
   - b. Hemichordates
   - c. both a and b
   - d. neither is correct

5. A shift from a reliance on _____ to reliance on _____ was pivotal in the evolution of all vertebrates.
   - a. the notochord; a backbone
   - b. filter feeding; jaws
   - c. gills; lungs
   - d. all are correct

6. The first vertebrates were _____ .
   - a. bony fishes
   - b. jawless fishes
   - c. jawed fishes
   - d. both a and b

7. The bony fishes include _____ .
   - a. ray-finned fishes
   - b. lobe-finned fishes
   - c. lungfishes
   - d. all are correct

8. Of all existing vertebrates, _____ are the most diverse.
   - a. cartilaginous fishes
   - b. bony fishes
   - c. amphibians
   - d. reptiles
   - e. birds
   - f. mammals

9. A four-chambered heart is characteristic of _____ .
   - a. bony fishes
   - b. amphibians
   - c. birds
   - d. mammals
   - e. both b and c
   - f. both c and d

10. The only amphibians to entirely escape dependency on aquatic habitats are _____ .
    - a. salamanders
    - b. desert toads
    - c. caecilians
    - d. none is correct

11. Adaptations that permitted reptiles to escape dependency on aquatic habitats were _____ .
    - a. tough skin
    - b. internal fertilization
    - c. good kidneys
    - d. amniote eggs
    - e. both b and d
    - f. all are correct

12. _____ have highly efficient circulatory and respiratory systems, and a complex nervous system and sensory organs.
    - a. Reptiles
    - b. Birds
    - c. Mammals
    - d. all are correct

13. Birds use feathers in _____ .
    - a. flight
    - b. heat conservation
    - c. social functions
    - d. all are correct

14. Various mammals _____ .
    - a. hatch
    - b. complete embryonic development in pouches
    - c. complete embryonic development in the uterus
    - d. both b and c
    - e. all are correct

15. Match the organisms with the appropriate features.
    - _____ jawless fishes
    - _____ cartilaginous fishes
    - _____ bony fishes
    - _____ amphibians
    - _____ reptiles
    - _____ birds
    - _____ mammals
      - a. complex three-part brain, thick skin or hair
      - b. respiration by skin and lungs
      - c. include coelocanths
      - d. include hagfishes
      - e. include sharks and rays
      - f. complex social behavior, feathers
      - g. first with amniote eggs

Selected Key Terms

amniote egg *458*	lancelet *448*
amphibian *456*	lobe-finned fish *455*
bird *462*	lung *451*
bony fish *454*	mammal *464*
cartilaginous fish *454*	nerve cord *446*
cerebral cortex *458*	notochord *446*
chordate *446*	ostracoderm *452*
dentition *464*	pharynx *446*
feather *462*	placenta *466*
fin *450*	placoderm *453*
gill *451*	reptile *458*
gill slit *447*	scale *454*
hagfish *453*	swim bladder *452*
hemichordate *448*	therapsid *464*
jaw *450*	tunicate *447*
lamprey *453*	vertebra *450*

Readings

Carroll, R. L. 1988. *Vertebrate Paleontology and Evolution.* New York: Freeman.

Hoffman, E. 1990. "Paradox of the Platypus." *International Wildlife* 20(1):18–21.

Romer, A. S., and T. S. Parsons. 1986. *The Vertebrate Body.* Sixth edition. Philadelphia: Saunders.

Welty, J., and L. Baptista. 1988. *The Life of Birds.* Fourth edition. New York: Saunders.

28 HUMAN EVOLUTION: A CASE STUDY

The Cave at Lascaux and the Hands of Gargas

Half a century ago, on a warm autumn day, four boys out for a romp stumbled into a cave near Lascaux, a town in the Perigord region of France. What they discovered inside that intricately tunneled cave stunned the world.

Magnificent sketches, engravings, and paintings swept out across the cave walls (Figure 28.1). The red, yellow, purple, and brown pigments were as vivid as if they had been painted only recently. Yet radioisotope measurements revealed that they were 17,000 to 20,000 years old. The prehistoric artists worked deep within the cave, where sunlight could not fade the images and winds and water could not wear them away. By the

light of crude oil lamps, they captured the graceful, dynamic lines of bison, stags, stallions, ibexes, lions, a rhinoceros, and a heifer, now known as the Great Black Cow.

Caves throughout southern France, northern Spain, and Africa hold treasures from even earlier times. About 25,000 years ago, for example, prehistoric peoples carefully committed more than 150 imprints and outlines of their hands to walls in the cave of Gargas, in the Pyrenees.

Who were the people who did this? From their fossilized remains, we know that they were anatomically like us. From the way they planned and executed their

Figure 28.1 Part of the human cultural heritage—prehistoric cave paintings, a unique outcome of a long history of biological evolution.

art, we sense a level of abstract thinking that is unique to humans.

The quality of "humanness" did not materialize out of thin air. The story of the human species began more the 60 million years ago, with the origin of primates in tropical forests. In turn, the primate story began more than 250 million years ago, with the origin of mammals. The story extends back to the origin of animals at some time before 750,000,000 years ago—and so on back in time, to the origin of the first living cells.

As we poke about the branches of our family tree, keep this greater evolutionary story in mind. *Our "uniquely" human traits emerged through modification of traits that had already evolved in ancestral forms.* At each branch in the family tree, certain mutations produced workable changes in traits, which proved useful in prevailing environments.

From this perspective, "ancient" cave paintings are the legacy of individuals who departed only yesterday, so to speak. The artists of Lascaux and Gargas are not remote from us. They *are* us.

KEY CONCEPTS

1. The primate branch of the mammalian lineage includes prosimians, tarsioids, and anthropoids, which include the monkeys, apes, humans, and ancestral humanlike forms. Apes and humans, which share common ancestry, are hominoids. Only humans and their humanlike ancestors are further classified as hominids.

2. Unlike other primates, many early hominids did not become restrictively specialized in a single habitat. They remained flexible and became adapted to a wide range of challenges in complex, unpredictable environments.

3. The capacity to make generalized responses that work in different environments is an outcome of several trends that occurred in certain primate lineages. They involved changes in bones, muscles, teeth, sensory systems, and the brain itself.

4. Ancestral, four-legged primates underwent skeletal modifications that led to an upright stance, which in turn freed forelimbs and hands for new functions. Modifications in handbones and muscles led to a capacity to hold, carry, use, and make objects.

5. Teeth became less specialized, and this allowed a greater variety of food sources to be tapped.

6. Reorganization of the skull and eye sockets led to increased reliance on daytime vision.

7. Brain tissue increased in volume and became more complex. The evolution of the brain was interlocked with cultural evolution.

In the preceding chapters, we likened the history of life to a great tree. Each branch of the tree represents a line of descent, and the branch points represent a divergence leading to new species. There are more than 47,000 existing species of fishes, amphibians, reptiles, birds, and mammals on the vertebrate branch of that tree. When we turn to the evolution of any one of those species—as we do here—it helps to keep a key point in mind. At each crossroad leading to a new species, complex traits were already in place and functioning—*and new traits emerged only through modification of traits that already were in place.* The evolution of the human species speaks eloquently of this characteristic of life.

28.1 PRIMATE CLASSIFICATION

The order **Primates** includes prosimians, tarsioids, and anthropoids. Figure 28.2*a* through *c* shows a few of these distinctive mammals, and Figure 28.2*d* shows their presumed evolutionary relationships.

Prosimians are the oldest lineage (*pro*, before; *simian*, ape). For millions of years, prosimians dominated the trees in North America, Europe, and Asia. That was before monkeys and apes evolved and almost displaced them entirely. Tarsioids, represented by tarsiers of southeastern Asia, are small primates with features that place them between prosimians and anthropoids.

Monkeys, apes, and humans are **anthropoids**. In structural details and biochemistry, apes are far more similar to humans than to monkeys. That is why apes and all species of the human lineage are classified together, as **hominoids**. A divergence from their common ancestor began many millions of years ago. All species on the separate evolutionary road leading to humans are classified as the **hominids**.

Figure 28.2 Representative primates. (**a**) Gibbons have limbs and a body adapted for swinging arm over arm through the trees. (**b**) Monkeys are quadrupedal (four-legged) climbers, leapers, and runners, as this spider monkey demonstrates. (**c**) Tarsiers are vertical clingers and leapers. (**d**) Evolutionary tree for the primates. *Green* boxes show the two highest groupings (the prosimians, and the tarsioids and anthropoids). *Tan* boxes show the major groups of living primates. A few representative members are named. Monkeys, apes, and humans are all anthropoids. The only hominids are humans and now-extinct species of their lineage.

a

b

c

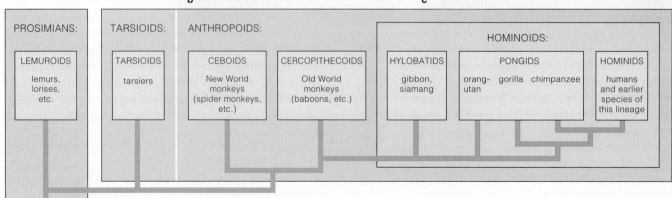

PROSIMIANS:	TARSIOIDS:	ANTHROPOIDS:		HOMINOIDS:		
LEMUROIDS lemurs, lorises, etc.	TARSIOIDS tarsiers	CEBOIDS New World monkeys (spider monkeys, etc.)	CERCOPITHECOIDS Old World monkeys (baboons, etc.)	HYLOBATIDS gibbon, siamang	PONGIDS orang-utan gorilla chimpanzee	HOMINIDS humans and earlier species of this lineage

d

tree-dwelling, rodentlike primate of Paleocene

28.2 FROM PRIMATE TO HUMAN: KEY EVOLUTIONARY TRENDS

Most primates live in tropical or subtropical forests, woodlands, or **savannas** (open grasslands with a few stands of trees). Like their ancient ancestors, the vast majority are tree dwellers. Yet no one feature sets "the primates" apart from other mammals. Each primate lineage evolved in a distinct way and has its own defining traits. Five trends define the lineage we are concerned with here. They were set in motion when primates first started adapting to life in the trees, and they contributed to the emergence of modern humans.

1. Skeletal changes led to upright walking, which freed the hands for new functions.

2. Changes in bones and muscles led to refined hand movements.

3. There was less reliance on the sense of smell and more on daytime vision.

4. Changes led to fewer, less specialized teeth.

5. Brain elaboration and changes in the skull led to speech. These developments became interlocked with each other and with cultural evolution.

Upright Walking

Of all primates, only humans can stride freely on two legs for a long time. Their habitual two-legged gait is called **bipedalism**. By contrast, monkeys are adapted to life in the trees. Their skeleton permits rapid climbing, leaping, and running along branches. Their armbones and legbones are about the same length (Figure 28.2). This means monkeys can run palms-down. Try this yourself and see what happens.

Unlike monkeys, apes hang onto overhead branches and use their long arms to carry some body weight. The arms often support body weight when an ape is on the ground. Because of the way their shoulder blades are positioned, apes can swivel the arms freely above the head when the body is erect or semi-erect.

Compared with monkeys and apes, humans have a shorter, S-shaped, and somewhat flexible backbone. The position and shape of their backbone, shoulder blades, and pelvic girdle are the basis of bipedalism (Figure 28.3). These skeletal traits emerged not long after the divergence that led to the hominids.

Figure 28.3 Comparison of the skeletal organization and stance of a monkey, ape (gorilla), and human. Their modes of locomotion resulted from modifications of the basic mammalian plan. The quadrupedal monkeys climb and leap; apes climb and swing by forelimbs. Both modes are suited for life in the trees. Humans are two-legged striders. The drawings are not to the same scale.

monkey   gorilla   human

Precision Grips and Power Grips

The first mammals spread their toes apart to help support the body as they walked or ran on four legs. Primates still spread their toes or fingers. Many also make cupping motions, as when monkeys bring food to the mouth. Two other hand movements developed in ancient tree-dwelling primates. Through alterations in handbones, fingers could be wrapped around objects (*prehensile* movements), and the thumb and tip of each finger could touch (*opposable* movements).

Hands began to be freed from load-bearing functions when early primates lived in the trees. Later, when hominids were evolving, refinements in hand movements led to the precision grip and power grip:

These hand positions gave early humans the capacity to make and use tools. They were a foundation for unique technologies and cultural development.

Enhanced Daytime Vision

Early primates had an eye on each side of the head. Later ones had forward-directed eyes, an arrangement that is better for sampling shapes and movements in three dimensions. Through additional modifications, the eyes were able to respond to variations in color and light intensity (dim to bright). These visual stimuli are typical of life in the trees.

Teeth for All Occasions

Monkeys have rectangular jaws and long canines (page 464). Humans have a bow-shaped jaw and smaller teeth of about the same length. Jaws and teeth became modified on the road from early primates to humans. There was a shift from eating insects, then fruit and leaves, and on to a mixed diet.

Better Brains, Bodacious Behavior

Living on tree branches favored shifts in reproductive and social behavior. Imagine the advantages of single births over litters, for example, or of clinging longer to the mother. In many lineages, parents started to invest more in fewer offspring. They formed strong bonds with their young, maternal care became intense, and the learning period grew longer (Figure 28.4).

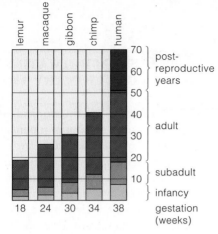

Figure 28.4 Trend toward longer life spans and longer periods of infant dependency among existing primates.

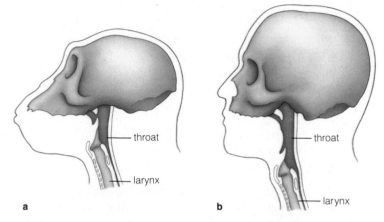

Figure 28.5 Structural basis of speech. (**a**) Like modern chimpanzees, early humans had a skull with a flattened base. Their larynx (a tube leading to the lungs) was not far below the skull, so the throat volume was small. (**b**) In modern humans, the skull's base angles down sharply as it develops. This moves the larynx down also. The result is an increase in the volume of the throat—the area in which sounds are produced.

And now brain regions concerned with encoding and processing information started expanding greatly. New behavior stimulated development of those regions—which in turn stimulated more new behavior. And so brain modifications and behavioral complexity became highly interlocked. The interlocking is most evident in the parallel evolution of the human brain and culture. **Culture** is the sum total of behavior patterns of a social group, passed between generations by learning and by symbolic behavior—especially language. The capacity for language arose among ancestral humans. And it arose through changes in the skull and expansion of parts of the brain (Figure 28.5).

These features emerged along the evolutionary road leading to humans: upright walking, refined hand movements, generalized dentition, refined vision, and the interlocked elaboration of brain regions and cultural behavior.

28.3 PRIMATE ORIGINS

Primates evolved from ancestral mammals more than 60 million years ago, during the Paleocene. Fossils indicate that the first ones resembled small rodents or tree shrews (Figures 28.6 and 28.7). Like tree shrews, they probably had huge appetites and foraged at night for insects, seeds, buds, and eggs beneath the trees of tropical forests. They had a long snout and a good sense of smell, suitable for detecting food or predators. They could claw their way up through the shrubbery, although not with much speed or grace.

Between 54 and 38 million years ago (the Eocene), some primates were staying in the trees. Fossils provide tangible evidence of increased brain size, a shorter snout, enhanced daytime vision, and refined grasping movements. How did these traits evolve?

Consider the trees. Trees offered food and safety from ground-dwelling predators. They also were a habitat of uncompromising selection. Imagine dappled sunlight, boughs swaying in the wind, colorful fruit tucked among the leaves, perhaps predatory birds. A long, odor-sensitive snout would not have been of much use up in the trees, where air currents disperse odors. But a brain that could assess movement, depth, shape, and color would have been a definite plus. So would a brain that worked fast when its owner was running, swinging, and leaping (especially!) from branch to branch. Distance, body weight, winds, and suitability of the destination had to be estimated, and any adjustments for miscalculations had to be quick.

By the dawn of the Oligocene, 35 million years ago, tree-dwelling anthropoids had evolved in the forests. The ancestors of monkeys, apes, and humans were among them. Some forms lived above swamps that were infested with predatory reptiles. Perhaps that is why they rarely ventured to the ground—and why it

Figure 28.7 A night-foraging tree shrew of Indonesia.

was imperative to think fast and grip strongly. Slip-ups were always possible; a surprising number of primates still fall out of the trees.

During the Miocene, which extended from 25 million to 5 million years ago, continents began to assume their current positions (Figure 21.16). Climates were becoming cooler and drier. An adaptive radiation of apelike forms—the first hominoids—took place, and by 13 million years ago, ape populations were scattered through Africa, Europe, and southern Asia. Among them were chimpanzee-sized **dryopiths** (Figure 28.6). Most Miocene apes became extinct around that time. However, fossils and biochemical studies point to three divergences that occurred between 10 million and 5 million years ago. Two branchings gave rise to gorillas and chimpanzees. The third gave rise to early hominids—including the ancestors of humans.

---

1. Small, rodentlike mammals started moving into arboreal habitats about 60 million years ago. The founders of the primate lineage were among them.

2. After 25 million years, during Miocene times, their descendants included the anthropoid ancestors of monkeys, apes, and the hominids.

3. A divergence led to the first hominids—the earliest ancestors of humans. It began between 10 and 5 million years ago, before the Miocene-Pliocene boundary.

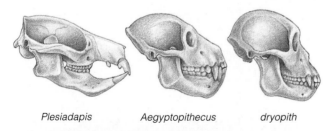

Plesiadapis   Aegyptopithecus   dryopith

Figure 28.6 Comparison of skull shape and teeth of some extinct primates. *Plesiadapis*, a Paleocene primate, had rodentlike teeth. *Aegyptopithecus*, an Oligocene anthropoid, probably predates the divergence leading to Old World monkeys and the apes. An adaptive radiation that began in the Miocene produced many dryopiths and other apelike forms. The drawings are not to the same scale. In size, *Plesiadapis* was like a tiny tree shrew. *Aegyptopithecus* was monkey-sized, and the dryopiths, chimpanzee-sized.

28.4 THE FIRST HOMINIDS

Most of the earliest known hominids lived in the East African Rift Valley, where a 3,200-kilometer (2,000-mile) fracture in the earth's crust runs from the Red Sea down into Mozambique. For the past 20 million years, volcanoes have intermittently spewed, the valley floor has buckled upward and downward, and lakewaters have collected in the lowland basins.

At the Miocene-Pliocene boundary, the long-term shift to a cooler, drier climate triggered the breakup of the once-vast tropical forests, with their bounty of soft fruits and insects. Now there were scattered woodlands and open, grassy plains. There were pronounced seasonal variations in food availability. The African savanna had emerged.

The survivors in these challenging environments were able to locate and exploit new kinds of food and hide from new kinds of predators. This must have been a "bushy" time of evolution, with branchings and radiations into new adaptive zones. Why? Fossil hunters have found and accurately dated the fragmented remains of a variety of humanlike forms.

We don't have enough fossils to discern how all the forms were related (Figure 28.8). Even so, many had three features in common. *First*, they were upright walkers; their hands were freed for new functions. *Second*, modifications in their teeth, jawbones, and jaw muscles allowed them to vary their diet. *Third*, their brain must have been more elaborate than that of their forerunners. At the least, they had to be able to think ahead, to plan when and where to get foods when seasonal supplies ran out. All three features emerged through modifications of traits that can be observed among the other primates. *They were based on the primate heritage.*

We call the first known hominids **australopiths** (southern apes). They fall into two broad categories:

1. Gracile forms (slightly built), called *Australopithecus ramidus*, *A. afarensis*, and *A. Africanus*.

2. Robust forms (muscular, heavily built), including *A. boisei* and *A. robustus*.

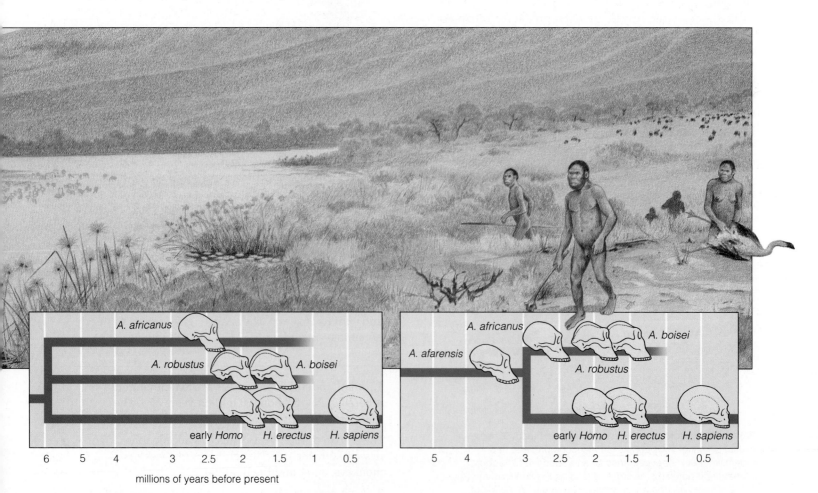

millions of years before present

Figure 28.8 Reconstruction of one of the environments in which the first humanlike forms (early hominids) evolved. Their evolutionary connections with one another and with later hominids are not understood. Several phylogenetic trees have been proposed, including the two shown here for australopiths and for species on the road to modern humans.

With an average volume of 400 cubic centimeters, the australopith brain was a long way from Einstein's. But these hominids were fine striders, like later ones. Legbone fragments, 4 million years old, have muscle attachment sites similar to yours. Figures 28.9 and 28.10 show the reconstructed form and skeleton of a female dubbed Lucy. Her thighbones angled inward, so her body weight was centered directly beneath the pelvis—a sure sign of bipedalism. Thighbones angle outward in apes, which have a waddling, four-legged gait. Most telling, the australopiths left footprints.

Also like later hominids, some australopiths had a bowed jaw. Different types ate different things. *A. robustus* had the strong jaw muscles and the grinding platform typical of plant eaters. *A. boisei*, with its huge, heavily cusped molars, may have been adapted to chewing dry seeds, nuts, and other tough plant material. The cheek teeth of *A. africanus* served as a grinding platform for plants—but the relatively large incisors were more typical of carnivores.

Australopiths, the earliest hominids, were small brained and bipedal, apelike in some ways and humanlike in others.

Australopiths endured for at least 3 million years in the African savanna. The ancestors of humans may have been among these transitional forms.

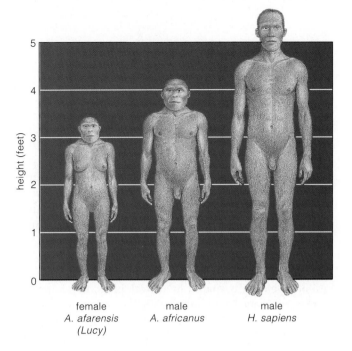

female
A. afarensis
(Lucy)

male
A. africanus

male
H. sapiens

Figure 28.9 Size and stature of two australopiths, compared with a modern human. In 1994, an Ethiopian research team led by Tim White of the University of California at Berkeley found several fossils of a different australopith, *A. ramidus*. This form existed 4.4 million years ago, which puts it close to the presumed time of divergence from the last common ancestor of apes and humans. At that time, tropical forests were giving way to woodlands and grasslands. *A. ramidus* was chimplike in some respects. Yet its flattened face, smaller canines, and other features seem to place it on the evolutionary road that led to humans.

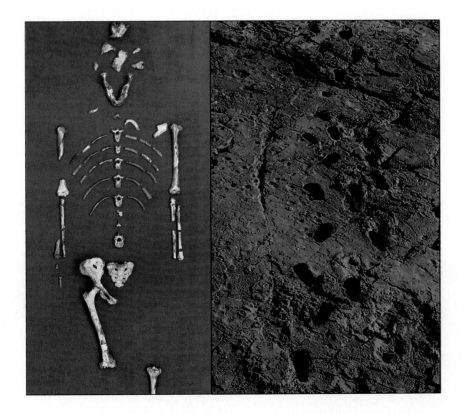

Following the path produces, at least for me, a kind of poignant time wrench. At one point, and you need not be an expert tracker to discern this, the traveler stops, pauses, turns to the left to glance at some possible threat or irregularity, then continues to the north. This motion, so intensely human, transcends time. Three million seven hundred thousand years ago, a remote ancestor—just as you or I—experienced a moment of doubt.

—Mary Leakey

Figure 28.10 Evidence of early hominids. The photograph at far left shows the remains of Lucy, one of the earliest known australopiths. The limb bone density indicates she had strong muscles. The adjacent photograph shows footprints made in soft, damp volcanic ash 3.7 million years ago at Laetoli, Tanzania, as discovered by Mary Leakey. The arch, big toe, and heel marks were made by bipedal hominids.

28.5 ON THE ROAD TO MODERN HUMANS

Early *Homo*

By 2.5 million years ago, some hominids were making stone tools. Maybe they were using sticks and other perishable items before then, much like modern apes do, but we have no way of knowing. We also don't know *which* species actually made stone tools. Maybe it was **early *Homo*.** (This is only one of its names. In other taxonomic schemes the same form is called "late australopith" or *H. habilis*.) Early *Homo* may have been on or near the road leading to modern humans.

Some skull fragments of early *Homo* are dated at 2.4 million years. Compared to australopiths, this hominid had a smaller face, more generalized teeth, and a larger brain (Figure 28.11). It ate plants and may have scavenged the remains of small animals.

Early *Homo* may have started down a toolmaking road by picking up rocks to crack open animal bones and expose the soft, edible marrow. Perhaps it started using sharp-edged flakes, formed naturally as rocks tumbled through a river or down a hill. Flakes could scrape flesh from animal bones. And then some hominid started *shaping* stone implements. The earliest known "manufactured" tools were crudely chipped pebbles, first discovered by Mary Leakey at Olduvai Gorge (Figure 28.12). This African gorge cuts through a sequence of sedimentary deposits. The crudest tools are in the deepest layers. They may have been used in different tasks, such as digging the ground for roots and tubers, smashing bones for marrow, and poking insects out from tree bark. In more recent layers we find more complex tools. They signify that hominids were getting serious about exploiting a major food source—the abundant game of open grasslands.

Figure 28.11 Comparison of skull shapes of early hominids with that of an anatomically modern human (*Homo sapiens*). The drawings are not all to the same scale. White indicates reconstructed portions of missing bone tissue.

From *Homo erectus* to *H. sapiens*

Where modern human populations arose is a hotly debated topic. By one model, they evolved in Africa, from populations of an early human species called ***Homo erectus***, then migrated into Asia, Europe, and other regions starting about 900,000 years ago. By another model, *H. erectus* migrated out of Africa, then distinctive subpopulations of modern humans arose later, as an outcome of genetic divergence in different geographic regions (Figure 28.13). In 1994, fossil hunters in Southeast Asia and in the former Soviet republic of Georgia unearthed *H. erectus* specimens that are 1.8 million and 1.6 million years old. The finds lend support to the second model.

H. erectus individuals had a long, chinless face, a thick-walled skull, a heavy browridge, and other archaic traits (Figure 28.11). Like some modern humans, they also were tall, rather narrow hipped, and long legged. They were suitably built for travel. In cold regions they became stockier, with bones adapted to muscular stress. Life must have been challenging, and strenuous. More than once, vast glaciers advanced and retreated in northern Europe, Asia, and North America. Some glaciers were 2 miles high!

This was the time of cultural lift-off for the human species. From southern Africa to England, *H. erectus* populations were using the same kinds of hand axes and other tools designed to pound, scrape, shred, cut, and whittle. *H. erectus* also learned to control fire.

Fully modern humans, ***Homo sapiens***, evolved 300,000 to 200,000 years ago. Older populations of *H. erectus* coexisted with them, then disappeared gradually. Early *H. sapiens* had smaller teeth and jaws. Many individuals had a chin. They had thinner facial bones, a larger brain, and a rounder, higher skull. Certain aspects of their skeletal organization indicate they had the capacity for complex human language (Figure 28.5).

One group of early humans, the Neandertals, lived in southern France, central Europe, and the Near East about 130,000 years ago. Although large brained, they

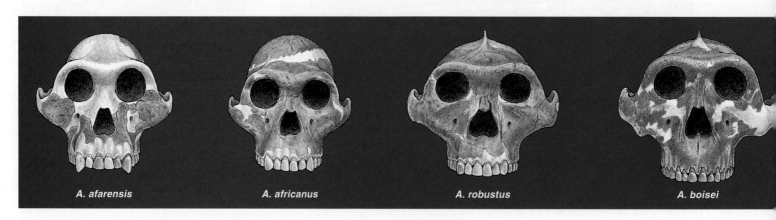

A. afarensis   *A. africanus*   *A. robustus*   *A. boisei*

Figure 28.12 A sampling of the 37,000+ stone tools from Olduvai Gorge. *First row, top to bottom:* The stone ball may herald a transition to aggressive tool use. It resembles Argentine bolas, which are strung on lengths of hide. Bolas are thrown at animals to entangle the legs and bring them down. Below the ball are three choppers. The last two are more advanced forms that have a joint as well as a sharp edge. *Second row:* A hand ax and a cleaver.

had heavy facial bones, often a large browridge, and a skull base flatter than that of *H. erectus*. Neandertals lived at the edge of forests, in caves, rock shelters, and open-air camps. Apparently they hunted or gathered food at their doorstep, so to speak. They did not learn how to exploit the abundant game herds. They disappeared 35,000 or 40,000 years ago, when anatomically modern humans arose. The more recent groups learned to store and share food among themselves. These newly evolved humans developed spectacularly rich cultures. They were the creators of the exquisite tools and artistic treasures at Lascaux and elsewhere.

From 40,000 years ago to the present, human evolution has been almost entirely cultural rather than biological, and so we leave the story. From the biological perspective, however, we can make these concluding remarks: Humans spread throughout the world by rapidly devising the cultural means to deal with a broad range of environments. Compared with their predecessors, modern humans developed rich and varied cultures, moving from "stone-age" technology to the age of "high tech." Yet hunters and gatherers persist in parts of the world, attesting to the great plasticity and depth of human adaptations.

Cultural evolution has outpaced the biological evolution of the only remaining human species. Humans everywhere rely on cultural innovation to adapt rapidly to a broad range of environmental challenges.

0.2%   0.1%   0% = genetic distance

NEW GUINEA, AUSTRALIA

PACIFIC ISLANDS

SOUTHEAST ASIA

ARCTIC, NORTHEAST ASIA

NORTH, SOUTH AMERICA

NORTHEAST ASIA

EUROPE, MIDDLE EAST

AFRICA

Figure 28.13 (*Above*) Proposed family tree for modern human populations that are native to different geographic regions. The branch points reflect divergences that began after the far-reaching migrations of *Homo erectus* populations. Data are based on biochemical and immunological comparisons (page 308). The first major divergence seems to have occurred after some *H. erectus* populations spread from Africa into Europe and others spread into Southeast Asia.

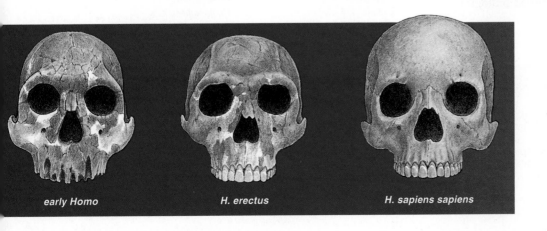

early Homo   *H. erectus*   *H. sapiens sapiens*

SUMMARY

1. Like other mammals, humans and other primates have an internal skeleton, a complex brain and sensory organs within a skull, mammary glands (in females), and distinctive teeth. Adults nourish, protect, and serve as behavioral models for the young.

2. Primates include prosimians (lemurs and related forms) and anthropoids (including tarsioids, monkeys, apes, and humans). Only apes and humans are hominoids. Only anatomically modern humans (*H. sapiens*) and others of their lineage (from *A. afarensis* to *H. erectus*) are further classified as hominids.

3. Unlike most primates, humans are not restrictively specialized. They remain flexible and adapt to a wide range of challenges in diverse environments. This capacity resulted from five evolutionary trends that occurred among certain primate lineages:

a. From a four-legged gait to bipedalism.

b. Increased manipulative skills owing to changes in the hands, which began to be freed from load-bearing functions among tree-dwelling primates.

c. Less reliance on the sense of smell and more on enhanced daytime vision.

d. From specialized to omnivorous eating habits.

e. Increases in brain complexity and behavior.

4. The first primates—small, rodentlike mammals—evolved more than 60 million years ago. Anthropoids, including the ancestors of monkeys, apes, and humans, had evolved by 35 million years ago. The earliest known fossils of hominids (*A. ramidus*) date from approximately 4.4 million years ago.

5. About 2.4 million years ago, early *Homo* evolved. It may have been the first stone toolmaker. *Homo erectus*, the presumed ancestor of modern human populations, not only had evolved by 1.8 million years ago, some populations of this species had already moved out of Africa, into Asia and possibly Europe by then. Anatomically modern humans (*H. sapiens*) arose between 300,000 and 200,000 years ago. Starting about 40,000 years ago, cultural evolution has outstripped biological evolution of the human form.

Review Questions

1. What is the difference between "hominoid" and "hominid"? Are we hominoids, hominids, or both? *472*

2. What are the general evolutionary trends that occurred among the primates as a group? What way of life apparently was the foundation for these trends? *473–474*

3. What conditions seem to have been responsible for the great adaptive radiation of apelike forms during the Miocene? *475*

Self-Quiz *(Answers in Appendix IV)*

1. Primates include _____ .
   a. lemurs   d. humans
   b. monkeys   e. all of the above
   c. apes

2. Behavioral flexibility can be observed among _____ .
   a. mammals   c. humans
   b. primates   d. all of the above

3. _____ are hominoids; _____ are hominids.
   a. Lemurs and monkeys; apes and humans
   b. Apes and humans; australopiths and humans
   c. Monkeys; apes and australopiths
   d. Monkeys, apes, and humans; apes only

4. Hominids showed plasticity. This means they _____ .
   a. could adapt to a wide range of demands in complex environments
   b. were adapted for a narrow range of demands in complex environments
   c. had flexible bones that cracked easily
   d. were limber enough to swing through the trees

5. The oldest known primates date from the _____ .
   a. Miocene   c. Oligocene
   b. Paleocene   d. Pliocene

6. The first known hominids were the _____ .
   a. *Homos*   c. cercopiths
   b. dryopiths   d. australopiths

7. The earliest known stone tools are _____ years old.
   a. 8 million   c. 4.2 million
   b. 6 million   d. 2.5 million

8. Match the primates with their descriptions.
   ____ prosimian   a. monkeys, apes, and humans
   ____ tarsioid   b. lemurs
   ____ anthropoid   c. humans and recent ancestors
   ____ hominoid   d. apes and humans
   ____ hominid   e. tarsiers

Selected Key Terms

anthropoid *472*   hominid *472*
australopith *476*   hominoid *472*
bipedalism *473*   *Homo erectus* *478*
culture *474*   *Homo sapiens* *478*
dryopith *475*   primate *472*
early *Homo* *478*   savanna *473*

Readings

Stringer, C. December 1990. "The Emergence of Modern Humans." *Scientific American* 263:98–104.

Weiss, M., and A. Mann. 1990. *Human Biology and Behavior.* Fifth edition. New York: Harper Collins.

FACING PAGE: *A flowering plant (Prunus) busily doing what it does best: producing flowers for the fine art of reproduction.*

APPENDIX I
A Brief Classification Scheme

The following classification scheme is a composite of several that are used in microbiology, botany, and zoology. The major groupings are more or less agreed upon. There is not always agreement on what to call a given grouping or where it fits in the overall hierarchy. There are several reasons for this.

First, the fossil record varies in its quality and completeness. Therefore, the relationship of one group to others is sometimes open to interpretation. Comparative studies at the molecular level are firming up the picture, but this work is still under way.

Second, since the time of Linnaeus, classification schemes have been based on perceived morphological similarities and differences among organisms. Although some original interpretations are now open to question, we are so used to thinking about organisms in certain ways that reclassification proceeds slowly. Traditionally, for example, birds and reptiles are separate classes (Reptilia and Aves). Yet there are compelling arguments for grouping lizards and snakes as one class, and crocodilians, dinosaurs, and birds as another.

Finally, microbiologists, mycologists, botanists, zoologists, and other researchers have inherited a wealth of literature, based on classification schemes peculiar to their fields. Most see no good reason to give up established terminology and so disrupt access to the past. Until very recently, botanists were using *division*, and zoologists, *phylum* for groupings that are equivalent in the hierarchy. Opinions are polarized with respect to an entire kingdom (the Protista), certain members of which could just as easily be called single-celled plants, fungi, or animals. Indeed, the term protozoan is a holdover from earlier schemes that ranked the amoebas and some other single cells as simple animals.

Given the problems, why do we bother imposing artificial frameworks on the history of life? We do this for the same reason that a writer might decide to break up the history of civilization into several volumes, many chapters, and many more paragraphs. Both efforts are attempts to impart structure to what might otherwise be an overwhelming body of information.

Bear in mind, we include this classification scheme mainly for your reference purposes. It is by no means complete. Numerous existing and extinct organisms of the so-called lesser phyla are not represented. Our strategy is to focus mainly on organisms mentioned in the text. A few examples of organisms also are listed under the entries.

SUPERKINGDOM PROKARYOTA. Prokaryotes. Single-celled organisms with DNA concentrated in a cytoplasmic region, not in a membrane-bound nucleus.

KINGDOM MONERA. Bacteria, either single cells or simple associations of cells. Both autotrophs and heterotrophs (refer to Table 22.2). *Bergey's Manual of Systematic Bacteriology,* the authoritative reference in the field, calls this "a time of taxonomic transition." It groups bacteria mainly on the basis of form, physiology, and behavior, not on phylogeny. The scheme presented here does reflect the growing evidence of evolutionary relationships for at least some bacterial groups.

SUBKINGDOM ARCHAEBACTERIA. Methanogens, halophiles, thermophiles. Strict anaerobes, distinct from other bacteria in cell wall, membrane lipids, ribosomes, and RNA sequences. *Methanobacterium, Halobacterium, Sulfolobus.*

SUBKINGDOM EUBACTERIA. Gram-negative and gram-positive forms. Peptidoglycan in cell wall. Photosynthetic autotrophs, chemosynthetic autotrophs, and heterotrophs.

PHYLUM GRACILICUTES. Typical Gram-negative, thin wall. Autotrophs (photosynthetic and chemosynthetic) and heterotrophs. *Anabaena* and other cyanobacteria. *Escherichia, Pseudomonas, Neisseria, Myxococcus.*
PHYLUM FIRMICUTES. Typical Gram-positive, thick wall. Heterotrophs. *Bacillus, Staphylococcus, Streptococcus, Clostridium, Actinomycetes.*
PHYLUM TENERICUTES. Gram-negative, wall absent. Heterotrophs (saprobes, pathogens). *Mycoplasma.*

SUPERKINGDOM EUKARYOTA. Eukaryotes (single-celled and multicelled organisms. Cells typically have a nucleus (enclosing the DNA) and other membrane-bound organelles.

KINGDOM PROTISTA. Mostly single-celled eukaryotes, some colonial forms. Diverse autotrophs and heterotrophs. Many lineages apparently related evolutionarily to certain plants, fungi, and possibly animals.

PHYLUM CHYTRIDIOMYCOTA. Chytrids. Heterotrophs; saprobic decomposers, parasites. *Chytridium.*
PHYLUM OOMYCOTA. Water molds. Heterotrophs. Decomposers, some parasites. *Saprolegnia, Phytophthora, Plasmopara.*
PHYLUM ACRASIOMYCOTA. Cellular slime molds. Heterotrophs with amoeboid and spore-bearing stages. *Dictyostelium.*
PHYLUM MYXOMYCOTA. Plasmodial slime molds. Heterotrophs with amoeboid and spore-bearing stages. *Physarum.*
PHYLUM SARCODINA. Amoeboid protozoans. Heterotrophs. Soft-bodied; some shelled. Amoebas, foraminiferans, radiolarians, heliozoans. *Amoeba, Entomoeba.*
PHYLUM CILIOPHORA. Ciliated protozoans. Heterotrophs. Distinctive arrays of cilia, used as motile structures. *Paramecium,* hypotrichs.
PHYLUM MASTIGOPHORA. Flagellated protozoans. Some free-living, many internal parasites; all with one to several flagella. *Trypanosoma, Trichomonas, Giardia.*

SPOROZOANS.   Parasitic protozoans, many intracellular. "Sporozoans" is the common name with no formal taxonomic status. *Plasmodium, Toxoplasma.*

PHYLUM EUGLENOPHYTA.   Euglenoids. Mostly heterotrophs, some photosynthetic types. Flagellated. *Euglena.*

PHYLUM CHRYSOPHYTA.   Golden algae, yellow-green algae, diatoms. Photosynthetic. Some flagellated, others not. *Mischococcus, Synura, Vaucheria.*

PHYLUM PYRRHOPHYTA.   Dinoflagellates. Photosynthetic, mostly, but some heterotrophs. *Gymnodinium breve.*

PHYLUM RHODOPHYTA.   Red algae. Photosynthetic, nearly all marine, some freshwater. *Porphyra. Bonnemaisonia, Euchema.*

PHYLUM PHAEOPHYTA.   Brown algae. Photosynthetic, nearly all temperate or marine waters. *Macrocystis, Fucus, Sargassum, Ectocarpus, Postelsia.*

PHYLUM CHLOROPHYTA.   Green algae. Photosynthetic. Most freshwater, some marine or terrestrial. *Chlamydomonas, Spirogyra, Ulva, Volvox, Codium, Halimeda.*

KINGDOM FUNGI.   Mostly multicelled eukaryotes. Heterotrophs (mostly saprobes, some parasites). Major decomposers of nearly all communities. Reliance on extracellular digestion of organic matter and absorption of nutrients by individual cells.

PHYLUM ZYGOMYCOTA.   Zygomycetes. All produce nonmotile spores. Bread molds, related forms. *Rhizopus, Philobolus.*

PHYLUM ASCOMYCOTA.   Ascomycetes. Sac fungi. Most yeasts and molds; morels, truffles. *Saccharomycetes, Morchella Neurospora, Sarcoscypha. Claviceps, Ophiostoma.*

PHYLUM BASIDIOMYCOTA.   Basidiomycetes. Club fungi. Mushrooms, shelf fungi, stinkhorns. *Agaricus, Amanita, Puccinia, Ustilago.*

IMPERFECT FUNGI.   Sexual spores absent or undetected. The group has no formal taxonomic status. If better understood, a given species might be grouped with sac fungi or club fungi. *Verticillium, Candida, Microsporum, Histoplasma.*

LICHENS.   Mutualistic interactions between a fungus and a cyanobacterium, green alga, or both. *Usnea, Cladonia.*

KINGDOM PLANTAE.   Nearly all multicelled eukaryotes. Photosynthetic autotrophs, except for a few parasitic types.

PHYLUM RHYNIOPHYTA.   Earliest known vascular plants; extinct. *Cooksonia, Rhynia.*

PHYLUM TRIMEROPHYTA.   Trimerophytes. *Psilophyton.*

PHYLUM PROGYMNOSPERMOPHYTA. Progymnosperms. Ancestral to early seed-bearing plants; extinct. *Archaeopteris.*

PHYLUM CHAROPHYTA.   Stoneworts.

PHYLUM BRYOPHYTA.   Liverworts, hornworts, mosses. *Marchantia, Polytrichum, Sphagnum.*

PHYLUM PSILOPHYTA.   Whisk ferns. *Psilotum.*

PHYLUM LYCOPHYTA.   Lycophytes, club mosses. *Lycopodium, Selaginella.*

PHYLUM SPHENOPHYTA.   Horsetails. *Equisetum.*

PHYLUM PTEROPHYTA.   Ferns.

PHYLUM PTERIDOSPERMOPHYTA.   Seed ferns. Fernlike gymnosperms; extinct.

PHYLUM CYCADOPHYTA.   Cycads. *Zamia.*

PHYLUM GINKGOPHYTA.   Ginkgo. *Ginkgo.*

PHYLUM GNETOPHYTA.   Gnetophytes. *Ephedra, Welwitschia.*

PHYLUM CONIFEROPHYTA.   Conifers.
   Family Pinaceae. Pines, firs, spruces, hemlock, larches, Douglas firs, true cedars. *Pinus.*
   Family Cupressaceae. Junipers, cypresses. *Juniperus.*
   Family Taxodiaceae. Bald cypress, redwoods, Sierra bigtree, dawn redwood. *Sequoia.*
   Family Taxaceae. Yews.

PHYLUM ANTHOPHYTA.   Flowering plants.
   Class Dicotyledonae. Dicotyledons (dicots). Some families of several different orders are listed:
      Family Nymphaeaceae. Water lilies.
      Family Papaveraceae. Poppies.
      Family Brassicaceae. Mustards, cabbages, radishes.
      Family Malvaceae. Mallows, cotton, okra, hibiscus.
      Family Solanaceae. Potatoes, eggplant, petunias.
      Family Salicaceae. Willows, poplars.
      Family Rosaceae. Roses, apples, almonds, strawberries.
      Family Fabaceae. Peas, beans, lupines, mesquite.
      Family Cactaceae. Cacti.
      Family Euphorbiaceae. Spurges, poinsettia.
      Family Cucurbitaceae. Gourds, melons, cucumbers, squashes.
      Family Apiaceae. Parsleys, carrots, poison hemlock.
      Family Aceraceae. Maples.
      Family Asteraceae. Composites. Chrysanthemums, sunflowers, lettuces, dandelions.
   Class Monocotyledonae. Monocotyledons (monocots). Some families of several different orders are listed:
      Family Liliaceae. Lilies, hyacinths, tulips, onions, garlic.
      Family Iridaceae. Irises, gladioli, crocuses.
      Family Orchidaceae. Orchids.
      Family Arecaceae. Date palms, coconut palms.
      Family Cyperaceae. Sedges.
      Family Poaceae. Grasses, bamboos, corn, wheat, sugarcane.
      Family Bromeliaceae. Bromeliads, pineapples, Spanish moss.

KINGDOM ANIMALIA.   Multicelled eukaryotes. Heterotrophs (herbivores, carnivores, omnivores, parasites, detritivores).

PHYLUM PLACOZOA.   Small, organless marine animal. *Trichoplax.*

PHYLUM MESOZOA.   Ciliated, wormlike parasites, about the same level of complexity as *Trichoplax.*

PHYLUM PORIFERA.   Sponges.

PHYLUM CNIDARIA.
   Class Hydrozoa. Hydrozoans. *Hydra, Obelia, Physalia.*
   Class Scyphozoa. Jellyfishes. *Aurelia.*
   Class Anthozoa. Sea anemones, corals. *Telesto.*

PHYLUM CTENOPHORA.   Comb jellies. *Pleurobrachia.*

PHYLUM PLATYHELMINTHES.   Flatworms.
   Class Turbellaria. Triclads (planarians), polyclads. *Dugesia.*
   Class Trematoda. Flukes. *Schistosoma.*
   Class Cestoda. Tapeworms. *Taenia.*

PHYLUM NEMERTEA.   Ribbon worms.

PHYLUM NEMATODA.   Roundworms. *Ascaris, Trichinella.*

PHYLUM ROTIFERA.   Rotifers.

PHYLUM MOLLUSCA.   Mollusks.
   Class Polyplacophora. Chitons.
   Class Gastropoda. Snails (periwinkles, whelks, limpets, abalones, cowries, conches, nudibranchs, tree snails, garden snails), sea slugs, land slugs.
   Class Bivalvia. Clams, mussels, scallops, cockles, oysters, shipworms.
   Class Cephalopoda. Squids, octopuses, cuttlefish, nautiluses. *Loligo.*

PHYLUM BRYOZOA.   Bryozoans (moss animals).

PHYLUM BRACHIOPODA.   Lampshells.

PHYLUM ANNELIDA.   Segmented worms.
   Class Polychaeta. Mostly marine worms.
   Class Oligochaeta. Mostly freshwater and terrestrial worms, but many marine. *Lumbricus* (earthworms).
   Class Hirudinea. Leeches.

PHYLUM TARDIGRADA.   Water bears.

PHYLUM ONYCHOPHORA.   Onychophorans. *Peripatus.*

PHYLUM ARTHROPODA.
   Subphylum Trilobita. Trilobites; extinct.
   Subphylum Chelicerata. Chelicerates. Horseshoe crabs, spiders, scorpions, ticks, mites.
   Subphylum Crustacea. Shrimps, crayfishes, lobsters, crabs, barnacles, copepods, isopods (sowbugs).
   Subphylum Uniramia.
      Superclass Myriapoda. Centipedes, millipedes.
      Superclass Insecta.
         Order Ephemeroptera. Mayflies.
         Order Odonata. Dragonflies, damselflies.

Order Orthoptera. Grasshoppers, crickets, katydids.
Order Dermaptera. Earwigs.
Order Blattodea. Cockroaches.
Order Mantodea. Mantids.
Order Isoptera. Termites.
Order Mallophaga. Biting lice.
Order Anoplura. Sucking lice.
Order Homoptera. Cicadas, aphids, leafhoppers, spittlebugs.
Order Hemiptera. Bugs.
Order Coleoptera. Beetles.
Order Diptera. Flies.
Order Mecoptera. Scorpion flies. *Harpobittacus.*
Order Siphonaptera. Fleas.
Order Lepidoptera. Butterflies, moths.
Order Hymenoptera. Wasps, bees, ants.

PHYLUM ECHINODERMATA. Echinoderms.
Class Asteroidea. Sea stars.
Class Ophiuroidea. Brittle stars.
Class Echinoidea. Sea urchins, heart urchins, sand dollars.
Class Holothuroidea. Sea cucumbers.
Class Crinoidea. Feather stars, sea lilies.
Class Concentricycloidea. Sea daisies.

PHYLUM HEMICHORDATA. Acorn worms.

PHYLUM CHORDATA. Chordates.
Subphylum Urochordata. Tunicates, related forms.
Subphylum Cephalochordata. Lancelets.
Subphylum Vertebrata. Vertebrates.
Class Agnatha. Jawless vertebrates (lampreys, hagfishes).
Class Placodermi. Jawed, heavily armored fishes; extinct.
Class Chondrichthyes. Cartilaginous fishes (sharks, rays, skates, chimaeras).
Class Osteichthyes. Bony fishes.
Subclass Dipnoi. Lungfishes.
Subclass Crossopterygii. Coelacanths, related forms.
Subclass Actinopterygii. Ray-finned fishes.
Order Acipenseriformes. Sturgeons, paddlefishes.
Order Salmoniformes. Salmon, trout.
Order Atheriniformes. Killifishes, guppies.
Order Gasterosteiformes. Seahorses.
Order Perciformes. Perches, wrasses, barracudas, tunas, freshwater bass, mackerels.
Order Lophiiformes. Angler fishes.
Class Amphibia. Mostly tetrapods; embryo enclosed in amnion.
Order Caudata. Salamanders.
Order Anura. Frogs, toads.
Order Apoda. Apodans (caecilians).
Class Reptilia. Skin with scales, embryo enclosed in amnion.
Subclass Anapsida. Turtles, tortoises.
Subclass Lepidosaura. *Sphenodon*, lizards, snakes.
Subclass Archosaura. Dinosaurs (extinct), crocodiles, alligators.
Class Aves. Birds. (In more recent schemes, dinosaurs, crocodilians, and birds are grouped in the same category.)

Order Struthioniformes. Ostriches.
Order Sphenisciformes. Penguins.
Order Procellariiformes. Albatrosses, petrels.
Order Ciconiiformes. Herons, bitterns, storks, flamingoes.
Order Anseriformes. Swans, geese, ducks.
Order Falconiformes. Eagles, hawks, vultures, falcons.
Order Galliformes. Ptarmigan, turkeys, domestic fowl.
Order Columbiformes. Pigeons, doves.
Order Strigiformes. Owls.
Order Apodiformes. Swifts, hummingbirds.
Order Passeriformes. Sparrows, jays, finches, crows, robins, starlings, wrens.
Class Mammalia. Skin with hair; young nourished by milk-secreting glands of adult.
Subclass Prototheria. Egg-laying mammals (duckbilled platypus, spiny anteaters).
Subclass Metatheria. Pouched mammals or marsupials (opossums, kangaroos, wombats).
Subclass Eutheria. Placental mammals.
Order Insectivora. Tree shrews, moles, hedgehogs.
Order Scandentia. Insectivorous tree shrews.
Order Chiroptera. Bats.
Order Primates.
Suborder Strepsirhini (prosimians). Lemurs, lorises.
Suborder Haplorhini (tarsioids and anthropoids).
Infraorder Tarsiiformes. Tarsiers.
Infraorder Platyrrhini (New World monkeys).
Family Cebidae. Spider monkeys, howler monkeys, capuchin.
Infraorder Catarrhini (Old World monkeys and hominoids).
Superfamily Cercopithecoidea. Baboons, macaques, langurs.
Superfamily Hominoidea. Apes and humans.
Family Hylobatidae. Gibbons.
Family Pongidae. Chimpanzees, gorillas, orangutans.
Family Hominidae. Humans and most recent ancestors of humans.
Order Carnivora. Carnivores.
Suborder Feloidea. Cats, civets, mongooses, hyenas.
Suborder Canoidea. Dogs, weasels, skunks, otters, raccoons, pandas, bears.
Order Proboscidea. Elephants; mammoths (extinct).
Order Sirenia. Sea cows (manatees, dugongs).
Order Perissodactyla. Odd-toed ungulates (horses, tapirs, rhinos).
Order Artiodactyla. Even-toed ungulates (camels, deer, bison, sheep, goats, antelopes, giraffes).
Order Edentata. Anteaters, tree sloths, armadillos.
Order Tubulidentata. African aardvarks.
Order Cetacea. Whales, porpoises.
Order Rodentia. Most gnawing animals (squirrels, rats, mice, guinea pigs, porcupines).

APPENDIX II
Units of Measure

Metric-English Conversions

Length

English		Metric
inch	=	2.54 centimeters
foot	=	0.30 meter
yard	=	0.91 meter
mile (5,280 feet)	=	1.61 kilometer

To convert	multiply by	to obtain
inches	2.54	centimeters
feet	30.00	centimeters
centimeters	0.39	inches
millimeters	0.039	inches

Weight

English		Metric
grain	=	64.80 milligrams
ounce	=	28.35 grams
pound	=	453.60 grams
ton (short) (2,000 pounds)	=	0.91 metric ton

To convert	multiply by	to obtain
ounces	28.3	grams
pounds	453.6	grams
pounds	0.45	kilograms
grams	0.035	ounces
kilograms	2.2	pounds

Volume

English		Metric
cubic inch	=	16.39 cubic centimeters
cubic foot	=	0.03 cubic meter
cubic yard	=	0.765 cubic meters
ounce	=	0.03 liter
pint	=	0.47 liter
quart	=	0.95 liter
gallon	=	3.79 liters

To convert	multiply by	to obtain
fluid ounces	30.00	milliliters
quart	0.95	liters
milliliters	0.03	fluid ounces
liters	1.06	quarts

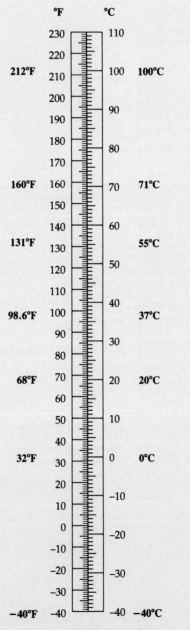

To convert temperature scales:

Fahrenheit to Celsius: $°C = 5/9 (°F - 32)$

Celsius to Fahrenheit: $°F = 9/5 (°C) + 32$

APPENDIX III
Answers to Self-Quizzes

Chapter 21
1. e
2. c
3. b
4. c
5. d
6. c
7. d
8. d
9. b, d, e, a, c

Chapter 22
1. c
2. c
3. b
4. c
5. d
6. d
7. e
8. d
9. c
10. d, a, b, e, c

Chapter 23
1. d
2. b
3. d
4. a
5. c
6. a
7. a
8. d
9. b

Chapter 24
1. c
2. d
3. b
4. c
5. b
6. e
7. a
8. a
9. c
10. d, f, c, e, a, b

Chapter 25
1. b
2. d
3. b
4. c
5. d
6. c
7. b
8. d
9. b
10. c, e, g, h, f, a, b, d

Chapter 26
1. body symmetry, cephalization, type of gut, type of body cavity, segmentation
2. b
3. b
4. c
5. a
6. a
7. b
8. b and c
9. c
10. d, f, c, g, e, i, j, h, a, b

Chapter 27
1. d
2. e
3. f
4. d
5. a
6. b
7. d
8. b
9. c
10. d
11. f
12. d
13. d
14. e
15. d, e, c, b, g, f, a

Chapter 28
1. e
2. c
3. b
4. a
5. b
6. d
7. d
8. b, e, a, d, c

A GLOSSARY OF BIOLOGICAL TERMS

ABO blood typing Method of characterizing an individual's blood according to whether one or both of two protein markers, A and B, are present at the surface of red blood cells. The O signifies that neither marker is present.

abortion Spontaneous or induced expulsion of the embryo or fetus from the uterus.

abscisic acid (ab-SISS-ik) Plant hormone that promotes stomatal closure, bud dormancy, and seed dormancy.

abscission (ab-SIH-zhun) [L. *abscindere*, to cut off] The dropping of leaves, flowers, fruits, or other plant parts due to hormonal action.

absorption Of complex animals, the movement of nutrients, fluid, and ions across the gut lining and into the internal environment.

accessory pigment A light-trapping pigment that contributes to photosynthesis by extending the range of usable wavelengths beyond those absorbed by the chlorophylls.

acid [L. *acidus*, sour] A substance that releases hydrogen ions (H^+) in water.

acid rain The falling to earth of snow or rain that contains sulfur and nitrogen oxides. Also called wet acid deposition (as opposed to dry acid deposition, or the falling to earth of airborne particles of sulfur and nitrogen oxides).

acoelomate (ay-SEE-luh-mate) Type of animal that has no fluid-filled cavity between the gut and body wall.

acoustical signal Sounds that are used as a communication signal.

actin (AK-tin) A globular contractile protein. In muscle cells, actin interacts with another protein, myosin, to bring about contraction.

action potential An abrupt, brief reversal in the steady voltage difference across the plasma membrane (that is, the resting membrane potential) of a neuron and some other cells.

activation energy The minimum amount of collision energy required to bring reactant molecules to an activated condition (the transition state) at which a reaction will proceed spontaneously. Enzymes enhance reaction rates by lowering the activation energy (they put substrates on a precise collision course).

active site A crevice on the surface of an enzyme molecule where a specific reaction is catalyzed.

active transport The pumping of one or more specific solutes through a transport protein that spans the lipid bilayer of a cell membrane. Most often, the solute is transported against its concentration gradient. The protein is activated by an energy boost, as from ATP.

adaptation [L. *adaptare*, to fit] In evolutionary biology, the process of becoming adapted (or more adapted) to a given set of environmental conditions. Of sensory neurons, a decrease in the frequency of action potentials (or their cessation) even when a stimulus is maintained at constant strength.

adaptive behavior A behavior that promotes the propagation of an individual's genes and that tends to increase in frequency in a population over time.

adaptive radiation A burst of speciation events, with lineages branching away from one another as they partition the existing environment or invade new ones.

adaptive trait Any aspect of form, function, or behavior that helps an organism survive and reproduce under a given set of environmental conditions.

adaptive zone A way of life, such as "catching insects in the air at night." A lineage must have physical, ecological, and evolutionary access to an adaptive zone to become a successful occupant of it.

adenine (AH-de-neen) A purine; a nitrogen-containing base found in nucleotides.

adenosine diphosphate (ah-DEN-uh-seen die-FOSS-fate) ADP, a molecule involved in cellular energy transfers; typically formed by hydrolysis of ATP.

adenosine phosphates Any of several relatively small molecules, some of which function as chemical messengers within and between cells, and others that function as energy carriers.

adenosine triphosphate *See* ATP.

ADH Antidiuretic hormone. Produced by the hypothalamus and released by the posterior pituitary, it stimulates reabsorption in the kidneys and so reduces urine volume.

adipose tissue A type of connective tissue having an abundance of fat-storing cells and blood vessels for transporting fats.

ADP Adenosine diphosphate. A nucleotide coenzyme that accepts unbound phosphate or a phosphate group to become ATP.

ADP/ATP cycle In cells, a mechanism of ATP renewal. When ATP donates a phosphate group to other molecules (and so energizes them), it reverts to ADP, then forms again by phosphorylation of ADP.

adrenal cortex (ah-DREE-nul) Outer portion of the adrenal gland; its hormones have roles in metabolism, inflammation, maintaining extracellular fluid volume, and other functions.

adrenal medulla Inner region of the adrenal gland; its hormones help control blood circulation and carbohydrate metabolism.

aerobic respiration (air-OH-bik) [Gk. *aer*, air, + *bios*, life] The main energy-releasing metabolic pathway of ATP formation, in which oxygen is the final acceptor of electrons stripped from glucose or some other organic compound. The pathway proceeds from glycolysis through the Krebs cycle and electron transport phosphorylation. A typical net yield is 36 ATP for each glucose molecule.

age structure Of a population, the number of individuals in each of several or many age categories.

agglutination (ah-glue-tin-AY-shun) Clumping together of foreign cells that have invaded the body (as pathogens or in tissue grafts or transplants). Clumping is induced by cross-linking between antibody molecules that have already latched onto antigen at the surface of the foreign cells.

aging A range of processes, including the breakdown of cell structure and function, by which the body gradually deteriorates. All organisms showing extensive cell differentiation undergo aging.

AIDS Acquired immunodeficiency syndrome. A set of chronic disorders following infection by the human immunodeficiency virus (HIV), which destroys key cells of the immune system.

alcoholic fermentation Anaerobic pathway of ATP formation in which pyruvate from glycolysis is broken down to acetaldehyde, which accepts electrons from NADH to become ethanol, and NAD^+ is regenerated. Its net yield is two ATP.

aldosterone (al-DOSS-tuh-rohn) Hormone secreted by the adrenal cortex that helps regulate sodium reabsorption.

allantois (ah-LAN-twahz) [Gk. *allas*, sausage] Of vertebrates, one of four extraembryonic membranes that form during embryonic development. It functions in respiration and storage of metabolic wastes in reptiles, birds, and some mammals. In humans, it functions in early blood formation and development of the urinary bladder.

allele (uh-LEEL) For a given location on a chromosome, one of two or more slightly different molecular forms of a gene that code for different versions of the same trait.

allele frequency Of a given gene locus, the relative abundances of each kind of allele carried by the individuals of a population.

allergy An immune response made against a normally harmless substance.

allopatric speciation [Gk. *allos*, different, + *patria*, native land] Speciation that follows geographic isolation of populations of the same species.

allosteric control (AL-oh-STARE-ik) Control over a metabolic reaction or pathway that operates through the binding of a specific substance at a control site on a specific enzyme.

alpine tundra A type of biome that exists at high elevations in mountains throughout the world.

altruistic behavior (al-true-ISS-tik) Self-sacrificing behavior; the individual behaves in a way that helps others but decreases its own chances of reproductive success.

alveolar sac (al-VEE-uh-lar) Any of the pouch-like clusters of alveoli in the lungs; the major sites of gas exchange.

alveolus (ahl-VEE-uh-lus), plural **alveoli** [L. *alveus*, small cavity] Any of the many cup-shaped, thin-walled outpouchings of respiratory bronchioles. A site where oxygen diffuses from air in the lungs to the blood, and carbon dioxide diffuses from blood to the lungs.

amino acid (uh-MEE-no) A small organic molecule having a hydrogen atom, an amino group, an acid group, and an R group covalently bonded to a central carbon atom. The subunit of polypeptide chains, which represent the primary structure of proteins.

ammonification (uh-moan-ih-fih-KAY-shun) Together with decomposition, a process by which certain bacteria and fungi break down nitrogen-containing wastes and remains of other organisms.

amnion (AM-nee-on) Of land vertebrates, one of four extraembryonic membranes. It becomes a fluid-filled sac in which the embryo (and fetus) can grow, move freely, and be protected from sudden temperature shifts and impacts.

amniote egg A type of egg, often with a leathery or calcified shell, that contains extraembryonic membranes, including the amnion. An adaptation that figured in the vertebrate invasion of land.

amphibian A type of vertebrate somewhere between fishes and reptiles in body plan and reproductive mode; salamanders, frogs and toads, and caecilians are the existing groups.

anaerobic pathway (an-uh-ROW-bik) [Gk. *an*, without, + *aer*, air] Metabolic pathway in which a substance other than oxygen serves as the final acceptor of electrons that have been stripped from substrates.

analogous structures Body parts, once different in separate lineages, that were put to comparable uses in similar environments and that came to resemble one another in form and function. They are evidence of morphological convergence.

anaphase (AN-uh-faze) The stage at which microtubules of a spindle apparatus separate sister chromatids of each chromosome and move them to opposite spindle poles. During anaphase I of *meiosis*, the two members of each pair of homologous chromosomes separate. During anaphase II, sister chromatids of each chromosome separate.

aneuploidy (AN-yoo-ploy-dee) A change in the chromosome number following inheritance of one extra or one less chromosome.

angiosperm (AN-gee-oh-spurm) [Gk. *angeion*, vessel, and *spermia*, seed] A flowering plant.

animal A heterotroph that eats or absorbs nutrients from other organisms; is multicelled, usually with tissues arranged in organs and organ systems; is usually motile during at least part of the life cycle; and goes through a period of embryonic development.

annelid The type of invertebrate classified as a segmented worm; an oligochaete (such as an earthworm), leech, or polychaete.

annual plant A flowering plant that completes its life cycle in one growing season.

anther [Gk. *anthos*, flower] In flowering plants, the pollen-bearing part of the male reproductive structure (stamen).

antibiotic A normal metabolic product of certain microorganisms that kills or inhibits the growth of other microorganisms.

antibody [Gk. *anti*, against] Any of a variety of Y-shaped receptor molecules with binding sites for specific antigens. Only B cells produce antibodies, then position them at their surface or secrete them.

anticodon In a tRNA molecule, a sequence of three nucleotide bases that can pair with an mRNA codon.

antigen (AN-tih-jen) [Gk. *anti*, against, + *genos*, race, kind] Any molecular configuration that is recognized as foreign to the body and that triggers an immune response. Most antibodies are protein molecules at the surface of infectious agents or tumor cells.

antigen-presenting cell A macrophage or some other white blood cell that engulfs and digests antigen, then displays it with certain MHC molecules at its surface. Recognition of antigen-MHC complexes by T and B cells triggers an immune response.

aorta (ay-OR-tah) [Gk. *airein*, to lift, heave] Main artery of systemic circulation; carries oxygenated blood away from the heart to all body regions except the lungs.

apical dominance The inhibitory influence of a terminal bud on the growth of lateral buds.

apical meristem (AY-pih-kul MARE-ih-stem) [L. *apex*, top, + Gk. *meristos*, divisible] Of most plants, a mass of self-perpetuating cells responsible for primary growth (elongation) at root and shoot tips.

appendicular skeleton (ap-en-DIK-yoo-lahr) In vertebrates, bones of the limbs, hips, and shoulders.

appendix A slender projection from the cup-shaped pouch at the start of the colon.

archaebacteria One of three great prokaryotic lineages that arose early in the history of life; now represented by methanogens, halophiles, and thermophiles.

arctic tundra A type of biome that lies between the polar ice cap and boreal forests of North America, Europe, and Asia.

arteriole (ar-TEER-ee-ole) Any of the blood vessels between arteries and capillaries. They are control points where the volume of blood delivered to different body regions can be adjusted.

artery Any of the large-diameter blood vessels that conduct oxygen-poor blood to the lungs and oxygen-enriched blood to all body tissues. Their thick, muscular wall allows them to smooth out pulsations in blood pressure caused by heart contractions.

arthropod An invertebrate having a hardened exoskeleton, specialized segments, and jointed appendages. Spiders, crabs, and insects are examples.

asexual reproduction Mode of reproduction by which offspring arise from a single parent, and inherit the genes of that parent only.

atmosphere A region of gases, airborne particles, and water vapor enveloping the earth; 80 percent of its mass is distributed within seventeen miles of the earth's surface.

atmospheric cycle A biogeochemical cycle in which the atmosphere is the largest reservoir of an element. The carbon and nitrogen cycles are examples.

atom The smallest unit of matter that is unique to a particular element.

atomic number The number of protons in the nucleus of each atom of an element; it differs for each element.

ATP Adenosine triphosphate (ah-DEN-uh-seen try-FOSS-fate). A nucleotide composed of adenine, ribose, and three phosphate groups. As the main energy carrier in cells, it directly or indirectly delivers energy to or picks up energy from nearly all metabolic pathways.

atrium (AYE-tree-um) Of the human heart, one of two chambers that receive blood. *Compare* ventricle.

australopith (OHSS-trah-low-pith) [L. *australis*, southern, + Gk. *pithekos*, ape] Any of the earliest known species of hominids, that is, the first species of the evolutionary branch leading to humans.

autoimmune response Misdirected immune response in which lymphocytes mount an attack against normal body cells.

autonomic nervous system (auto-NOM-ik) Those nerves leading from the central nervous system to the smooth muscle, cardiac muscle, and glands of internal organs and structures, that is, to the visceral portion of the body.

autosomal dominant inheritance Condition arising from the presence of a dominant allele on an autosome (not a sex chromosome). The allele is always expressed to some extent, even in heterozygotes.

autosomal recessive inheritance Condition arising from a recessive allele on an autosome (not a sex chromosome). Only recessive homozygotes show the resulting phenotype.

autosome Any of the chromosomes that are of the same number and kind in both males and females of the species.

autotroph (AH-toe-trofe) [Gk. *autos*, self, + *trophos*, feeder] An organism able to build its own large organic molecules by using carbon dioxide and energy from the physical environment. Photosynthetic autotrophs use sunlight energy; chemosynthetic autotrophs extract energy from chemical reactions involving inorganic substances. *Compare* heterotroph.

auxin (AWK-sin) Any of a class of growth-regulating hormones in plants; auxins promote stem elongation as one effect.

axial skeleton (AX-ee-uhl) In vertebrates, the skull, backbone, ribs, and breastbone (sternum).

axon Of a neuron, a long, cylindrical extension from the cell body, with finely branched endings. Action potentials move rapidly, without alteration, along an axon; their arrival at axon endings may trigger the release of neurotransmitter molecules that influence an adjacent cell.

B lymphocyte, or **B cell** The only white blood cell that produces antibodies, then positions them at the cell surface or secretes them as weapons in immune responses.

bacterial conjugation The transfer of plasmid DNA from one bacterial cell to another.

bacterial flagellum Of many bacterial cells, a whiplike motile structure that does not contain a core of microtubules.

bacteriophage (bak-TEER-ee-oh-fahj) [Gk. *baktērion*, small staff, rod, + *phagein*, to eat] Category of viruses that infect bacterial cells.

balanced polymorphism The maintenance of two or more forms of a trait in fairly stable proportions over generations.

Barr body In the cells of female mammals, a condensed X chromosome that was inactivated during early embryonic development.

basal body A centriole which, after having given rise to the microtubules of a flagellum or cilium, remains attached to its base in the cytoplasm.

base A substance that releases OH⁻ in water.

base pair A pair of hydrogen-bonded nucleotide bases in two strands of nucleic acids. In a DNA double helix, adenine pairs with thymine, and guanine with cytosine. When an mRNA strand forms on a DNA strand during transcription, uracil (U) pairs with the DNA's adenine.

base sequence The particular order in which one nucleotide base follows the next in a strand of DNA or RNA. The order differs to some extent for each kind of organism.

basophil Fast-acting white blood cells that secrete histamine and other substances during inflammation.

behavior, animal A response to external and internal stimuli, following integration of sensory, neural, endocrine, and effector components. Because behavior has a genetic basis, it is subject to natural selection and commonly can be modified through experience.

benthic province All of the sediments and rocky formations of the ocean bottom; begins with the continental shelf and extends down through deep-sea trenches.

biennial A flowering plant that lives through two growing seasons.

bilateral symmetry Body plan in which the left and right halves of the animal are mirror-images of each other.

binary fission Of bacteria, a mode of asexual reproduction in which the parent cell replicates its single chromosome, then divides into two genetically identical daughter cells.

biogeochemical cycle The movement of an element such as carbon or nitrogen from the environment to organisms, then back to the environment.

biogeographic realm [Gk. *bios*, life, + *geographein*, to describe the surface of the earth] Any of six major land regions, each having distinguishing types of plants and animals and generally retaining its identity because of climate and geographic barriers to gene flow.

biological clocks Internal time-measuring mechanisms that have roles in adjusting an organism's daily activities, seasonal activities, or both in response to environmental cues.

biological magnification The increasing concentration of a nondegradable or slowly degradable substance in body tissues as it is passed along food chains.

biological systematics Branch of biology that assesses patterns of diversity based on information from taxonomy, phylogenetic reconstruction, and classification.

bioluminescence A flashing of light that emanates from an organism when excited electrons of luciferins, or highly fluorescent substances, return to a lower energy level.

biomass The combined weight of all the organisms at a particular trophic (feeding) level in an ecosystem.

biome A broad, vegetational subdivision of a biogeographic realm shaped by climate, topography, and composition of regional soils.

biosphere [Gk. *bios*, life, + *sphaira*, globe] All regions of the earth's waters, crust, and atmosphere in which organisms live.

biosynthetic pathway A metabolic pathway in which small molecules are assembled into lipids, proteins, and other large organic molecules.

biotic potential Of a population, the maximum rate of increase per individual under ideal conditions.

bipedalism A habitual standing and walking on two feet, as by ostriches and humans.

bird A type of vertebrate, the only one having feathers, with strong resemblances and evolutionary connections to reptiles.

blastocyst (BLASS-tuh-sist) [Gk. *blastos*, sprout, + *kystis*, pouch] In mammalian development, a blastula stage consisting of a hollow ball of surface cells and an inner cell mass.

blastula (BLASS-chew-lah) An embryonic stage consisting of a ball of cells produced by cleavage.

blood A fluid connective tissue composed of water, solutes, and formed elements (blood cells and platelets); it carries substances to and from cells and helps maintain an internal environment that is favorable for cell activities.

blood pressure Fluid pressure, generated by heart contractions, that keeps blood circulating.

blood-brain barrier Set of mechanisms that helps control which blood-borne substances reach neurons in the brain.

bone The mineral-hardened connective tissue of bones.

bones In vertebrate skeletons, organs that function in movement and locomotion, protection of other organs, mineral storage, and (in some bones) blood cell production.

bottleneck An extreme case of genetic drift. A catastrophic decline in population size leads to a random shift in the allele frequencies among survivors. Because the population must rebuild from so few individuals, severely limited genetic variation may be an outcome.

Bowman's capsule The cup-shaped portion of a nephron that receives water and solutes filtered from blood.

brain Of most nervous systems, the most complex integrating center; it receives, processes, and integrates sensory input and issues coordinated commands for response.

brainstem The vertebrate midbrain, pons, and medulla oblongata, the core of which contains the reticular formation that helps govern activity of the nervous system as a whole.

bronchiole Of most vertebrates, a component of the finely branched bronchial tree inside each lung.

bronchus, plural **bronchi** (BRONG-CUSS, BRONG-kee) [Gk. *bronchos*, windpipe] Tubelike branchings of the trachea that lead into the lungs of most vertebrates.

brown alga A type of aquatic plant, found in nearly all marine habitats, that has an abundance of xanthophyll pigments.

bryophyte A nonvascular land plant that requires free water to complete its life cycle.

bud An undeveloped shoot of mostly meristematic tissue; often covered and protected by scales (modified leaves).

buffer A substance that can combine with hydrogen ions, release them, or both. Buffers help stabilize the pH of blood and other fluids.

bulk Of human digestion, a volume of fiber and other undigested material that absorption processes in the colon cannot decrease.

bulk flow In response to a pressure gradient, a movement of more than one kind of molecule in the same direction in the same medium (as in blood, sap, or air).

C4 pathway Of many plants, a pathway of photosynthesis in which carbon dioxide is fixed twice, in two different cell types. Carbon dioxide accumulates in the leaf and helps counter photorespiration. The first compound formed is the 4-carbon oxaloacetate.

Calvin-Benson cycle Cyclic reactions that are the "synthesis" part of the light-independent reactions of photosynthesis. In land plants, RuBP, or some other compound to which carbon has been affixed, undergoes rearrangements that lead to formation of a sugar phosphate and to regeneration of the RuBP. The cycle runs on ATP and NADPH from light-dependent reactions.

CAM plant A plant that conserves water by opening stomata only at night, when it fixes carbon dioxide by way of a C4 pathway.

cambium (KAM-bee-um), plural **cambia** In vascular plants, one of two types of meristems that are responsible for secondary growth (increases in stem and root diameter). Vascular cambium gives rise to secondary xylem and phloem; cork cambium gives rise to periderm.

camouflage An outcome of an organism's form, patterning, color, or behavior that helps it blend with its surroundings and escape detection.

cancer A type of malignant tumor, the cells of which show profound abnormalities in the plasma membrane and cytoplasm, abnormal growth and division, and weakened capacity for adhesion within the parent tissue (leading to metastasis). Unless eradicated, cancer is lethal.

canine A pointed tooth that functions in piercing food.

capillary [L. *capillus*, hair] A thin-walled blood vessel that functions in the exchange of gases and other substances between blood and interstitial fluid.

capillary bed A diffusion zone, consisting of numerous capillaries, where substances are exchanged between blood and interstitial fluid.

carbohydrate [L. *carbo*, charcoal, + *hydro*, water] A simple sugar or large molecule composed of sugar units. All cells use carbohydrates as structural materials, energy stores, and transportable forms of energy. The three classes of carbohydrates include: monosaccharides, oligosaccharides, or polysaccharides.

carbon cycle A biogeochemical cycle in which carbon moves from its largest reservoir in the atmosphere, through oceans and organisms, then back to the atmosphere.

carbon dioxide fixation First step of the light-independent reactions of photosynthesis. Carbon (from carbon dioxide) becomes affixed to a carbon compound (such as RuBP) that can enter the Calvin-Benson cycle.

carcinogen (kar-SIN-uh-jen) An environmental agent or substance, such as ultraviolet radiation, that can trigger cancer.

cardiac cycle (KAR-dee-ak) [Gk. *kardia*, heart, + *kyklos*, circle] The sequence of muscle contraction and relaxation constituting one heartbeat.

cardiac pacemaker Sinoatrial (SA) node; the basis of the normal rate of heartbeat. The self-excitatory cardiac muscle cells that spontaneously generate rhythmic waves of excitation over the heart chambers.

cardiovascular system Of most animals, an organ system that is composed of blood, one or more hearts, and blood vessels and that functions in the rapid transport of substances to and from cells.

carnivore [L. *caro*, *carnis*, flesh, + *vovare*, to devour] An animal that eats other animals; a type of heterotroph.

carotenoids (kare-OTT-en-oyds) Light-sensitive, accessory pigments that transfer absorbed energy to chlorophylls. They absorb violet and blue wavelengths but transmit red, orange, and yellow.

carpel (KAR-pul) The female reproductive part of a flower; sometimes called a pistil. The lower portion of a single carpel (or of a structure composed of two or more carpels) is an ovary, where eggs develop and are fertilized and where seeds mature. The upper portion has a stigma (a pollen-capturing surface tissue) and often a style (a slender extension of the ovary wall).

carrier protein Type of transport protein that binds specific substances and changes shape in ways that shunt the substances across a plasma membrane. Some carrier proteins function passively, others require an energy input.

carrying capacity The maximum number of individuals in a population (or species) that can be sustained indefinitely by a given environment.

cartilage A type of connective tissue with solid yet pliable intercellular material that resists compression.

Casparian strip In the exodermis and endodermis of roots, a waxy band that acts as an impermeable barrier between the walls of abutting cells.

cDNA Any DNA molecule copied from a mature mRNA transcript by way of reverse transcription.

cell [L. *cella*, small room] The smallest living unit; an organized unit that can survive and reproduce on its own, given DNA instructions and suitable environmental conditions, including appropriate sources of energy and raw materials.

cell count The number of cells of a given type in a microliter of blood.

cell cycle Events during which a cell increases in mass, roughly doubles its number of cytoplasmic components, duplicates its DNA, then undergoes nuclear and cytoplasmic division. It extends from the time a new cell is produced until it completes its own division.

cell differentiation The developmental process in which different cell types activate and suppress a fraction of their genes in different ways and so become specialized in composition, structure, and function. Regulatory proteins, enzymes, hormonal signals, and control sites built into DNA interact to bring about this selective gene expression.

cell junction Of multicelled organisms, a point of contact that physically links two cells or that provides functional links between their cytoplasm.

cell plate Of a plant cell undergoing cytoplasmic division, a disklike structure that forms from remnants of the spindle; it becomes a crosswall that partitions the cytoplasm.

cell theory A theory in biology, the key points of which are that (1) all organisms are composed of one or more cells, (2) the cell is the smallest unit that still retains a capacity for independent life, and (3) all cells arise from preexisting cells.

cell wall A rigid or semirigid wall outside the plasma membrane that supports a cell and imparts shape to it; a cellular feature of plants, fungi, protistans, and most bacteria.

central nervous system The brain and spinal cord of vertebrates.

central vacuole Of mature, living plant cells, a fluid-filled organelle that stores amino acids, sugars, ions, and toxic wastes. Its enlargement during growth causes the increases in surface area that improve nutrient uptake.

centriole (SEN-tree-ohl) A cylinder of triplet microtubules that gives rise to the microtubules of cilia and flagella.

centromere (SEN-troh-meer) [Gk. *kentron*, center, + *meros*, a part] A small, constricted region of a chromosome having attachment sites for microtubules that help move the chromosome during nuclear division.

cephalization (sef-ah-lah-ZAY-shun) [Gk. *kephalikos*, head] During the evolution of bilateral animals, the concentration of sensory structures and nerve cells in the head.

cerebellum (ser-ah-BELL-um) [L. diminutive of *cerebrum*, brain] Hindbrain region with reflex centers for maintaining posture and refining limb movements.

cerebral cortex Thin surface layer of the cerebral hemispheres. Some regions of the cortex receive sensory input, others integrate information and coordinate appropriate motor responses.

cerebrospinal fluid Clear extracellular fluid that surrounds and cushions the brain and spinal cord.

cerebrum (suh-REE-bruhm) Part of the vertebrate forebrain that originally integrated olfactory input and selected motor responses to it. In mammals, it evolved into the most complex integrating center.

channel protein Type of transport protein that serves as a pore through which ions or other water-soluble substances move across the plasma membrane. Some channels remain open, while others are gated and open and close in controlled ways.

chemical bond A union between the electron structures of two or more atoms or ions.

chemical synapse (SIN-aps) [Gk. *synapsis*, union] A small gap, the synaptic cleft, that separates two neurons (or a neuron and a muscle cell or gland cell) and that is bridged by neurotransmitter molecules released from the presynaptic neuron.

chemiosmotic theory (kim-ee-OZ-MOT-ik) Theory that an electrochemical gradient across a cell membrane drives ATP formation. Metabolic reactions cause hydrogen ions (H^+) to accumulate in a compartment formed by the membrane. The combined force of the resulting concentration and electric gradients propels hydrogen ions down the gradient, through channel proteins. Through enzyme action at these proteins, ADP and inorganic phosphate combine to form ATP.

chemoreceptor (KEE-moe-ree-sep-tur) Sensory receptor that detects chemical energy (ions or molecules) dissolved in the surrounding fluid.

chemosynthetic autotroph (KEE-moe-sin-THET-ik) One of a few kinds of bacteria able to synthesize its own organic molecules using carbon dioxide as the carbon source and certain inorganic substances (such as sulfur) as the energy source.

chlorofluorocarbon (KLORE-oh-FLOOR-oh-car-bun), or **CFC** One of a variety of odorless, invisible compounds of chlorine, fluorine, and carbon, widely used in commercial products, that are contributing to the destruction of the ozone layer above the earth's surface.

chlorophylls (KLOR-uh-fills) [Gk. *chloros*, green, + *phyllon*, leaf] Light-sensitive pigment molecules that absorb violet-to-blue and red wavelengths but that transmit green. Certain chlorophylls donate the electrons required for photosynthesis.

chloroplast (KLOR-uh-plast) An organelle that specializes in photosynthesis in plants and certain protistans.

chordate An animal having a notochord, a dorsal hollow nerve cord, a pharynx, and gill slits in the pharynx wall for at least part of its life cycle.

chorion (CORE-ee-on) Of placental mammals, one of four extraembryonic membranes; it becomes a major component of the placenta. Absorptive structures (villi) that develop at its surface are crucial for the transfer of substances between the embryo and mother.

chromatid Of a duplicated eukaryotic chromosome, one of two DNA molecules and its associated proteins. One chromatid remains attached to its "sister" chromatid at the centromere until they are separated from each other during a nuclear division; then each is a separate chromosome.

chromosome (CROW-moe-some) [Gk. *chroma*, color, + *soma*, body] Of eukaryotes, a DNA molecule with many associated proteins. A bacterial chromosome does not have a comparable profusion of proteins associated with the DNA.

chromosome number Of eukaryotic species, the number of each type of chromosome in all cells except dividing germ cells or gametes.

chytrid A type of single-celled fungus of muddy and aquatic habitats.

ciliated protozoan One of four major groups of protozoans.

cilium (SILL-ee-um), plural **cilia** [L. *cilium*, eyelid] Of eukaryotic cells, a short, hairlike projection that contains a regular array of microtubules. Cilia serve as motile structures, help create currents of fluids, or are part of sensory structures. They typically are more profuse than flagella.

circadian rhythm (ser-KAYD-ee-un) [L. *circa*, about, + *dies*, day] Of many organisms, a cycle of physiological events that is completed every 24 hours or so, even when environmental conditions remain constant.

circulatory system Of multicelled animals, an organ system consisting of a muscular pump (heart, most often), blood vessels, and blood; the system transports materials to and from cells and often helps stabilize body temperature and pH.

cladistics An approach to biological systematics in which organisms are grouped according to similarities that are derived from a common ancestor.

cladogram Branching diagram that represents patterns of relative relationships between organisms based on discrete morphological, physiological, and behavioral traits that vary among taxa being studied.

classification system A way of organizing and retrieving information about species.

cleavage Stage of animal development when mitotic cell divisions convert a zygote to a ball of cells, the blastula.

cleavage furrow Of an animal cell undergoing cytoplasmic division, a shallow, ringlike depression that forms at the cell surface as contractile microfilaments pull the plasma membrane inward. It defines where the cytoplasm will be cut in two.

climate Prevailing weather conditions for an ecosystem, including temperature, humidity, wind speed, cloud cover, and rainfall.

climax community Following primary or secondary succession, the array of species that remains more or less steady under prevailing conditions.

clonal selection theory Theory that lymphocytes activated by a specific antigen rapidly multiply and differentiate into huge subpopulations of cells, all having the parent cell's specificity against that antigen.

cloned DNA Multiple, identical copies of DNA fragments that have been inserted into plasmids or some other cloning vector.

club fungus One of a highly diverse group of multicelled fungi, the reproductive structures of which have microscopic, club-shaped cells that produce and bear spores.

cnidarian A radially symmetrical invertebrate, usually marine, that has tissues (not organs), and nematocysts. Two body forms (medusae and polyps) are common. Jellyfishes, corals, and sea anemones are examples.

coal A nonrenewable source of energy that formed more than 280 million years ago from submerged, undecayed plant remains that were buried in sediments, compressed, then compacted further by heat and pressure.

codominance Condition in which a pair of nonidentical alleles are both expressed even though they specify two different phenotypes.

codon One of a series of base triplets in an mRNA molecule, most of which code for a sequence of amino acids of a specific polypeptide chain. (Of sixty-four codons, sixty-one specify different amino acids and three of these also serve as start signals for translation; one other serves only as a stop signal for translation.)

coelum (SEE-lum) [Gk. *koilos*, hollow] Of many animals, a type of body cavity located between the gut and body wall and having a distinctive lining (peritoneum).

coenzyme A type of nucleotide that transfers hydrogen atoms and electrons from one reaction site to another. NAD$^+$ is an example.

coevolution The joint evolution of two or more closely interacting species; when one species evolves, the change affects selection pressures operating between the two species, so the other also evolves.

cofactor A metal ion or coenzyme; it helps catalyze a reaction or serves briefly as an agent that transfers electrons, atoms, or functional groups from one substrate to another.

cohesion Condition in which molecular bonds resist rupturing when under tension.

cohesion theory of water transport Theory that water moves up through vascular plants due to hydrogen bonding among water molecules confined inside the xylem pipelines. The collective cohesive strength of those bonds allows water to be pulled up as columns in response to transpiration (evaporation from leaves).

collenchyma One of the simple tissues of flowering plants; lends flexible support to primary tissues, such as those of lengthening stems.

colon (CO-lun) The large intestine.

commensalism [L. *com*, together, + *mensa*, table] Two-species interaction in which one species benefits significantly while the other is neither helped nor harmed to any notable extent.

communication signal Of social animals, an action or cue sent by one member of a species (the signaler) that can change the behavior of another member (the signal receiver).

community The populations of all species occupying a habitat; also applied to groups of organisms with similar life-styles in a habitat (such as the bird community).

companion cell A specialized parenchyma cell that helps load dissolved organic compounds into the conducting cells of the phloem.

comparative morphology [Gk. *morph*, form] Anatomical comparisons of major lineages.

competitive exclusion Theory that populations of two species competing for a limited resource cannot coexist indefinitely in the same habitat; the population better adapted to exploit the resource will enjoy a competitive (hence reproductive) edge and will eventually exclude the other population from the habitat.

complement system A set of about twenty proteins circulating in blood plasma with roles in nonspecific defenses and in immune responses. Some induce lysis of pathogens, others promote inflammation, and others stimulate phagocytes to engulf pathogens.

compound A substance in which the relative proportions of two or more elements never vary. Organic compounds have a backbone of carbon atoms arranged as a chain or ring structure. The simpler, inorganic compounds do not have comparable backbones.

concentration gradient A difference in the number of molecules (or ions) of a substance between two adjacent regions, as in a volume of fluid.

condensation reaction Enzyme-mediated reaction leading to the covalent linkage of small molecules and, often, the formation of water as a by-product.

cone cell In the vertebrate eye, a type of photoreceptor that responds to intense light and contributes to sharp daytime vision and color perception.

conifer A type of plant belonging to the dominant group of gymnosperms; mostly evergreen, woody trees and shrubs with pollen- and seed-bearing cones.

connective tissue proper A category of animal tissues, all having mostly the same components but in different proportions. These tissues contain fibroblasts and other cells, the secretions of which form fibers (of collagen and elastin) and a ground substance (of modified polysaccharides).

consumers [L. *consumere*, to take completely] Of ecosystems, heterotrophic organisms that obtain energy and raw materials by feeding on the tissues of other organisms. Herbivores, carnivores, omnivores, and parasites are examples.

continuous variation A more or less continuous range of small differences in a given trait among all the individuals of a population.

contractile vacuole (kun-TRAK-till VAK-you-ohl) [L. *contractus*, to draw together] In some protistans, a membranous chamber that takes up excess water in the cell body, then contracts, expelling the water outside the cell through a pore.

control group In a scientific experiment, a group used to evaluate possible side effects of a test involving an experimental group. Ideally, the control group should differ from the experimental group only with respect to the variable being studied.

convergence, morphological *See* morphological convergence.

cork cambium A type of lateral meristem that produces a tough, corky replacement for epidermis on parts of woody plants showing extensive secondary growth.

corpus callosum (CORE-pus ka-LOW-sum) A band of axons (200 million in humans) that functionally link two cerebral hemispheres.

corpus luteum (CORE-pus LOO-tee-um) A glandular structure; it develops from cells of a ruptured ovarian follicle and secretes progesterone and some estrogen, both of which maintain the lining of the uterus (endometrium).

cortex [L. *cortex*, bark] In general, a rindlike layer; the kidney cortex is an example. In vascular plants, ground tissue that makes up most of the primary plant body, supports plant parts, and stores food.

cotyledon A seed leaf, which develops as part of a plant embryo; cotyledons provide nourishment for the germinating seedling.

courtship display Social behavior by which individuals assess and respond to sexual overtures of potential partners.

covalent bond (koe-VAY-lunt) [L. *con*, together, + *valere*, to be strong] A sharing of one or more electrons between atoms or groups of atoms. When electrons are shared equally, the bond is nonpolar. When electrons are shared unequally, the bond is polar—slightly positive at one end and slightly negative at the other.

cross-bridge formation Of muscle cells, the interaction between actin and myosin filaments that is the basis of contraction.

crossing over During prophase I of meiosis, an interaction between a pair of homologous chromosomes. Their nonsister chromatids break at the same place along their length and exchange corresponding segments at the break points. Crossing over breaks up old combinations of alleles and puts new ones together in chromosomes.

culture The sum total of behavior patterns of a social group, passed between generations by learning and by symbolic behavior, especially language.

cuticle (KEW-tih-kull) A body covering. Of land plants, a covering of waxes and lipid-rich cutin deposited on the outer surface of epidermal cell walls. Of annelids, a thin, flexible surface coat. Of arthropods, a hardened yet lightweight covering with protein and chitin components that functions as an external skeleton.

cycad A type of gymnosperm of the tropics and subtropics; slow growing, with massive, cone-shaped structures that bear ovules or pollen.

cyclic AMP (SIK-lik) Cyclic adenosine monophosphate. A nucleotide that has roles in intercellular communication, as when it serves as a second messenger (a cytoplasmic mediator of a cell's response to signaling molecules).

cyclic pathway of ATP formation Photosynthetic pathway in which excited electrons move from a photosystem to an electron transport system, and back to the photosystem. The electron flow contributes to the formation of ATP from ADP and inorganic phosphate.

cyst Of some microorganisms, a walled, resting structure that forms during the life cycle.

cytochrome (SIGH-toe-krome) [Gk. *kytos*, hollow vessel, + *chrōma*, color] Iron-containing protein molecule; a component of electron transport systems used in photosynthesis and aerobic respiration.

cytokinesis (SIGH-toe-kih-NEE-sis) [Gk. *kinesis*, motion] Cytoplasmic division; the splitting of a parental cell into daughter cells.

cytokinin (SIGH-tow-KY-nin) Any of the class of plant hormones that stimulate cell division, promote leaf expansion, and retard leaf aging.

cytomembrane system [Gk. *kytos*, hollow vessel] Organelles, functioning as a system to modify, package, and distribute newly formed proteins and lipids. Endoplasmic reticulum, Golgi bodies, lysosomes, and a variety of vesicles are its components.

cytoplasm (SIGH-toe-plaz-um) [Gk. *plassein*, to mold] All cellular parts, particles, and semifluid substances enclosed by the plasma membrane except for the region of DNA (which in eukaryotes, is the nucleus).

cytosine (SIGH-toe-seen) A pyrimidine; one of the nitrogen-containing bases in nucleotides.

cytoskeleton Of eukaryotic cells, an internal "skeleton." Its microtubules and other components structurally support the cell, organize and move its internal components. The cytoskeleton also helps free-living cells move through their environment.

cytotoxic T cell A T lymphocyte that eliminates infected body cells or tumor cells with a single hit of toxins and perforins.

decomposers [L. *de-*, down, away, + *companere*, to put together] Of ecosystems, heterotrophs that obtain energy by chemically breaking down the remains, products, or wastes of other organisms. Their activities help cycle nutrients back to producers. Certain fungi and bacteria are examples.

deforestation The removal of all trees from a large tract of land, such as the Amazon Basin or the Pacific Northwest.

degradative pathway A metabolic pathway by which molecules are broken down in stepwise reactions that lead to products of lower energy.

deletion A change in a chromosome's structure after one of its regions is lost as a result of irradiation, viral attack, chemical action, or some other factor.

demographic transition model Model of human population growth in which changes in the growth pattern correspond to different stages of economic development. These are a preindustrial stage, when birth and death rates are both high, a transitional stage, an industrial stage, and a postindustrial stage, when the death rate exceeds the birth rate.

denaturation (deh-NAY-chur-AY-shun) Of any molecule, the loss of three-dimensional shape following disruption of hydrogen bonds and other weak bonds.

dendrite (DEN-drite) [Gk. *dendron*, tree] A short, slender extension from the cell body of a neuron.

denitrification (DEE-nite-rih-fih-KAY-shun) The conversion of nitrate or nitrite by certain bacteria to gaseous nitrogen (N_2) and a small amount of nitrous oxide (N_2O).

density-dependent controls Factors such as predation, parasitism, disease, and competition for resources, which limit population growth by reducing the birth rate, increasing the rates of death and dispersal, or all of these.

density-independent controls Factors such as storms or floods that increase a population's death rate more or less independently of its density.

dentition (den-TIH-shun) The type, size, and number of an animal's teeth.

dermal tissue system Of vascular plants, the tissues that cover and protect the plant surfaces.

dermis The layer of skin underlying the epidermis, consisting mostly of dense connective tissue.

desert A type of biome that exists where the potential for evaporation greatly exceeds rainfall and vegetation cover is limited.

desertification (dez-urt-ih-fih-KAY-shun) The conversion of grasslands, rain-fed cropland, or irrigated cropland to desertlike conditions, with a drop in agricultural productivity of 10 percent or more.

detrital food web Of most ecosystems, the flow of energy mainly from plants through detritivores and decomposers.

detritivores (dih-TRY-tih-vorez) [L. *detritus*; after *deterere*, to wear down] Of ecosystems, heterotrophs that consume dead or decomposing particles of organic matter. Earthworms, crabs, and nematodes are examples.

deuterostome (DUE-ter-oh-stome) [Gk. *deuteros*, second, + *stoma*, mouth] Any of the bilateral animals, including echinoderms and chordates, in which the first indentation in the early embryo develops into the anus.

diaphragm (DIE-uh-fram) [Gk. *diaphragma*, to partition] Muscular partition between the thoracic and abdominal cavities, the contraction and relaxation of which contribute to breathing. Also, a contraceptive device used temporarily to prevent sperm from entering the uterus during sexual intercourse.

dicot (DIE-kot) [Gk. *di*, two, + *kotylēdōn*, cup-shaped vessel] Short for dicotyledon; class of flowering plants characterized generally by seeds having embryos with two cotyledons (seed leaves), net-veined leaves, and floral parts arranged in fours, fives, or multiples of these.

differentiation *See* cell differentiation.

diffusion Net movement of like molecules (or ions) down their concentration gradient. In the absence of other forces, molecular motion and random collisions cause their net outward movement from one region into a neighboring region where they are less concentrated (because collisions are more frequent where the molecules are most crowded together).

digestive system An internal tube or cavity from which ingested food is absorbed into the internal environment; often divided into regions specialized for food transport, processing, and storage.

dihybrid cross An experimental cross in which offspring inherit two gene pairs, each consisting of two nonidentical alleles.

dinoflagellate A photosynthetic or heterotrophic protistan, often flagellated, that is a component of plankton.

diploid number (DIP-loyd) For many sexually reproducing species, the chromosome number of somatic cells and of germ cells prior to meiosis. Such cells have two chromosomes of each type (that is, pairs of homologous chromosomes). *Compare* haploid number.

directional selection Of a population, a shift in allele frequencies in a steady, consistent direction in response to a new environment or to a directional change in the old one. The outcome is that forms of traits at one end of the range of phenotypic variation become more common than the intermediate forms.

disaccharide (die-SAK-uh-ride) [Gk. *di*, two, + *sakcharon*, sugar] A type of simple carbohydrate, of the class called oligosaccharides; two monosaccharides covalently bonded.

disruptive selection Of a population, a shift in allele frequencies to forms of traits at both ends of a range of phenotypic variation and away from intermediate forms.

distal tubule The tubular section of a nephron most distant from the glomerulus; a major site of water and sodium reabsorption.

divergence Accumulation of differences in allele frequencies between populations that have become reproductively isolated from one another.

divergence, morphological *See* morphological divergence.

diversity, organismic Sum total of variations in form, function, and behavior that have accumulated in different lineages. Those variations generally are adaptive to prevailing conditions or were adaptive to conditions that existed in the past.

DNA Deoxyribonucleic acid (dee-OX-ee-RYE-bow-new-CLAY-ik). For all cells (and many viruses), the molecule of inheritance. A category of nucleic acids, each usually consisting of two nucleotide strands twisted together helically and held together by hydrogen bonds. The nucleotide sequence encodes the instructions for assembling proteins, and, ultimately, new individuals of a particular species.

DNA-DNA hybridization *See* nucleic acid hybridization.

DNA fingerprint Of each individual, a unique array of RFLPs, resulting from the DNA sequences inherited (in a Mendelian pattern) from each parent.

DNA library A collection of DNA fragments produced by restriction enzymes and incorporated into plasmids.

DNA ligase (LYE-gaze) Enzyme that seals together the new base-pairings during DNA replication; also used by recombinant DNA technologists to seal base-pairings between DNA fragments and cut plasmid DNA.

DNA polymerase (poe-LIM-uh-raze) Enzyme that assembles a new strand on a parent DNA strand during replication; also takes part in DNA repair.

DNA probe A short DNA sequence that has been assembled from radioactively labeled nucleotides and that can base-pair with part of a gene under investigation.

DNA repair Following an alteration in the base sequence of a DNA strand, a process that restores the original sequence, as carried out by DNA polymerases, DNA ligases, and other enzymes.

DNA replication Of cells, the process by which the hereditary material is duplicated for distribution to daughter nuclei. An example is the duplication of eukaryotic chromosomes during interphase, prior to mitosis.

dominance hierarchy Form of social organization in which some members of the group have adopted a subordinate status to others.

dominant allele In a diploid cell, an allele that masks the expression of its partner on the homologous chromosome.

dormancy [L. *dormire*, to sleep] Of plants, the temporary, hormone-mediated cessation of growth under conditions that might appear to be quite suitable for growth.

double fertilization Of flowering plants only, the fusion of one sperm nucleus with the egg nucleus (to produce a zygote), *and* fusion of a second sperm nucleus with the two nuclei of the endosperm mother cell, which gives rise to triploid (3n) nutritive tissue.

doubling time The length of time it takes for a population to double in size.

drug addiction Chemical dependence on a drug, following habituation and tolerance of

it; the drug takes on an "essential" biochemical role in the body.

dry shrubland A type of biome that exists where annual rainfall is less than 25 to 60 centimeters and where short, woody, multi-branched shrubs predominate; chaparral is an example.

dry woodland A type of biome that exists when annual rainfall is about 40 to 100 centimeters; there may be tall trees, but these do not form a dense canopy.

dryopith A type of hominoid, one of the first to appear during the Miocene about the time of the divergences that led to gorillas, chimpanzees, and humans.

duplication A change in a chromosome's structure resulting in the repeated appearance of the same gene sequence.

early *Homo* A type of early hominid that may have been the maker of stone tools that date from about 2.5 million years ago.

echinoderm A type of invertebrate that has calcified spines, needles, or plates on the body wall. It is radially symmetrical but with some bilateral features. Sea stars and sea urchins are examples.

ecology [Gk. *oikos*, home, + *logos*, reason] Study of the interactions of organisms with one another and with their physical and chemical environment.

ecosystem [Gk. *oikos*, home] An array of organisms and their physical environment, all of which interact through a flow of energy and a cycling of materials.

ecosystem modeling Analytical method of predicting unforeseen effects of disturbances to an ecosystem, based on computer programs and models.

ectoderm [Gk. *ecto*, outside, + *derma*, skin] Of animal embryos, the outermost primary tissue layer (germ layer) that gives rise to the outer layer of the integument and to tissues of the nervous system.

effector Of homeostatic systems, a muscle (or gland) that responds to signals from an integrator (such as the brain) by producing movement (or chemical change) that helps adjust the body to changing conditions.

effector cell Of the differentiated subpopulations of lymphocytes that form during an immune response, the type of cell that engages and destroys the antigen-bearing agent that triggered the response.

egg A type of mature female gamete; also called an ovum.

El Niño A recurring, massive displacement to the east of warm surface waters of the western equatorial Pacific, which in turn displaces the cooler waters of the Humboldt Current off the coast of Peru.

electron Negatively charged unit of matter, with both particulate and wavelike properties, that occupies one of the orbitals around the atomic nucleus. Atoms can gain, lose, or share electrons with other atoms.

electron transport phosphorylation (FOSS-for-ih-LAY-shun) Final stage of aerobic respiration, in which ATP forms after hydrogen ions and electrons (from the Krebs cycle) are sent through a transport system that gives up the electrons to oxygen.

electron transport system An organized array of enzymes and cofactors, bound in a cell membrane, that accept and donate electrons in sequence. When such systems operate, hydrogen ions (H^+) flow across the membrane, and the flow drives ATP formation and other reactions.

element Any substance that cannot be decomposed into substances with different properties.

embryo (EM-bree-oh) [Gk. *en*, in, + probably *bryein*, to swell] Of animals generally, the stage formed by way of cleavage, gastrulation, and other early developmental events. Of seed plants, the young sporophyte, from the first cell divisions after fertilization until germination.

embryo sac The female gametophyte of flowering plants.

emulsification Of chyme in the small intestine, a suspension of droplets of fat coated with bile salts.

end product A substance present at the end of a metabolic pathway.

endangered species A species poised at the brink of extinction, owing to the extremely small size and severely limited genetic diversity of its remaining populations.

endergonic reaction (en-dur-GONE-ik) Chemical reaction showing a net gain in energy.

endocrine gland Ductless gland that secretes hormones into interstitial fluid, after which they are distributed by way of the bloodstream.

endocrine system System of cells, tissues, and organs that is functionally linked to the nervous system and that exerts control by way of its hormones and other chemical secretions.

endocytosis (EN-doe-sigh-TOE-sis) Movement of a substance into cells; the substance becomes enclosed by a patch of plasma membrane that sinks into the cytoplasm, then forms a vesicle around it. Phagocytic cells also engulf pathogens or prey in this manner.

endoderm [Gk. *endon*, within, + *derma*, skin] Of animal embryos, the inner primary tissue layer, or germ layer, that gives rise to the inner lining of the gut and organs derived from it.

endodermis A sheetlike wrapping of single cells around the vascular cylinder of a root; it functions in controlling the uptake of water and dissolved nutrients. An impermeable barrier (Casparian strip) prevents water from passing between the walls of abutting endodermal cells.

endometrium (EN-doh-MEET-ree-um) [Gk. *metrios*, of the womb] Inner lining of the uterus, consisting of connective tissues, glands, and blood vessels.

endoplasmic reticulum or **ER** (EN-doe-PLAZ-mik reh-TIK-yoo-lum) An organelle that begins at the nucleus and curves through the cytoplasm. In rough ER (which has many ribosomes on its cytoplasmic side), many new polypeptide chains acquire specialized side chains. In many cells, smooth ER (with no

attached ribosomes) is the main site of lipid synthesis.

endoskeleton [Gk. *endon*, within, + *skleros*, hard, stiff] In chordates, the internal framework of bone, cartilage, or both. Together with skeletal muscle, supports and protects other body parts, helps maintain posture, and moves the body.

endosperm (EN-doe-sperm) Nutritive tissue that surrounds and serves as food for a flowering plant embryo and, later, for the germinating seedling.

endospore Of certain bacteria, a resistant body that forms around DNA and some cytoplasm; it germinates and gives rise to new bacterial cells when conditions become favorable.

endosymbiosis A permanent, mutually beneficial interdependency between two species, one of which resides permanently inside the other's body.

energy The capacity to do work.

energy carrier A molecule that delivers energy from one metabolic reaction site to another. ATP is the most widely travelled of these; it readily donates energy to nearly all metabolic reactions.

energy flow pyramid A pyramid-shaped representation of an ecosystem's trophic structure, illustrating the energy losses at each transfer to a different trophic level.

entropy (EN-trow-pee) A measure of the degree of disorder in a system (how much energy has become so disorganized and dispersed, usually as heat, that it is no longer readily available to do work).

enzyme (EN-zime) One of a class of proteins that greatly speed up (catalyze) reactions between specific substances, usually at their functional groups. The substances that each type of enzyme acts upon are called its substrates.

eosinophil Fast-acting, phagocytic white blood cell that takes part in inflammation but not in immune responses.

epidermis The outermost tissue layer of a multicelled plant or animal.

epiglottis A flaplike structure at the start of the larynx, the position of which directs the movement of air into the trachea or food into the esophagus.

epistasis (eh-PISS-tih-sis) A type of gene interaction, whereby two alleles of a gene influence the expression of alleles of a different gene.

epithelium (EP-ih-THEE-lee-um) An animal tissue consisting of one or more layers of adhering cells that covers the body's external surfaces and lines its internal cavities and tubes. Epithelium has one free surface; the opposite surface rests on a basement membrane between it and an underlying connective tissue. Epidermis or skin is an example.

equilibrium, dynamic [Gk. *aequus*, equal, + *libra*, balance] The point at which a chemical reaction runs forward as fast as it runs in reverse; thus the concentrations of reactant molecules and product molecules show no net change.

erythrocyte (eh-RITH-row-site) [Gk. *erythros*, red, + *kytos*, vessel] Red blood cell.

esophagus (ee-SOF-uh-gus) Tubular portion of a digestive system that receives swallowed food and leads to the stomach.

essential amino acid Any of eight amino acids that certain animals cannot synthesize for themselves and must obtain from food.

essential fatty acid Any of the fatty acids that certain animals cannot synthesize for themselves and must obtain from food.

estrogen (ESS-trow-jen) A sex hormone that helps oocytes mature, induces changes in the uterine lining during the menstrual cycle and pregnancy, and maintains secondary sexual traits; also influences bodily growth and development.

estrus (ESS-truss) [Gk. *oistrus*, frenzy] For mammals generally, the cyclic period of a female's sexual receptivity to the male.

estuary (EST-you-ehr-ee) A partly enclosed coastal region where seawater mixes with freshwater from rivers, streams, and runoff from the surrounding land.

ethylene (ETH-il-een) Plant hormone that stimulates fruit ripening and triggers abscission.

eubacteria The subkingdom of all bacterial species except the archaebacteria; one of the three great prokaryotic lineages that arose early in the history of life.

euglenoid A type of flagellated protistan, most of which are photosynthesizers in stagnant or freshwater ponds.

eukaryotic cell (yoo-CARRY-oh-tic) [Gk. *eu*, good, + *karyon*, kernel] A type of cell that has a "true nucleus" and other distinguishing membrane-bound organelles. *Compare* prokaryotic cell.

eutrophication Nutrient enrichment of a body of water, such as a lake, that typically results in reduced transparency and a phytoplankton-dominated community.

evaporation [L. *e-*, out, + *vapor*, steam] Conversion of a substance from the liquid to the gaseous state; some or all of its molecules leave in the form of vapor.

evolution, biological [L. *evolutio*, act of unrolling] Change within a line of descent over time. A population is evolving when some forms of a trait are becoming more or less common, relative to the other kinds of traits. The shifts are evidence of changes in the relative abundances of alleles for that trait, as brought about by mutation, natural selection, genetic drift, and gene flow.

evolutionary tree A treelike diagram in which branches represent separate lines of descent from a common ancestor.

excitatory postsynaptic potential or **EPSP** One of two competing signals at an input zone of a neuron; a graded potential that brings the neuron's plasma membrane closer to threshold.

excretion Any of several processes by which excess water, excess or harmful solutes, or waste materials leave the body by way of the urinary system or certain glands.

exergonic reaction (EX-ur-GONE-ik) A chemical reaction that shows a net loss in energy.

exocrine gland (EK-suh-krin) [Gk. *es*, out of, + *krinein*, to separate] Glandular structure that secretes products, usually through ducts or tubes, to a free epithelial surface.

exocytosis (EK-so-sigh-TOE-sis) Movement of a substance out of a cell by means of a transport vesicle, the membrane of which fuses with the plasma membrane, so that the vesicle's contents are released outside.

exodermis Layer of cells just inside the root epidermis of most flowering plants; helps control the uptake of water and solutes.

exon Of eukaryotic cells, any of the nucleotide sequences of a pre-mRNA molecule that are spliced together to form the mature mRNA transcript and are ultimately translated into protein.

exoskeleton [Gk. *exo*, out, + *skleros*, hard, stiff] An external skeleton, as in arthropods.

experiment A test in which some phenomenon in the natural world is manipulated in controlled ways to gain insight into its function, structure, operation, or behavior.

exploitation competition Interaction in which both species have equal access to a required resource but differ in how fast or efficiently they exploit it.

exponential growth (EX-po-NEN-shul) Pattern of population growth in which greater and greater numbers of individuals are produced during the successive doubling times; the pattern that emerges when the per capita birth rate remains even slightly above the per capita death rate, putting aside the effects of immigration and emigration.

extinction, background A steady rate of species turnover that characterizes lineages through most of their histories.

extinction, mass An abrupt increase in the rate at which major taxa disappear, with several taxa being affected simultaneously.

extracellular fluid In animals generally, all the fluid not inside cells; includes plasma (the liquid portion of blood) and interstitial fluid (which occupies the spaces between cells and tissues).

extracellular matrix A material, largely secreted, that helps hold many animal tissues together in certain shapes; it consists of fibrous proteins and other components in a ground substance.

FAD Flavin adenine dinucleotide, a nucleotide coenzyme. When delivering electrons and unbound protons (H^+) from one reaction to another, it is abbreviated $FADH_2$.

fall overturn The vertical mixing of a body of water in autumn. Its upper layer cools, increases in density, and sinks; dissolved oxygen moves down and nutrients from bottom sediments are brought to the surface.

family pedigree A chart of genetic relationships of the individuals in a family through successive generations.

fat A lipid with a glycerol head and one, two, or three fatty acid tails. The tails of saturated fats have only single bonds between carbon atoms and hydrogen atoms attached to all other bonding sites. Tails of unsaturated fats additionally have one or more double bonds between certain carbon atoms.

fatty acid A long, flexible hydrocarbon chain with a —COOH group at one end.

feedback inhibition Of cells, a control mechanism by which the production (or secretion) of a substance triggers a change in some activity that in turn shuts down further production of the substance.

fermentation [L. *fermentum*, yeast] A type of anaerobic pathway of ATP formation, it starts with glycolysis, ends when electrons are transferred back to one of the breakdown products or intermediates, and regenerates the NAD^+ required for the reaction. Its net yield is two ATP per glucose molecule degraded.

fern One of the seedless vascular plants, mostly of wet, humid habitats; requires free water to complete its life cycle.

fertilization [L. *fertilis*, to carry, to bear] Fusion of a sperm nucleus with the nucleus of an egg, which thereupon becomes a zygote.

fever A body temperature higher than a set point that is preestablished in the brain region governing temperature.

fibrous root system Of most monocots, all the lateral branchings of adventitious roots, which arose earlier from the young stem.

filtration Of urine formation, the process by which blood pressure forces water and solutes out of glomerular capillaries and into the cupped portion of a nephron wall (Bowman's capsule).

fin Of fishes generally, an appendage that helps propel, stabilize, and guide the body through water.

first law of thermodynamics [Gk. *therme*, heat, + *dynamikos*, powerful] Law stating that the total amount of energy in the universe remains constant. Energy cannot be created and existing energy cannot be destroyed. It can only be converted from one form to another.

fish An aquatic animal of the most ancient vertebrate lineage; jawless fishes (such as lampreys and hagfishes), cartilaginous fishes (such as sharks), and bony fishes (such as coelacanths and salmon) are the three existing groups.

fixed action pattern An instinctive response that is triggered by a well-defined, simple stimulus and that is performed in its entirety once it has begun.

flagellated protozoan A member of one of four major groups of protozoans, many of which cause serious diseases.

flagellum (fluh-JELL-um), plural **flagella** [L. whip] Tail-like motile structure of many free-living eukaryotic cells; it has a distinctive 9 + 2 array of microtubules.

flatworm A type of invertebrate having bilateral symmetry, a flattened body, and a saclike gut; a turbellarian, fluke, or tapeworm.

flower The reproductive structure that distinguishes angiosperms from other seed plants and often attracts pollinators.

fluid mosaic model Model of membrane structure in which proteins are embedded in a lipid bilayer or attached to one of its surfaces. The lipid molecules give the membrane its basic structure, impermeability to water-soluble molecules, and (through packing variations and movements) fluidity. Proteins carry out most membrane functions, such as transport, enzyme action, and reception of signals or substances.

follicle (FOLL-ih-kul) In a mammalian ovary, a primary oocyte (immature egg) together with the surrounding layer of cells.

food chain A straight-line sequence of who eats whom in an ecosystem.

food web A network of cross-connecting, interlinked food chains, encompassing primary producers and an array of consumers, detritivores, and decomposers.

forebrain Brain region that includes the cerebrum and cerebral cortex, the olfactory lobes, and the hypothalamus.

forest A type of biome where tall trees grow together closely enough to form a fairly continuous canopy over a broad region.

fossil Recognizable evidence of an organism that lived in the distant past. Most fossils are skeletons, shells, leaves, seeds, and tracks that were buried in rock layers before they decomposed.

fossil fuel Coal, petroleum, or natural gas; a nonrenewable source of energy formed in sediments by the compression of carbon-containing plant remains over hundreds of millions of years.

founder effect An extreme case of genetic drift. By chance, a few individuals that leave a population and establish a new one carry fewer (or more) alleles for certain traits. Increased variation between the two populations is one outcome. Limited genetic variability in the new population is another.

free radical A highly reactive, unbound molecular fragment with the wrong number of electrons.

fruit [L. after *frui*, to enjoy] Of flowering plants, the expanded and ripened ovary of one or more carpels, sometimes with accessory structures incorporated.

FSH Follicle-stimulating hormone. Produced and secreted by the anterior lobe of the pituitary gland, this hormone has roles in the reproductive functions of both males and females.

functional group An atom or group of atoms that is covalently bonded to the carbon backbone of an organic compound and that influences its behavior.

Fungi The kingdom of fungi.

fungus A eukaryotic heterotroph that uses extracellular digestion and absorption; it secretes enzymes able to break down an external food source into molecules small enough to be absorbed by its cells. Saprobic types feed on nonliving organic matter; parasitic types feed on living organisms. Fungi as a group are major decomposers.

gall bladder Organ of the digestive system that stores bile secreted from the liver.

gamete (GAM-eet) [Gk. *gametes*, husband, and *gamete*, wife] A haploid cell that functions in sexual reproduction. Sperm and eggs are examples.

gamete formation Generally, the formation of gametes by way of meiosis. Of animals, the first stage of development, in which sperm or eggs form and mature within reproductive tissues of parents.

gametophyte (gam-EET-oh-fite) [Gk. *phyton*, plant] The haploid, multicelled, gamete-producing phase in the life cycle of most plants.

ganglion (GANG-lee-un), plural **ganglia** [Gk. *ganglion*, a swelling] A distinct clustering of cell bodies of neurons in regions other than the brain or spinal cord.

gastrulation (gas-tru-LAY-shun) Of animals, the stage of embryonic development in which cells become arranged into two or three primary tissue layers (germ layers); in humans, the layers are an inner endoderm, an intermediate mesoderm, and a surface ectoderm.

gene [short for German *pangan*, after Gk. *pan*, all + *genes*, to be born] A unit of information about a heritable trait that is passed on from parents to offspring. Each gene has a specific location on a chromosome.

gene flow A microevolutionary process; a physical movement of alleles out of a population as individuals leave (emigrate) or enter (immigrate), the outcome being changes in allele frequencies.

gene frequency More precisely, allele frequency: the relative abundances of all the different alleles for a trait that are carried by the individuals of a population.

gene locus A given gene's particular location on a chromosome.

gene mutation [L. *mutatus*, a change] Change in DNA due to the deletion, addition, or substitution of one to several bases in the nucleotide sequence.

gene pair In diploid cells, the two alleles at a given locus on a pair of homologous chromosomes.

gene pool Sum total of all genotypes in a population. More accurately, allele pool.

gene therapy Generally, the transfer of one or more normal genes into the body cells of an organism in order to correct a genetic defect.

genetic code [After L. *genesis*, to be born] The correspondence between nucleotide triplets in DNA (then in mRNA) and specific sequences of amino acids in the resulting polypeptide chains; the basic language of protein synthesis.

genetic disorder An inherited condition that results in mild to severe medical problems.

genetic drift A microevolutionary process; a change in allele frequencies over the generations due to chance events alone.

genetic engineering Altering the information content of DNA through use of recombinant DNA technology.

genetic equilibrium Hypothetical state of a population in which allele frequencies for a trait remain stable through the generations; a reference point for measuring rates of evolutionary change.

genetic recombination Presence of a new combination of alleles in a DNA molecule compared to the parental genotype; the result of processes such as crossing over at meiosis, chromosome rearrangements, gene mutation, and recombinant DNA technology.

genome All the DNA in a haploid number of chromosomes of a given species.

genotype (JEEN-oh-type) Genetic constitution of an individual. Can mean a single gene pair or the sum total of the individual's genes. *Compare* phenotype.

genus, plural **genera** (JEEN-US, JEN-er-ah) [L. *genus*, race, origin] A taxon into which all species exhibiting certain phenotypic similarities and evolutionary relationship are grouped.

geologic time scale A time scale for earth history, the subdivisions of which have been refined by radioisotope dating work.

germ cell Of animals, one of a cell lineage set aside for sexual reproduction; germ cells give rise to gametes. *Compare* somatic cell.

germ layer Of animal embryos, one of two or three primary tissue layers that form during gastrulation and that gives rise to certain tissues of the adult body. *Compare* ectoderm; endoderm; mesoderm.

germination (jur-min-AY-shun) Generally, the resumption of growth following a rest stage; of seed plants, the time at which an embryo sporophyte breaks through its seed coat and resumes growth.

gibberellin (JIB-er-ELL-un) Any of a class of plant hormones that promote stem elongation.

gill A respiratory organ, typically with a moist, thin vascularized layer of epidermis that functions in gas exchange.

ginkgo A type of gymnosperm with fan-shaped leaves and fleshy coated seeds; now represented by a single species of deciduous trees.

gland A secretory cell or multicelled structure derived from epithelium and often connected to it.

glomerular capillaries The set of blood capillaries inside Bowman's capsule of the nephron.

glomerulus (glow-MARE-you-luss) [L. *glomus*, ball] The first portion of the nephron, where water and solutes are filtered from blood.

glucagon (GLUE-kuh-gone) Hormone that stimulates conversion of glycogen and amino acids to glucose; secreted by alpha cells of the pancreas when the flow of glucose decreases.

glyceride (GLISS-er-eyed) One of the molecules, commonly called fats and oils, that has one, two, or three fatty acid tails attached to a glycerol backbone. They are the body's most abundant lipids and its richest source of energy.

glycerol (GLISS-er-oh) [Gk. *glykys*, sweet, + L. *oleum*, oil] A three-carbon molecule with three hydroxyl groups attached; together with fatty acids, a component of fats and oils.

glycogen (GLY-kuh-jen) In animals, a storage polysaccharide that is a main food reserve; can be readily broken down into glucose subunits.

glycolysis (gly-CALL-ih-sis) [Gk. *glykys*, sweet, + *lysis*, loosening or breaking apart] Initial reactions of both aerobic and anaerobic pathways by which glucose (or some other organic compound) is partially broken down to pyruvate, with a net yield of two ATP. Gly-

colysis proceeds in the cytoplasm of all cells, and oxygen has no role in it.

gnetophyte A type of gymnosperm limited to deserts and tropics.

Golgi body (GOHL-gee) Organelle in which newly synthesized polypeptide chains as well as lipids are modified and packaged in vesicles for export or for transport to specific locations within the cytoplasm.

gonad (GO-nad) Primary reproductive organ in which gametes are produced.

graded potential Of neurons, a local signal that slightly changes the voltage difference across a small patch of the plasma membrane. Such signals vary in magnitude, depending on the stimulus. With prolonged or intense stimulation, they may spread to a trigger zone of the membrane and initiate an action potential.

granum, plural **grana** Within many chloroplasts, any of the stacks of flattened, membranous compartments with chlorophyll and other light-trapping pigments and reaction sites for ATP formation.

grassland A type of biome with flat or rolling land, 25 to 100 centimeters of annual rainfall, warm summers, and often grazing and periodic fires that regenerate the dominant species.

gravitropism (GRAV-ih-TROPE-izm) [L. *gravis*, heavy, + Gk. *trepein*, to turn] The tendency of a plant to grow directionally in response to the earth's gravitational force.

gray matter Of vertebrates, the dendrites, neuron cell bodies, and neuroglial cells of the spinal cord and cerebral cortex.

grazing food web Of most ecosystems, the flow of energy from plants to herbivores, then through an array of carnivores.

green alga One of a group or division of aquatic plants with an abundance of chlorophylls a and b; early members of its lineage may have given rise to the bryophytes and vascular plants.

green revolution In developing countries, the use of improved crop varieties, modern agricultural practices (including massive inputs of fertilizers and pesticides), and equipment to increase crop yields.

greenhouse effect Warming of the lower atmosphere due to the presence of greenhouse gases—carbon dioxide, methane, nitrous oxide, ozone, water vapor, and chlorofluorocarbons.

ground meristem (MARE-ih-stem) [Gk. *meristos*, divisible] Of vascular plants, a primary meristem that produces the ground tissue system, hence the bulk of the plant body.

ground substance Of certain animal tissues, the intercellular material made up of cell secretions and other noncellular components.

ground tissue system Tissues that make up the bulk of the vascular plant body; parenchyma is the most common of these.

guanine A nitrogen-containing base; present in one of the four nucleotide building blocks of DNA and RNA.

guard cell Either of two adjacent cells having roles in the movement of gases and water vapor across leaf or stem epidermis. An opening (stoma) forms when both cells swell with water and move apart; it closes when they lose water and collapse against each other.

gut A body region where food is digested and absorbed; of complete digestive systems, the gastrointestinal tract (the portions from the stomach onward).

gymnosperm (JIM-noe-sperm) [Gk. *gymnos*, naked, + *sperma*, seed] A plant that bears seeds at exposed surfaces of reproductive structures, such as cone scales. Pine trees are examples.

habitat [L. *habitare*, to live in] The type of place where an organism normally lives, characterized by physical features, chemical features, and the presence of certain other species.

hair cell Type of mechanoreceptor that may give rise to action potentials when bent or tilted.

halophile A type of archaebacterium that lives in extremely salty habitats.

haploid number (HAP-loyd) The chromosome number of a gamete which, as an outcome of meiosis, is only half that of the parent germ cell (it has only one of each pair of homologous chromosomes). *Compare* diploid number.

HCG Human chorionic gonadotropin. A hormone that helps maintain the lining of the uterus during the menstrual cycle and during the first trimester of pregnancy.

heart Muscular pump that keeps blood circulating through the animal body.

helper T cell One of the T lymphocytes; when activated, it produces and secretes interleukins that promote formation of huge populations of effector and memory cells for immune responses.

hemoglobin (HEEM-oh-glow-bin) [Gk. *haima*, blood, + L. *globus*, ball] Iron-containing, oxygen-transporting protein that gives red blood cells their color.

hemostasis (HEE-mow-STAY-sis) [Gk. *haima*, blood, + *stasis*, standing] Stopping of blood loss from a damaged blood vessel through coagulation, blood vessel spasm, platelet plug formation, and other mechanisms.

herbivore [L. *herba*, grass, + *vovare*, to devour] Plant-eating animal.

heterocyst (HET-er-oh-sist) Of some filamentous cyanobacteria, a type of thick-walled, nitrogen-fixing cell that forms when nitrogen is scarce.

heterotroph (HET-er-oh-trofe) [Gk. *heteros*, other, + *trophos*, feeder] Organism that cannot synthesize its own organic compounds and must obtain nourishment by feeding on autotrophs, each other, or organic wastes. Animals, fungi, many protistans, and most bacteria are heterotrophs. *Compare* autotroph.

heterozygous condition (HET-er-oh-ZYE-guss) [Gk. *zygoun*, join together] For a given trait, having nonidentical alleles at a particular locus on a pair of homologous chromosomes.

hindbrain One of the three divisions of the vertebrate brain; the medulla oblongata, cerebellum, and pons; includes reflex centers for respiration, blood circulation, and other basic functions; also coordinates motor responses and many complex reflexes.

histone Any of a class of proteins that are intimately associated with DNA and that are largely responsible for its structural (and possibly functional) organization in eukaryotic chromosomes.

homeostasis (HOE-me-oh-STAY-sis) [Gk. *homo*, same, + *stasis*, standing] Of multicelled organisms, a physiological state in which the physical and chemical conditions of the internal environment are being maintained within tolerable ranges.

homeostatic feedback loop An interaction in which an organ (or structure) stimulates or inhibits the output of another organ, then shuts down or increases this activity when it detects that the output has exceeded or fallen below a set point.

hominid [L. *homo*, man] All species on the evolutionary branch leading to modern humans. *Homo sapiens* is the only living representative.

hominoid Apes, humans, and their recent ancestors.

Homo erectus A hominid lineage that emerged between 1.5 million and 300,000 years ago and that may include the direct ancestors of modern humans.

Homo sapiens The hominid lineage of modern humans that emerged between 300,000 and 200,000 years ago.

homologous chromosome (huh-MOLL-uh-gus) [Gk. *homologia*, correspondence] Of sexually reproducing species, one of a pair of chromosomes that resemble each other in size, shape, and the genes they carry, and that line up with each other at meiosis I. The X and Y chromosomes differ in these respects but still function as homologues.

homologous structures The same body parts, modified in different ways, in different lines of descent from a common ancestor.

homozygous condition (HOE-moe-ZYE-guss) Having two identical alleles at a given locus (on a pair of homologous chromosomes).

homozygous dominant condition Having two dominant alleles at a given locus (on a pair of homologous chromosomes).

homozygous recessive condition Having two recessive alleles at a given gene locus (on a pair of homologous chromosomes).

hormone [Gk. *hormon*, to stir up, set in motion] Any of the signaling molecules secreted from endocrine glands, endocrine cells, and some neurons that the bloodstream distributes to nonadjacent target cells (any cell having receptors for that hormone).

horsetail One of the seedless vascular plants, which require free water to complete the life cycle; only one genus has survived to the present.

human genome project A basic research project in which researchers throughout the world are working together to sequence the estimated 3 billion nucleotides present in the DNA of human chromosomes.

hydrogen bond Type of chemical bond in which an atom of a molecule interacts weakly with a neighboring atom that is already taking part in a polar covalent bond.

hydrogen ion A free (unbound) proton; a hydrogen atom that has lost its electron and so bears a positive charge (H^+).

hydrologic cycle A biogeochemical cycle, driven by solar energy, in which water moves slowly through the atmosphere, on or through surface layers of land masses, to the ocean, and back again.

hydrolysis (high-DRAWL-ih-sis) [L. *hydro*, water, + Gk. *lysis*, loosening or breaking apart] Enzyme-mediated reaction in which covalent bonds break, splitting a molecule into two or more parts, and H^+ and OH^- (derived from a water molecule) become attached to the exposed bonding sites.

hydrophilic substance [Gk. *philos*, loving] A polar substance that is attracted to the polar water molecule and so dissolves easily in water. Sugars are examples.

hydrophobic substance [Gk. *phobos*, dreading] A nonpolar substance that is repelled by the polar water molecule and so does not readily dissolve in water. Oil is an example.

hydrosphere All liquid or frozen water on or near the earth's surface.

hydrothermal vent ecosystem A type of ecosystem that exists near fissures in the ocean floor and is based on chemosynthetic bacteria that use hydrogen sulfide as the energy source.

hypha (HIGH-fuh), plural **hyphae** [Gk. *hyphe*, web] Of fungi, a generally tube-shaped filament with chitin-reinforced walls and, often, reinforcing cross-walls; component of the mycelium.

hypodermis A subcutaneous layer having stored fat that helps insulate the body; although not part of skin, it anchors skin while allowing it some freedom of movement.

hypothalamus [Gk. *hypo*, under, + *thalamos*, inner chamber or possibly *tholos*, rotunda] Of vertebrate forebrains, a brain center that monitors visceral activities (such as salt-water balance, temperature control, and reproduction) and that influences related forms of behavior (as in hunger, thirst, and sex).

hypothesis A possible explanation of a specific phenomenon.

immune response A series of events by which B and T lymphocytes recognize a specific antigen, undergo repeated cell divisions that form huge lymphocyte populations, and differentiate into subpopulations of effector and memory cells. Effector cells engage and destroy antigen-bearing agents. Memory cells enter a resting phase and are activated during subsequent encounters with the same antigen.

immunization Various processes, including vaccination, that promote increased immunity against specific diseases.

immunoglobulins (Ig) Four classes of antibodies, each with binding sites for antigen and binding sites used in specialized tasks. Examples are IgM antibodies (first to be secreted during immune responses) and IgG antibodies (which activate complement proteins and neutralize many toxins).

implantation A process by which a blastocyst adheres to the endometrium and begins to establish connections by which the mother and embryo will exchange substances during pregnancy.

imprinting Category of learning in which an animal that has been exposed to specific key stimuli early in its behavioral development forms an association with the object.

incisor A tooth, shaped like a flat chisel or cone, used in nipping or cutting food.

incomplete dominance Of heterozygotes, the appearance of a version of a trait that is somewhere between the homozygous dominant and recessive conditions.

independent assortment Mendelian principle that each gene pair tends to assort into gametes independently of other gene pairs located on nonhomologous chromosomes.

indirect selection A theory in evolutionary biology that self-sacrificing individuals can indirectly pass on their genes by helping relatives survive and reproduce.

induced-fit model Model of enzyme action whereby a bound substrate induces changes in the shape of the enzyme's active site, resulting in a more precise molecular fit between the enzyme and its substrate.

industrial smog A type of gray-air smog that develops in industrialized regions when winters are cold and wet.

inflammation, acute In response to tissue damage or irritation, fast-acting phagocytes and plasma proteins, including complement proteins, leave the bloodstream, then defend and help repair the tissue. Proceeds during both nonspecific and specific (immune) defense responses.

inheritance The transmission, from parents to offspring, of structural and functional patterns that have a genetic basis and are characteristic of each species.

inhibiting hormone A signaling molecule produced and secreted by the hypothalamus that controls secretions by the anterior lobe of the pituitary gland.

inhibitor A substance that can bind with an enzyme and interfere with its functioning.

inhibitory postsynaptic potential, or **IPSP** Of neurons, one of two competing types of graded potentials at an input zone; tends to drive the resting membrane potential away from threshold.

instinctive behavior A complex, stereotyped response to a particular environmental cue that often is quite simple.

insulin Hormone that lowers the glucose level in blood; it is secreted from beta cells of the pancreas and stimulates cells to take up glucose; also promotes protein and fat synthesis and inhibits protein conversion to glucose.

integration, neural [L. *integrare*, to coordinate] Moment-by-moment summation of all excitatory and inhibitory synapses acting on a neuron; occurs at each level of synapsing in a nervous system.

integrator Of homeostatic systems, a control point where different bits of information are pulled together in the selection of a response. The brain is an example.

integument Of animals, a protective body covering such as skin. Of flowering plants, a protective layer around the developing ovule; when the ovule becomes a seed, its integument(s) harden and thicken into a seed coat.

integumentary exchange (in-teg-you-MEN-tuh-ree) Of some animals, a mode of respiration in which oxygen and carbon dioxide diffuse across a thin, vascularized layer of moist epidermis at the body surface.

interference competition Interaction in which one species may limit another species' access to some resource regardless of whether the resource is abundant or scarce.

interleukin One of a variety of communication signals, secreted by macrophages and by helper T cells, that drive immune responses.

intermediate compound A compound that forms between the start and the end of a metabolic pathway.

intermediate filament A cytoskeletal component that consists of different proteins in different types of animal cells.

interneuron Any of the neurons in the vertebrate brain and spinal cord that integrate information arriving from sensory neurons and that influence other neurons in turn.

internode In vascular plants, the stem region between two successive nodes.

interphase Of cell cycles, the time interval between nuclear divisions in which a cell increases its mass, roughly doubles the number of its cytoplasmic components, and finally duplicates its chromosomes (replicates its DNA). The interval is different for different species.

interspecific competition Two-species interaction in which both species can be harmed due to overlapping niches.

interstitial fluid (IN-ter-STISH-ul) [L. *interstitus*, to stand in the middle of something] In multicelled animals, that portion of the extracellular fluid occupying spaces between cells and tissues.

intertidal zone Generally, the area on a rocky or sandy shoreline that is above the low water mark and below the high water mark; organisms inhabiting it are alternately submerged, then exposed, by tides.

intervertebral disk One of a number of disk-shaped structures containing cartilage that serve as shock absorbers and flex points between bony segments of the vertebral column.

intraspecific competition Interaction among individuals of the same species that are competing for the same resources.

intron A noncoding portion of a newly formed mRNA molecule.

inversion A change in a chromosome's structure after a segment separated from it was then inserted at the same place, but in reverse. The reversal alters the position and order of the chromosome's genes.

invertebrate Animal without a backbone.

ion, negatively charged (EYE-on) An atom or a compound that has gained one or more electrons, and hence has acquired an overall negative charge.

ion, positively charged An atom or a compound that has lost one or more electrons, and hence has acquired an overall positive charge.

ionic bond An association between ions of opposite charge.

isotonic condition Equality in the relative concentrations of solutes in two fluids; for two fluids separated by a cell membrane, there is no net osmotic (water) movement across the membrane.

isotope (EYE-so-tope) For a given element, an atom with the same number of protons as the other atoms but with a different number of neutrons.

J-shaped curve A curve, obtained when population size is plotted against time, that is characteristic of unrestricted, exponential growth.

joint An area of contact or near-contact between bones.

karyotype (CARRY-oh-type) Of eukaryotic individuals (or species), the number of metaphase chromosomes in somatic cells and their defining characteristics.

keratin A tough, water-insoluble protein manufactured by most epidermal cells.

keratinization (care-AT-in-iz-AY-shun) Process by which keratin-producing epidermal cells of skin die and collect at the skin surface as keratinized "bags" that form a barrier against dehydration, bacteria, and many toxic substances.

kidney In vertebrates, one of a pair of organs that filter mineral ions, organic wastes, and other substances from the blood, and help regulate the volume and solute concentrations of extracellular fluid.

kilocalorie 1,000 calories of heat energy, or the amount of energy needed to raise the temperature of 1 kilogram of water by 1°C; the unit of measure for the caloric value of foods.

kinetochore A specialized group of proteins and DNA at the centromere of a chromosome that serves as an attachment point for several spindle microtubules during mitosis or meiosis. Each chromatid of a duplicated chromosome has its own kinetochore.

Krebs cycle Together with a few conversion steps that precede it, the stage of aerobic respiration in which pyruvate is completely broken down to carbon dioxide and water. Coenzymes accept the unbound protons (H^+) and electrons stripped from intermediates during the reactions and deliver them to the next stage.

lactate fermentation Anaerobic pathway of ATP formation in which pyruvate from glycolysis is converted to the three-carbon compound lactate, and NAD^+ (a coenzyme used in the reactions) is regenerated. Its net yield is two ATP.

lactation The production of milk by hormone-primed mammary glands.

lake A body of fresh water having littoral, limnetic, and profundal zones.

lancelet An invertebrate chordate having a body that tapers sharply at both ends, segmented muscles, and a full-length notochord.

large intestine The colon; a region of the gut that receives unabsorbed food residues from the small intestine and concentrates and stores feces until they are expelled from the body.

larva, plural **larvae** Of animals, a sexually immature, free-living stage between the embryo and the adult.

larynx (LARE-inks) A tubular airway that leads to the lungs. In humans, contains vocal cords, where sound waves used in speech are produced.

lateral meristem Of vascular plants, a type of meristem responsible for secondary growth; either vascular cambium or cork cambium.

lateral root Of taproot systems, a lateral branching from the first, primary root.

leaching The movement of soil water, with dissolved nutrients, out of a specified area.

leaf For most vascular plants, a structure having chlorophyll-containing tissue that is the major region of photosynthesis.

learned behavior The use of information gained from specific experiences to vary or change a response to stimuli.

lethal mutation A gene mutation that alters one or more traits in such a way that the individual inevitably dies.

LH Leutinizing hormone. Secreted by the anterior lobe of the pituitary gland, this hormone has roles in the reproductive functions of both males and females.

lichen (LY-kun) A symbiotic association between a fungus and a captive photosynthetic partner such as a green alga.

life cycle A recurring, genetically programmed frame of events in which individuals grow, develop, maintain themselves, and reproduce.

life table A tabulation of age-specific patterns of birth and death for a population.

ligament A strap of dense connective tissue that bridges a joint.

light-dependent reactions First stage of photosynthesis in which the energy of sunlight is trapped and converted to the chemical energy of ATP alone (by the cyclic pathway) or ATP and NADPH (by the noncyclic pathway).

light-independent reactions Second stage of photosynthesis, in which sugar phosphates form with the help of the ATP (and NADPH, in land plants) that were produced during the first stage. The sugar phosphates are used in other reactions by which starch, cellulose, and other end products of photosynthesis are assembled.

lignification Of mature land plants, a process by which lignin is deposited in secondary cell walls. The deposits impart strength and rigidity by anchoring cellulose strands in the walls, stabilize and protect other wall components, and form a waterproof barrier around the cellulose. Probably a key factor in the evolution of vascular plants.

lignin A substance that strengthens and waterproofs cell walls in certain tissues of vascular plants.

limbic system Brain regions that, along with the cerebral cortex, collectively govern emotions.

limiting factor Any essential resource that is in short supply and so limits population growth.

lineage (LIN-ee-age) A line of descent.

linkage The tendency of genes located on the same chromosome to end up in the same gamete. For any two of those genes, the probability that crossing over will disrupt the linkage is proportional to the distance separating them.

lipid A greasy or oily compound of mostly carbon and hydrogen that shows little tendency to dissolve in water, but that dissolves in nonpolar solvents (such as ether). Cells use lipids as energy stores and structural materials, especially in membranes.

lipid bilayer The structural basis of cell membranes, consisting of two layers of mostly phospholipid molecules. Hydrophilic heads force all fatty acid tails of the lipids to become sandwiched between the hydrophilic heads.

liver Glandular organ with roles in storing and interconverting carbohydrates, lipids, and proteins absorbed from the gut, maintaining blood; disposing of nitrogen-containing wastes; and other tasks.

local signaling molecules Secretions from cells in many different tissues that alter chemical conditions in the immediate vicinity where they are secreted, then are swiftly degraded.

locus (LOW-cuss) The specific location of a particular gene on a chromosome.

logistic population growth (low-JIS-tik) Pattern of population growth in which a low-density population slowly increases in size, goes through a rapid growth phase, then levels off once the carrying capacity is reached.

loop of Henle The hairpin-shaped, tubular region of a nephron that functions in reabsorption of water and solutes.

lung An internal respiratory surface in the shape of a cavity or sac.

lycophyte A type of seedless vascular plant of mostly wet or shade habitats; requires free water to complete its life cycle.

lymph (LIMF) [L. *lympha*, water] Tissue fluid that has moved into the vessels of the lymphatic system.

lymph capillary A small-diameter vessel of the lymph vascular system that has no pronounced entrance; tissue fluid moves inward by passing between overlapping endothelial cells at the vessel's tip.

lymph node A lymphoid organ that serves as a battleground of the immune system; each lymph node is packed with organized arrays of macrophages and lymphocytes that cleanse lymph of pathogens before it reaches the blood.

lymph vascular system [L. *lympha*, water, + *vasculum*, a small vessel] The vessels of the lymphatic system, which take up and transport excess tissue fluid and reclaimable solutes as well as fats absorbed from the digestive tract.

lymphatic system An organ system that supplements the circulatory system. Its vessels take up fluid and solutes from interstitial fluid and deliver them to the bloodstream; its lymphoid organs have roles in immunity.

lymphocyte Any of various white blood cells that take part in nonspecific and specific (immune) defense responses.

lymphoid organs The lymph nodes, spleen, thymus, tonsils, adenoids, and other organs with roles in immunity.

lysis [Gk. *lysis*, a loosening] Gross structural disruption of a plasma membrane that leads to cell death.

lysosome (LYE-so-sohm) The main organelle of digestion, with enzymes that can break down polysaccharides, proteins, nucleic acids, and some lipids.

lysozyme An infection-fighting enzyme that digests bacterial cell walls. Present in mucous membranes that line the body's surfaces.

lytic pathway During a viral infection, viral DNA or RNA quickly directs the host cell to produce the components necessary to produce new virus particles, which are released by lysis.

macroevolution The large-scale patterns, trends, and rates of change among groups of species.

macrophage One of the phagocytic white blood cells. It engulfs anything detected as foreign. Some also become antigen-presenting cells that serve as the trigger for immune responses by T and B lymphocytes. *Compare* antigen-presenting cell.

mammal A type of vertebrate; the only animal having offspring that are nourished by milk produced by mammary glands of females.

mass extinction An abrupt rise in extinction rates above the background level; a catastrophic, global event in which major taxa are wiped out simultaneously.

mass number The total number of protons and neutrons in an atom's nucleus. The relative masses of atoms are also called atomic weights.

maternal chromosome One of the chromosomes bearing the alleles that are inherited from a female parent.

mechanoreceptor Sensory cell or cell part that detects mechanical energy associated with changes in pressure, position, or acceleration.

medulla oblongata Part of the vertebrate brainstem with reflex centers for respiration, blood circulation, and other vital functions.

medusa (meh-DOO-sah) [Gk. *Medousa*, one of three sisters in Greek mythology having snake-entwined hair; this image probably evoked by the tentacles and oral arms extending from the medusa] Free-swimming, bell-shaped stage in cnidarian life cycles.

megaspore Of gymnosperms and flowering plants, a haploid spore that forms in the ovary; one of its cellular descendants develops into an egg.

meiosis (my-OH-sis) [Gk. *meioun*, to diminish] Two-stage nuclear division process in which the chromosome number of a germ cell is reduced by half, to the haploid number. (Each daughter nucleus ends up with one of each type of chromosome.) Meiosis is the basis of gamete formation and (in plants) of spore formation. *Compare* mitosis.

meltdown Events which, if unchecked, can blow apart a nuclear power plant, with a release of radioactive material into the environment.

membrane excitability A membrane property of any cell that can produce action potentials in response to appropriate stimulation.

memory The storage and retrieval of information about previous experiences; underlies the capacity for learning.

memory cell One of the subpopulations of cells that form during an immune response and that enters a resting phase, from which it is released during a secondary immune response.

memory lymphocyte Any of the various B or T lymphocytes of the immune system that are formed in response to invasion by a foreign agent and that circulate for some period, available to mount a rapid attack if the same type of invader reappears.

Mendel's theory of independent assortment Stated in modern terms, during meiosis, the gene pairs of homologous chromosomes tend to be stored independently of how gene pairs on other chromosomes are sorted for forthcoming gametes. The theory does not take into account the effects of gene linkage and crossing over.

Mendel's theory of segregation Stated in modern terms, diploid cells have two of each kind of gene (on pairs of homologous chromosomes), and the two segregate during meiosis so that they end up in different gametes.

menopause (MEN-uh-pozz) [L. *mensis*, month, + *pausa*, stop] End of the period of a human female's reproductive potential.

menstrual cycle The cyclic release of oocytes and priming of the endometrium (lining of the uterus) to receive a fertilized egg; the complete cycle averages about 28 days in female humans.

menstruation Periodic sloughing of the blood-enriched lining of the uterus when pregnancy does not occur.

mesoderm (MEH-so-derm) [Gk. *mesos*, middle, + *derm*, skin] In most animal embryos, a primary tissue layer (germ layer) between ectoderm and endoderm. Gives rise to muscle; organs of circulation, reproduction, and excretion; most of the internal skeleton (when present); and connective tissue layers of the gut and body covering.

mesophyll Of vascular plants, a type of parenchyma tissue with photosynthetic cells and an abundance of air spaces.

messenger RNA A linear sequence of ribonucleotides transcribed from DNA and translated into a polypeptide chain; the only type of RNA that carries protein-building instructions.

metabolic pathway One of many orderly sequences of enzyme-mediated reactions by which cells normally maintain, increase, or decrease the concentrations of substances.

Different pathways are linear or circular, and often they interconnect.

metabolism (meh-TAB-oh-lizm) [Gk. *meta*, change] All controlled, enzyme-mediated chemical reactions by which cells acquire and use energy. Through these reactions, cells synthesize, store, break apart, and eliminate substances in ways that contribute to growth, survival, and reproduction.

metamorphosis (met-uh-MOR-foe-sis) [Gk. *meta*, change, + *morphe*, form] Transformation of a larva into an adult form.

metaphase Of mitosis or meiosis II, the stage when each duplicated chromosome has become positioned at the midpoint of the microtubular spindle, with its two sister chromatids attached to microtubules from opposite spindle poles. Of meiosis I, the stage when all pairs of homologous chromosomes are positioned at the spindle's midpoint, with the two members of each pair attached to opposite spindle poles.

metazoan Any multicelled animal.

methanogen A type of archaebacterium that lives in oxygen-free habitats and that produces methane gas as a metabolic by-product.

MHC marker Any of a variety of proteins that are self-markers. Some occur on all body cells of an individual; others are unique to the macrophages and lymphocytes.

micelle formation Formation of a small droplet that consists of bile salts and products of fat digestion (fatty acids and monoglycerides) and that assists in their absorption from the small intestine.

microevolution Changes in allele frequencies brought about by mutation, genetic drift, gene flow, and natural selection.

microfilament [Gk. *mikros*, small, + L. *filum*, thread] In animal cells, one of a variety of cytoskeletal components. Actin and myosin filaments are examples.

microorganism An organism, usually single-celled, that is too small to be observed without the aid of a microscope.

microspore Of gymnosperms and flowering plants, a haploid spore, encased in a sculpted wall, that develops into a pollen grain.

microtubular spindle Of eukaryotic cells, a bipolar structure composed of organized arrays of microtubules that forms during nuclear division and that moves the chromosomes.

microtubule Hollow cylinder of mainly tubulin subunits; a cytoskeletal element with roles in cell shape, motion, and growth and in the structure of cilia and flagella.

microtubule organizing center, or **MTOC** Small mass of proteins and other substances in the cytoplasm; the number, type, and location of MTOCs determine the organization and orientation of microtubules.

microvillus (MY-crow-VILL-us) [L. *villus*, shaggy hair] A slender, cylindrical extension of the animal cell surface that functions in absorption or secretion.

midbrain Of vertebrates, a brain region that evolved as a coordination center for reflex responses to visual and auditory input; together with the pons and medulla oblongata,

part of the brainstem, which includes the reticular formation.

migration Of certain animals, a cyclic movement between two distant regions at times of year corresponding to seasonal change.

mimicry (MIM-ik-ree) Situation in which one species (the mimic) bears deceptive resemblance in color, form, and/or behavior to another species (the model) that enjoys some survival advantage.

mineral An inorganic substance required for the normal functioning of body cells.

mitochondrion (MY-toe-KON-dree-on), plural **mitochondria** Organelle that specializes in ATP formation; it is the site of the second and third stages of aerobic respiration, an oxygen-requiring pathway.

mitosis (my-TOE-sis) [Gk. *mitos*, thread] Type of nuclear division that maintains the parental chromosome number for daughter cells. It is the basis of bodily growth and, in many eukaryotic species, asexual reproduction.

molar One of the cheek teeth; a tooth with a platform having cusps (surface bumps) that help crush, grind, and shear food.

molecular clock With respect to the presumed regular accumulation of neutral mutations in highly conserved genes, a way of calculating the time of origin of one species or lineage relative to others.

molecule A unit of matter in which chemical bonding holds together two or more atoms of the same or different elements.

mollusk A type of invertebrate having a tissue fold (mantle) draped around a soft, fleshy body; snails, clams, and squids are examples.

molting The shedding of hair, feathers, horns, epidermis, or a shell (or some other exoskeleton) in a process of growth or periodic renewal.

Monera The kingdom of bacteria.

moneran A bacterium; a single-celled prokaryote.

monocot (MON-oh-kot) Short for monocotyledon; a flowering plant in which seeds have only one cotyledon, whose floral parts generally occur in threes (or multiples of three), and whose leaves typically are parallel-veined. *Compare* dicot.

monohybrid cross [Gk. *monos*, alone] An experimental cross in which offspring inherit a pair of nonidentical alleles for a single trait being studied, so that they are heterozygous.

monophyletic group A set of independently evolving lineages that share a common evolutionary heritage.

monosaccharide (MON-oh-SAK-ah-ride) [Gk. *monos*, alone, single, + *sakharon*, sugar] The simplest carbohydrate, with only one sugar unit. Glucose is an example.

monosomy Abnormal condition in which one chromosome of diploid cells has no homologue.

morphogenesis (MORE-foe-JEN-ih-sis) [Gk. *morphe*, form, + *genesis*, origin] Processes by which differentiated cells in an embryo become organized into tissues and organs, under genetic controls and environmental influences.

morphological convergence A macroevolutionary pattern of change in which separate lineages adopt similar lifestyles, put comparable body parts to similar uses, and in time resemble one another in structure and function. Analogous structures are evidence of this pattern.

morphological divergence A macroevolutionary pattern of change from a common ancestral form. Homologous structures are evidence of this pattern.

motor neuron A type of neuron; it delivers signals from the brain and spinal cord that can stimulate or inhibit the body's effectors (muscles, glands, or both).

mouth An oral cavity; in human digestion, the site where polysaccharide breakdown begins.

multicelled organism An organism that has differentiated cells arranged into tissues, organs, and often organ systems.

multiple allele system Three or more different molecular forms of the same gene (alleles) that exist in a population.

muscle fatigue A decline in tension of a muscle that has been kept in a state of tetanic contraction as a result of continuous, high-frequency stimulation.

muscle tension A mechanical force, exerted by a contracting muscle, that resists opposing forces such as gravity and the weight of objects being lifted.

muscle tissue Tissue having cells able to contract in response to stimulation, then passively lengthen and so return to their resting stage.

mutagen (MEW-tuh-jen) An environmental agent that can permanently modify the structure of a DNA molecule. Certain viruses and ultraviolet radiation are examples.

mutation [L. *mutatus*, a change, + *-ion*, result or a process or an act] A heritable change in the DNA. Generally, mutations are the source of all the different molecular versions of genes (alleles) and, ultimately, of life's diversity. *See also* lethal mutation; neutral mutation.

mutualism [L. *mutuus*, reciprocal] An interaction between two species that benefits both.

mycelium (my-SEE-lee-um), plural **mycelia** [Gk. *mykes*, fungus, mushroom, + *helos*, callus] A mesh of tiny, branching filaments (hyphae) that is the food-absorbing part of a multicelled fungus.

mycorrhiza (MY-coe-RISE-uh) "Fungus-root;" a symbiotic arrangement between fungal hyphae and the young roots of many vascular plants. The fungus obtains carbohydrates from the plant and in turn releases dissolved mineral ions to the plant roots.

myelin sheath Of many sensory and motor neurons, an axonal sheath that affects how fast action potentials travel; formed from the plasma membranes of Schwann cells that are wrapped repeatedly around the axon and are separated from each other by a small node.

myofibril (MY-oh-FY-brill) One of many thread-like structures inside a muscle cell; each is functionally divided into sarcomeres, the basic units of contraction.

myosin (MY-uh-sin) A type of protein with a head and long tail. In muscle cells, it interacts with actin, another protein, to bring about contraction.

NAD⁺ Nicotinamide adenine dinucleotide; a nucleotide coenzyme. When carrying electrons and unbound protons (H^+) between reaction sites, it is abbreviated NADH.

NADP⁺ Nicotinamide adenine dinucleotide phosphate; a phosphorylated nucleotide coenzyme. When carrying electrons and unbound protons (H^+) between reaction sites, it is abbreviated $NADPH_2$.

nasal cavity Of a respiratory system, the region where air is warmed, moistened, and filtered of airborne particles and dust.

natural selection A microevolutionary process; a difference in survival and reproduction among members of a population that vary in one or more traits.

negative feedback mechanism A homeostatic feedback mechanism in which an activity changes some condition in the internal environment and so triggers a response that reverses the changed condition.

nematocyst (NEM-ad-uh-sist) [Gk. *nema*, thread, + *kystis*, pouch] Of cnidarians only, a stinging capsule that assists in prey capture and possibly protection.

nephridium (neh-FRID-ee-um), plural **nephridia** Of earthworms and some other invertebrates, a system of regulating water and solute levels.

nephron (NEFF-ron) [Gk. *nephros*, kidney] Of the vertebrate kidney, a slender tubule in which water and solutes filtered from blood are selectively reabsorbed and in which urine forms.

nerve Cordlike communication line of the peripheral nervous system, composed of axons of sensory neurons, motor neurons, or both packed within connective tissue. In the brain and spinal cord, similar cord-like bundles are called nerve pathways or tracts.

nerve cord Of many animals, a cordlike communication line consisting of axons of neurons.

nerve impulse *See* action potential.

nerve net Cnidarian nervous system.

nervous system System of neurons oriented relative to one another in precise message-conducting and information-processing pathways.

nervous tissue A type of connective tissue composed of neurons.

net energy Of energy resources available to the human population, the amount of energy that is left over after subtracting the energy used to locate, extract, transport, store, and deliver energy to consumers.

net population growth rate per individual (r) Of population growth equations, a single variable in which birth and death rates, which are assumed to remain constant, are combined.

neuroglial cell (NUR-oh-GLEE-uhl) Of vertebrates, one of the cells that provide structural and metabolic support for neurons and that collectively represent about half the volume of the nervous system.

neuromodulator Type of signaling molecule that influences the effects of transmitter substances by enhancing or reducing membrane responses in target neurons.

neuromuscular junction Chemical synapses between axon terminals of a motor neuron and a muscle cell.

neuron A nerve cell; the basic unit of communication in nervous systems. Neurons collectively sense environmental change, integrate sensory inputs, then activate muscles or glands that initiate or carry out responses.

neurotransmitter Any of the class of signaling molecules that are secreted from neurons, act on immediately adjacent cells, and are then rapidly degraded or recycled.

neutral mutation A gene mutation that has neither harmful nor helpful effects on the individual's ability to survive and reproduce.

neutron Unit of matter, one or more of which occupies the atomic nucleus, that has mass but no electric charge.

neutrophil Fast-acting, phagocytic white blood cell that takes part in inflammatory responses against bacteria.

niche (NITCH) [L. *nidas*, nest] Of a species, the full range of physical and biological conditions under which its members can live and reproduce.

nitrification (nye-trih-fih-KAY-shun) A chemosynthetic process in which certain bacteria strip electrons from ammonia or ammonium present in soil. The end product, nitrite (NO_2^-), is broken down to nitrate (NO_3^-) by different bacteria.

nitrogen cycle Biogeochemical cycle in which the atmosphere is the largest reservoir of nitrogen.

nitrogen fixation Process by which a few kinds of bacteria convert gaseous nitrogen (N_2) to ammonia. This dissolves rapidly in their cytoplasm to form ammonium, which can be used in biosynthetic pathways.

NK cell Natural killer cell, possibly of the lymphocyte lineage, that reconnoiters and kills tumor cells and infected body cells.

nociceptor A receptor, such as a free nerve ending, that detects any stimulus causing tissue damage.

node In vascular plants, a point on a stem where one or more leaves are attached.

noncyclic pathway of ATP formation (non-SIK-lik) [L. *non*, not, + Gk. *kylos*, circle] Photosynthetic pathway in which excited electrons derived from water molecules flow through two photosystems and two transport chains, and ATP and NADPH form.

nondisjunction Failure of one or more chromosomes to separate properly during mitosis or meiosis.

nonsteroid hormone A type of water-soluble hormone, such as a protein hormone, that cannot cross the lipid bilayer of a target cell. These hormones enter the cell by receptor-mediated endocytosis, or they bind to receptors that activate membrane proteins or second messengers within the cell.

notochord (KNOW-toe-kord) Of chordates, a rod of stiffened tissue (not cartilage or bone) that serves as a supporting structure for the body.

nuclear envelope A double membrane (two lipid bilayers and associated proteins) that is the outermost portion of a cell nucleus.

nucleic acid (new-CLAY-ik) A long, single- or double-stranded chain of four different kinds of nucleotides joined one after the other at their phosphate groups. They differ in which nucleotide base follows the next in sequence. DNA and RNA are examples.

nucleic acid hybridization The base-pairing of nucleotide sequences from different sources, as used in genetics, genetic engineering, and studies of evolutionary relationship based on similarities and differences in the DNA or RNA of different species.

nucleoid Of bacteria, a region in which DNA is physically organized apart from other cytoplasmic components.

nucleolus (new-KLEE-oh-lus) [L. *nucleolus*, a little kernel] Within the nucleus of a nondividing cell, a site where the protein and RNA subunits of ribosomes are assembled.

nucleosome (NEW-klee-oh-sohm) Of eukaryotic chromosomes, one of many organizational units, each consisting of a small stretch of DNA looped twice around a "spool" of histone molecules, which another histone molecule stabilizes.

nucleotide (NEW-klee-oh-tide) A small organic compound having a five-carbon sugar (deoxyribose), nitrogen-containing base, and phosphate group. Nucleotides are the structural units of adenosine phosphates, nucleotide coenzymes, and nucleic acids.

nucleotide coenzyme A protein that transports hydrogen atoms (free protons) and electrons from one reaction site to another in cells.

nucleus (NEW-klee-us) [L. *nucleus*, a kernel] Of atoms, the central core of one or more positively charged protons and (in all but hydrogen) electrically neutral neutrons. In eukaryotic cells, a membranous organelle that physically isolates and organizes the DNA, out of the way of cytoplasmic machinery.

nutrition All those processes by which food is selectively ingested, digested, absorbed, and later converted to the body's own organic compounds.

obesity An excess of fat in the body's adipose tissues, caused by imbalances between caloric intake and energy output.

oligosaccharide A carbohydrate consisting of a short chain of two or more covalently bonded sugar units. One subclass, disaccharides, has two sugar units. *Compare* monosaccharide; polysaccharide.

omnivore [L. *omnis*, all, + *vovare*, to devour] An organism able to obtain energy from more than one source rather than being limited to one trophic level.

oncogene (ON-coe-jeen) Any gene having the potential to induce cancerous transformations in a cell.

oocyte An immature egg.

oogenesis (oo-oh-JEN-uh-sis) Formation of a female gamete, from a germ cell to a mature haploid ovum (egg).

operator A short base sequence between a promoter and the start of a gene; interacts with regulatory proteins.

operon Of transcription, a promoter-operator sequence that services more than a single gene. The lactose operon of *E. coli* is an example.

orbitals Volumes of space around the nucleus of an atom in which electrons are likely to be at any instant.

organ A structure of definite form and function that is composed of more than one tissue.

organ formation Stage of development in which primary tissue layers (germ layers) split into subpopulations of cells, and different lines of cells become unique in structure and function; foundation for growth and tissue specialization, when organs acquire specialized chemical and physical properties.

organ system Two or more organs that interact chemically, physically, or both in performing a common task.

organelle Of cells, an internal, membrane-bounded sac or compartment that has a specific, specialized metabolic function.

organic compound In biology, a compound assembled in cells and having a carbon backbone, often with carbon atoms arranged as a chain or ring structure.

osmosis (oss-MOE-sis) [Gk. *osmos*, act of pushing] Of cells, the tendency of water to move through channel proteins that span a membrane in response to a concentration gradient, fluid pressure, or both. Hydrogen bonds among water molecules prevent water *itself* from becoming more or less concentrated; but a gradient may exist when the water on either side of the membrane has more substances dissolved in it.

ovary (OH-vuh-ree) In female animals, the primary reproductive organ in which eggs form. In seed-bearing plants, the portion of the carpel where eggs develop, fertilization takes place, and seeds mature. A mature ovary (and sometimes other plant parts) is a fruit.

oviduct (OH-vih-dukt) Duct through which eggs travel from the ovary to the uterus. Formerly called Fallopian tube.

ovulation (AHV-you-LAY-shun) During each turn of the menstrual cycle, the release of a secondary oocyte (immature egg) from an ovary.

ovule (OHV-youl) [L. *ovum*, egg] Before fertilization in gymnosperms and angiosperms, a female gametophyte with egg cell, a surrounding tissue, and one or two protective layers (integuments). After fertilization, an ovule matures into a seed (an embryo sporophyte and food reserves encased in a hardened coat).

ovum (OH-vum) A mature female gamete (egg).

oxidation-reduction reaction An electron transfer from one atom or molecule to another. Often hydrogen is transferred along with the electron or electrons.

ozone hole A pronounced seasonal thinning of the ozone layer in the lower stratosphere above Antarctica.

pancreas (PAN-cree-us) Gland that secretes enzymes and bicarbonate into the small intestine during digestion, and that also secretes the hormones insulin and glucagon.

pancreatic islets Any of the two million clusters of endocrine cells in the pancreas, including alpha cells, beta cells, and delta cells.

parasite [Gk. *para*, alongside, + *sitos*, food] An organism that obtains nutrients directly from the tissues of a living host, which it lives on or in and may or may not kill.

parasitism A two-species interaction in which one species directly harms another that serves as its host.

parasitoid An insect larva that grows and develops inside a host organism (usually another insect), eventually consuming the soft tissues and killing it.

parasympathetic nerve Of the autonomic nervous system, any of the nerves carrying signals that tend to slow the body down overall and divert energy to basic tasks; also work continually in opposition with sympathetic nerves to bring about minor adjustments in internal organs.

parathyroid glands (PARE-uh-THY-royd) In vertebrates, endocrine glands embedded in the thyroid gland that secrete parathyroid hormone, which helps restore blood calcium levels.

parenchyma Most abundant of the simple tissues in flowering plant roots, stems, leaves, and other parts. Its cells function in photosynthesis, storage, secretion, and other tasks.

parthenogenesis Development of an embryo from an unfertilized egg.

passive immunity Temporary immunity conferred by deliberately introducing antibodies into the body.

passive transport Diffusion of a solute through a channel or carrier protein that spans the lipid bilayer of a cell membrane. Its passage does not require an energy input; the protein passively allows the solute to follow its concentration gradient.

paternal chromosome One of the chromosomes bearing alleles that are inherited from a male parent.

pathogen (PATH-oh-jen) [Gk. *pathos*, suffering, + *-genēs*, origin]. An infectious, disease-causing agent, such as a virus or bacterium.

pattern formation Of animals, mechanisms responsible for specialization and positioning of tissues during embryonic development.

PCR Polymerase chain reaction. A method used by recombinant DNA technologists to amplify the quantity of specific fragments of DNA.

peat An accumulation of saturated, undecayed remains of plants that have been compressed by sediments.

pedigree A chart of genetic connections among individuals, as constructed according to standardized methods.

pelagic province The entire volume of ocean water; subdivided into neritic zone (relatively shallow waters overlying continental shelves) and oceanic zone (water over ocean basins).

penis A male organ that deposits sperm into a female reproductive tract.

perennial [L. *per-*, throughout, + *annus*, year] A flowering plant that lives for three or more growing seasons.

perforin A type of protein, produced and secreted by cytotoxic cells, that destroys antigen-bearing targets.

pericycle (PARE-ih-sigh-kul) [Gk. *peri-*, around, + *kyklos*, circle] Of a root vascular cylinder, one or more layers just inside the endodermis that gives rise to lateral roots and contributes to secondary growth.

periderm Of vascular plants showing secondary growth, a protective covering that replaces epidermis.

peripheral nervous system (per-IF-ur-uhl) [Gk. *peripherein*, to carry around] Of vertebrates, the nerves leading into and out from the spinal cord and brain and the ganglia along those communication lines.

peristalsis (pare-ih-STAL-sis) A rhythmic contraction of muscles that moves food forward through the animal gut.

peritoneum A lining of the coelom that also covers and helps maintain the position of internal organs.

peritubular capillaries The set of blood capillaries that threads around the tubular parts of a nephron; they function in reabsorption of water and solutes back into the body and in secretion of hydrogen ions and some other substances in the forming urine.

permafrost A permanently frozen, water-impenetrable layer beneath the soil surface in arctic tundra.

peroxisome Enzyme-filled vesicle in which fatty acids and amino acids are digested first into hydrogen peroxide (which is toxic), then to harmless products.

PGA Phosphoglycerate (FOSS-foe-GLISS-er-ate). A key intermediate in glycolysis and in the Calvin-Benson cycle.

PGAL Phosphoglyceraldehyde. A key intermediate in glycolysis and in the Calvin-Benson cycle.

pH scale A scale used to measure the concentration of free hydrogen ions in blood, water, and other solutions; pH 0 is the most acidic, 14 the most basic, and 7, neutral.

phagocyte (FAG-uh-sight) [Gk. *phagein*, to eat, + *kytos*, hollow vessel] A macrophage or certain other white blood cells that engulf and destroy foreign agents.

phagocytosis (FAG-uh-sigh-TOE-sis) [Gk. *phagein*, to eat, + *kytos*, hollow vessel] Engulfment of foreign cells or substances by amoebas and some white blood cells by means of endocytosis.

pharynx (FARE-inks) A muscular tube by which food enters the gut; in land vertebrates, the dual entrance for the tubular part of the digestive tract and windpipe (trachea).

phenotype (FEE-no-type) [Gk. *phainein*, to show, + *typos*, image] Observable trait or traits of an individual; arises from interactions between genes, and between genes and the environment.

pheromone (FARE-oh-moan) [Gk. *phero*, to carry, + *-mone*, as in hormone] A type of signaling molecule secreted by exocrine glands that serves as a communication signal between individuals of the same species. Signaling pheromones elicit an immediate

behavioral response. Priming pheromones elicit a generalized physiological response.

phloem (FLOW-um) Of vascular plants, a tissue with living cells that interconnect and form the tubes through which sugars and other solutes are conducted.

phospholipid A type of lipid that is the main structural component of cell membranes. Each has a hydrophobic tail (of two fatty acids) and a hydrophilic head that incorporates glycerol and a phosphate group.

phosphorus cycle Movement of phosphorus from rock or soil through organisms, then back to soil.

phosphorylation (FOSS-for-ih-LAY-shun) The attachment of unbound (inorganic) phosphate to a molecule; also the transfer of a phosphate group from one molecule to another, as when ATP phosphorylates glucose.

photochemical smog A brown-air smog that develops over large cities when the surrounding land forms a natural basin.

photolysis (foe-TALL-ih-sis) [Gk. *photos*, light, + *-lysis*, breaking apart] A reaction sequence of the noncyclic pathway of photosynthesis, triggered by photon energy, in which water is split into oxygen, hydrogen, and electrons.

photoperiodism A biological response to a change in the relative length of daylight and darkness.

photoreceptor Light-sensitive sensory cell.

photosynthesis The trapping of sunlight energy and its conversion to chemical energy (ATP, NADPH, or both), followed by synthesis of sugar phosphates that become converted to sucrose, cellulose, starch, and other end products. It is the main biosynthetic pathway by which energy and carbon enter the web of life.

photosynthetic autotroph An organism able to synthesize all organic molecules it requires using carbon dioxide as the carbon source and sunlight as the energy source. All plants, some protistans, and a few bacteria are photosynthetic autotrophs.

photosystem One of the clusters of light-trapping pigments embedded in photosynthetic membranes. Photosystem I operates during the cyclic pathway; photosystem II operates during both the cyclic and noncyclic pathways.

phototropism [Gk. *photos*, light, + *trope*, turning, direction] Adjustment in the direction and rate of plant growth in response to light.

photovoltaic cell A device that converts sunlight energy into electricity.

phycobilins (FIE-koe-BY-lins) A class of light-sensitive, accessory pigments that transfer absorbed energy to chlorophylls. They are abundant in red algae and cyanobacteria.

phylogeny Evolutionary relationships among species, starting with most ancestral forms and including the branches leading to their descendants.

phytochrome Light-sensitive pigment molecule, the activation and inactivation of which triggers plant hormone activities governing leaf expansion, stem branching, stem length and often seed germination and flowering.

phytoplankton (FIE-toe-PLANK-tun) [Gk. *phyton*, plant, + *planktos*, wandering] A freshwater or marine community of floating or weakly swimming photosynthetic autotrophs, such as cyanobacteria, diatoms, and green algae.

pigment A light-absorbing molecule.

pineal gland (py-NEEL) A light-sensitive endocrine gland that secretes melatonin, a hormone that influences reproductive cycles and the development of reproductive organs.

pioneer species Typically small plants with short life cycles that are adapted to growing in exposed, often windy areas with intense sunlight, wide swings in air temperature, and soils deficient in nitrogen and other nutrients. By improving conditions in areas they colonize, pioneers invite their own replacement by other species.

pituitary gland Of endocrine systems, a gland that interacts with the hypothalamus to coordinate and control many physiological functions, including the activity of many other endocrine glands. Its posterior lobe stores and secretes hypothalamic hormones; the anterior lobe produces and secretes its own hormones.

placenta (play-SEN-tuh) Of the uterus, an organ composed of maternal tissues and extraembryonic membranes (the chorion especially); it delivers nutrients to the fetus and accepts wastes from it, yet allows the fetal circulatory system to develop separately from the mother's.

plankton [Gk. *planktos*, wandering] Any community of floating or weakly swimming organisms, mostly microscopic, living in freshwater and saltwater environments. *See* phytoplankton; zooplankton.

plant The type of eukaryotic organism, usually multicelled, that is a photosynthetic autotroph—it uses sunlight energy to drive the synthesis of all its required organic compounds from carbon dioxide, water, and mineral ions. Only a few nonphotosynthetic plants obtain nutrients by parasitism and other means.

Plantae The kingdom of plants.

plasma (PLAZ-muh) Liquid component of blood; consists of water, various proteins, ions, sugars, dissolved gases, and other substances.

plasma cell Of immune systems, any of the anitbody-secreting daughter cells of a rapidly dividing population of B cells.

plasma membrane Of cells, the outermost membrane. Its lipid bilayer structure and proteins carry out most functions, including transport across the membrane and reception of extracellular signals.

plasmid Of many bacteria, a small, circular molecule of extra DNA that carries only a few genes and replicates independently of the bacterial chromosome.

plasmodesma (PLAZ-moe-DEZ-muh) Of multicelled plants, a junction between linked walls of adjacent cells through which nutrients and other substances flow.

plasticity Of the human species, the ability to remain flexible and adapt to a wide range of environments.

plate tectonics Arrangement of the earth's outer layer (lithosphere) in slablike plates, all in motion and floating on a hot, plastic layer of the underlying mantle.

platelet (PLAYT-let) Any of the cell fragments in blood that release substances necessary for clot formation.

pleiotropy (PLEE-oh-troe-pee) [Gk. *pleon*, more, + *trope*, direction] A type of gene interaction in which a single gene exerts multiple effects on seemingly unrelated aspects of an individual's phenotype.

polar body Any of three cells that form during the meiotic cell division of an oocyte; the division also forms the mature egg, or ovum.

pollen grain [L. *pollen*, fine dust] Depending on the species, the immature or mature, sperm-bearing male gametophyte of gymnosperms and flowering plants.

pollen sac In anthers of flowers, any of the chambers in which pollen grains develop.

pollen tube A tube formed after a pollen grain germinates; grows down through carpel tissues and carries sperm to the ovule.

pollination Of flowering plants, the arrival of a pollen grain on the landing platform (stigma) of a carpel.

pollutant Any substance with which an ecosystem has had no prior evolutionary experience, in terms of kinds or amounts, and that can accumulate to disruptive or harmful levels. Can be naturally occurring or synthetic.

polymer (POH-lih-mur) [Gk. *polus*, many, + *meris*, part] A molecule composed of three to millions of small subunits that may or may not be identical.

polymerase chain reaction or **PCR** DNA amplification method; DNA containing a gene of interest is split into single strands, which enzymes (polymerases) copy; the enzymes also act on the accumulating copies, multiplying the gene sequence by the millions.

polymorphism (poly-MORE-fizz-um) [Gk. *polus*, many, + *morphe*, form] Of a population, the persistence through the generations of two or more forms of a trait, at a frequency greater than can be maintained by new mutations alone.

polyp (POH-lip) Vase-shaped, sedentary stage of cnidarian life cycles.

polypeptide chain Three or more amino acids joined by peptide bonds.

polyploidy (POL-ee-PLOYD-ee) A change in the chromosome number following inheritance of three or more of each type of chromosome.

polysaccharide [Gk. *polus*, many, + *sakcharon*, sugar] A straight or branched chain of hundreds of thousands of covalently linked sugar units, of the same or different kinds. The most common polysaccharides are cellulose, starch, and glycogen.

polysome Of protein synthesis, several ribosomes all translating the same messenger RNA molecule, one after the other.

population A group of individuals of the same species occupying a given area.

population density The number of individuals of a population that are living in a specified area or volume.

population distribution The general pattern of dispersion of individuals of a population throughout their habitat.

population size The number of individuals that make up the gene pool of a population.

positive feedback mechanism Homeostatic mechanism by which a chain of events is set in motion that intensifies a change from an original condition; after a limited time, the intensification reverses the change.

post-translational controls Of eukaryotes, controls that govern modification of newly formed polypeptide chains into functional enzymes and other proteins.

predation A two-species interaction in which one species (the predator) directly harms the other (its prey).

predator [L. *prehendere*, to grasp, seize] An organism that feeds on and may or may not kill other living organisms (its prey); unlike parasites, predators do not live on or in their prey.

prediction A claim about what you can expect to observe in nature if a theory or hypothesis is correct.

premolar One of the cheek teeth; a tooth having a platform with cusps (surface bumps) that can crush, grind, and shear food.

pressure flow theory Of vascular plants, a theory that organic compounds move through phloem because of gradients in solute concentrations and pressure between source regions (such as photosynthetically active leaves) and sink regions (such as growing plant parts).

primary growth Plant growth originating at root tips and shoot tips.

primary immune response Actions by white blood cells and their products elicited by a first-time encounter with an antigen; includes both antibody-mediated and cell-mediated responses.

primary productivity, gross Of ecosystems, the rate at which the producer organisms capture and store a given amount of energy during a specified interval.

primary productivity, net Of ecosystems, the rate of energy storage in the tissues of producers in excess of their rate of aerobic respiration.

primate The mammalian lineage that includes prosimians, tarsioids, and anthropoids (monkeys, apes, and humans).

probability With respect to any chance event, the most likely number of times it will turn out a certain way, divided by the total number of all possible outcomes.

procambium (pro-KAM-bee-um) Of vascular plants, a primary meristem that gives rise to the primary vascular tissues.

producers, primary Of ecosystems, the organisms that secure energy from the physical environment, as by photosynthesis or chemosynthesis.

progesterone (pro-JESS-tuh-rown) Female sex hormone secreted by the ovaries.

prokaryotic cell (pro-CARRY-oh-tic) [L. *pro*, before, + Gk. *karyon*, kernel] A bacterium; a single-celled organism that has no nucleus or any of the other membrane-bound organelles characteristic of eukaryotic cells.

promoter Of transcription, a base sequence that signals the start of a gene; the site where RNA polymerase initially binds.

prophase Of mitosis, the stage when each duplicated chromosome starts to condense, microtubules form a spindle apparatus, and the nuclear envelope starts to break up.

prophase I Of meiosis, the stage at which the microtubular spindle starts to form, the nuclear envelope starts to break up, and each duplicated chromosome also condenses and pairs with its homologous partner. At this time, their sister chromatids typically undergo crossing over and genetic recombination.

prophase II Of meiosis, a brief stage after interkinesis during which each chromosome still consists of two chromatids.

protein Large organic compound composed of one or more chains of amino acids held together by peptide bonds. Proteins have unique sequences of different kinds of amino acids in their polypeptide chains; such sequences are the basis of a protein's three-dimensional structure and chemical behavior.

Protista The kingdom of protistans.

protistan (pro-TISS-tun) [Gk. *protistos*, primal, very first] Single-celled eukaryote.

proto-oncogene A gene sequence similar to an oncogene but that codes for a protein required in normal cell function; may trigger cancer, generally when specific mutations alter its structure or function.

proton Positively charged particle, one or more of which is present in the atomic nucleus.

protostome (PRO-toe-stome) [Gk. *proto*, first, + *stoma*, mouth] A bilateral animal in which the first indentation in the early embryo develops into the mouth. Includes mollusks, annelids, and anthropods.

protozoan A type of protistan, some predatory and others parasitic; so named because they may resemble the single-celled heterotrophs that presumably gave rise to animals.

proximal tubule Of a nephron, the tubular region that receives water and solutes filtered from the blood.

pulmonary circuit Blood circulation route leading to and from the lungs.

Punnett-square method A way to predict the possible outcome of a mating or an experimental cross in simple diagrammatic form.

purine Nucleotide base having a double ring structure. Adenine and guanine are examples.

pyrimidine (phi-RIM-ih-deen) Nucleotide base having a single ring structure. Cytosine and thymine are examples.

pyruvate (PIE-roo-vate) A compound with a backbone of three carbon atoms. Two pyruvate molecules are the end products of glycolysis.

r Designates net population growth rate; the birth and death rates are assumed to remain constant and so are combined into this one variable for population growth equations.

radial symmetry Body plan having four or more roughly equivalent parts arranged around a central axis.

radioisotope An unstable atom that has dissimilar numbers of protons and neutrons and that spontaneously decays (emits electrons and energy) to a new, stable atom that is not radioactive.

rain shadow A reduction in rainfall on the leeward side of high mountains, resulting in arid or semiarid conditions.

reabsorption Of urine formation, the diffusion or active transport of water and usable solutes out of a nephron and into capillaries leading back to the general circulation; regulated by ADH and aldosterone.

receptor Of cells, a molecule at the surface of the plasma membrane or in the cytoplasm that binds molecules present in the extracellular environment. The binding triggers changes in cellular activities. Of nervous systems, a sensory cell or cell part that may be activated by a specific stimulus.

receptor protein Protein that binds a signaling molecule such as a hormone, then triggers alterations in cell behavior or metabolism.

recessive allele [L. *recedere*, to recede] In heterozygotes, an allele whose expression is fully or partially masked by expression of its partner; fully expressed only in the homozygous recessive condition.

recognition protein Protein at cell surface recognized by cells of like type; helps guide the ordering of cells into tissues during development and functions in cell-to-cell interactions.

recombinant technology Procedures by which DNA (genes) from different species may be isolated, cut, spliced together, and the new recombinant molecules multiplied in quantity in a population of rapidly dividing cells such as bacteria.

red blood cell Erythrocyte; an oxygen-transporting cell in blood.

red marrow A substance in the spongy tissue of many bones that serves as a major site of blood cell formation.

reflex [L. *reflectere*, to bend back] A simple, stereotyped movement elicited directly by sensory stimulation.

reflex pathway [L. *reflectere*, to bend back] Type of neural pathway in which signals from sensory neurons directly stimulate or inhibit motor neurons, without intervention by interneurons.

refractory period Of neurons, the period following an action potential at a given patch of membrane when sodium gates are shut and potassium gates are open, so that the patch is insensitive to stimulation.

regulatory protein A protein that enhances or suppresses the rate at which a gene is transcribed.

releasing hormone A hypothalamic signaling molecule that stimulates or slows down secretion by target cells in the anterior lobe of the pituitary gland.

repressor protein Regulatory protein that provides negative control of gene activity by

preventing RNA polymerase from binding to DNA.

reproduction In biology, processes by which a new generation of cells or multicelled individuals is produced. Sexual reproduction requires meiosis, formation of gametes, and fertilization. Asexual reproduction refers to the production of new individuals by any mode that does not involve gametes.

reproduction, sexual Mode of reproduction that begins with meiosis, proceeds through gamete formation, and ends at fertilization.

reproductive isolating mechanism Any aspect of structure, functioning, or behavior that restricts gene flow between two populations.

reproductive isolation An absence of gene flow between populations.

reproductive success The survival and production of the offspring of an individual.

reptile A type of carnivorous vertebrate; its ancestors were the first vertebrates to escape dependency of standing water, largely by means of internal fertilization and amniote eggs. They include turtles, crocodiles, lizards and snakes, and tuataras.

resource partitioning A community pattern in which similar species generally share the same kind of resource in different ways, in different areas, or at different times.

respiration [L. *respirare*, to breathe] In most animals, the overall exchange of oxygen from the environment for carbon dioxide wastes from cells by way of circulating blood. *Compare* aerobic respiration.

respiratory surface The surface, such as a thin epithelial layer, that gases diffuse across to enter and leave the animal body.

respiratory system An organ system that functions in respiration.

resting membrane potential Of neurons and other excitable cells that are not being stimulated, the steady voltage difference across the plasma membrane.

restriction enzymes Class of bacterial enzymes that cut apart foreign DNA injected into them, as by viruses; also used in recombinant DNA technology.

reticular formation Of the vertebrate brainstem, a major network of interneurons that helps govern activity of the whole nervous system.

reverse transcriptase Viral enzyme required for reverse transcription of mRNA into DNA; used in recombinant DNA technology.

reverse transcription Assembly of DNA on a single-stranded mRNA molecule by viral enzymes.

RFLPs Restriction fragment length polymorphisms. Of DNA samples from different individuals, slight but unique differences in the banding pattern of fragments of the DNA that have been cut with restriction enzymes.

Rh blood typing A method of characterizing red blood cells on the basis of a protein that serves as a self-marker at their surface; Rh^+ signifies its presence and Rh^-, its absence.

rhizoid Rootlike absorptive structure of some fungi and nonvascular plants.

ribosomal RNA (rRNA) Type of RNA molecule that combines with proteins to form ribosomes, on which the polypeptide chains of proteins are assembled.

ribosome In all cells, the structure at which amino acids are strung together in specified sequence to form the polypeptide chains of proteins. An intact ribosome consists of two subunits, each composed of ribosomal RNA and protein molecules.

RNA Ribonucleic acid. A category of single-stranded nucleic acids that function in processes by which genetic instructions are used to build proteins.

rod cell A vertebrate photoreceptor sensitive to very dim light and that contributes to coarse perception of movement.

root hair Of vascular plants, an extension of a specialized root epidermal cell; root hairs collectively enhance the surface area available for absorbing water and solutes.

root nodule A localized swelling on the roots of certain legumes and other plants that contain symbiotic, nitrogen-fixing bacteria.

roots Descending parts of a vascular plant that absorb water and nutrients, anchor aboveground parts, and usually store food.

rotifer An invertebrate common in food webs of lakes and ponds.

roundworm A type of parasitic or scavenging invertebrate with a bilateral, cylindrical, cuticle-covered body, usually tapered at both ends. Some cause diseases in humans.

RuBP Ribulose biphosphate. A compound with a backbone of five carbon atoms that is required for carbon fixation in the Calvin-Benson cycle of photosynthesis.

S-shaped curve A curve, obtained when population size is plotted against time, that is characteristic of logistic growth.

sac fungus A type of fungus, usually multicelled, with spores that develop inside the cells of reproductive structures shaped like globes, flasks, or dishes. Yeasts (single-celled) are also in this group.

salination A salt buildup in soil as a result of evaporation, poor drainage, and often the importation of mineral salts in irrigation water.

salivary gland Any of the glands that secrete saliva, a fluid that initially mixes with food in the mouth and starts the breakdown of starch.

salt An ionic compound formed when an acid reacts with a base.

saltatory conduction In myelinated neurons, rapid, node-to-node hopping of action potentials.

saprobe Heterotroph that obtains its nutrients from nonliving organic matter. Most fungi are saprobes.

sarcomere (SAR-koe-meer) Of vertebrate muscles, the basic unit of contraction; a region of myosin and actin filaments organized in parallel between two Z lines of a myofibril inside a muscle cell.

sarcoplasmic reticulum (sar-koe-PLAZ-mik reh-TIK-you-lum) In muscle cells, a membrane system that takes up, stores, and releases the calcium ions required for cross-bridge formation in sarcomeres, hence for contraction.

Schwann cells Specialized neuroglial cells that grow around neuron axons, forming a myelin sheath.

sclerenchyma One of the simple tissues of flowering plants, generally with cells having thick, lignin-impregnated walls. It supports mature plant parts and often protects seeds.

sea squirt An invertebrate chordate with a leathery or jellylike tunic and bilateral, free-swimming larvae that look like tadpoles. Ancient sea squirts may have resembled animals that gave rise to vertebrates.

second law of thermodynamics Law stating that the spontaneous direction of energy flow is from organized (high-quality) to less organized (low-quality) forms. With each conversion, some energy is randomly dispersed in a form, usually heat, that is not as readily available to do work.

second messenger A molecule inside a cell that mediates and generally triggers amplified response to a hormone.

secondary immune response Rapid, prolonged response by white blood cells, memory cells especially, to a previously encountered antigen.

secondary sexual trait A trait that is associated with maleness or femaleness but that does not play a direct role in reproduction.

secretion Generally, the release of a substance for use by the organism producing it. (Not the same as *excretion*, the expulsion of excess or waste material.) Of kidneys, a regulated stage in urine formation, in which ions and other substances move from capillaries into nephrons.

sedimentary cycle A biogeochemical cycle without a gaseous phase; the element moves from land to the seafloor, then returns only through long-term geological uplifting.

seed Of gymnosperms and flowering plants, a fully mature ovule (contains the plant embryo), with its integuments forming the seed coat.

segmentation Of earthworms and many other animals, a series of body units that may be externally similar to or quite different from one another.

segregation, Mendelian principle of [L. *se-*, apart, + *grex*, herd] The principle that diploid organisms inherit a pair of genes for each trait (on a pair of homologous chromosomes) and that the two genes segregate during meiosis and end up in separate gametes.

selective gene expression Of multicelled organisms, activation or suppression of a fraction of the genes in unique ways in different cells, leading to pronounced differences in structure and function among different cell lineages.

selfish behavior A behavior by which an individual protects or increases its own chance of producing offspring, regardless of the consequences of the group to which it belongs.

selfish herd A simple society held together by reproductive self-interest.

semen (SEE-mun) [L. *serere*, to sow] Sperm-bearing fluid expelled from a penis during male orgasm.

semiconservative replication [Gk. *hēmi*, half, + L. *conservare*, to keep] Reproduction of a DNA molecule when a complementary strand forms on each of the unzipping strands of an existing DNA double helix, the outcome being two "half-old, half-new" molecules.

senescence (sen-ESS-cents) [L. *senescere*, to grow old] Sum total of processes leading to the natural death of an organism or some of its parts.

sensation The conscious awareness of a stimulus.

sensory neuron Any of the nerve cells that act as sensory receptors, detecting specific stimuli (such as light energy) and relaying signals to the brain and spinal cord.

sensory system The "front door" of a nervous system; that portion of a nervous system that receives and sends on signals of specific changes in the external and internal environments.

sessile animal Animal that remains attached to a substrate during some stage (often the adult) of its life cycle.

sex chromosome Of most animals and some plants, a chromosome whose presence determines a new individual's gender. *Compare* autosomes.

sexual dimorphism Phenotypic differences between males and females of a species.

sexual reproduction Production of offspring from the union of gametes from two parents, by way of meiosis, gamete formation, and fertilization.

sexual selection A microevolutionary process; natural selection favoring a trait that gives the individual a competitive edge in reproductive success.

shifting cultivation The cutting and burning of trees, followed by tilling of ashes into the soil; once called slash-and-burn agriculture.

shoots The above-ground parts of vascular plants.

sieve tube member Of flowering plants, a cellular component of the interconnecting conducting tubes in phloem.

sink region In a vascular plant, any region using or stockpiling organic compounds for growth and development.

sister chromatids Of a duplicated chromosome, two DNA molecules (and associated proteins) that remain attached at their centromere only during nuclear division. Each ends up in a separate daughter nucleus.

skeletal muscle In vertebrates, an organ that contains hundreds to many thousands of muscle cells, arranged in bundles that are surrounded by connective tissue. The connective tissue extends beyond the muscle (as tendons that attach it to bone).

sliding filament model Model of muscle contraction, in which myosin filaments physically slide along and pull two sets of actin filaments toward the center of the sarcomere, which shortens. The sliding requires ATP energy and cross-bridge formation between the actin and myosin.

slime mold A type of heterotrophic protistan with a life cycle that includes free-living cells that at some point congregate and differentiate into spore-bearing structures.

small intestine Of vertebrates, the portion of the digestive system where digestion is completed and most nutrients absorbed.

smog, industrial Gray-colored air pollution that predominates in industrialized cities with cold, wet winters.

smog, photochemical Form of brown, smelly air pollution occurring in large cities with warm climates.

social behavior Cooperative, interdependent relationships among animals of the same species.

social parasite Animal that depends on the social behavior of another species to gain food, care for young, or some other factor to complete its life cycle.

sodium-potassium pump A transport protein spanning the lipid bilayer of the plasma membrane. When activated by ATP, its shape changes and it selectively transports sodium ions out of the cell and potassium ions in.

solute (SOL-yoot) [L. *solvere*, to loosen] Any substance dissolved in a solution. In water, this means spheres of hydration surround the charged parts of individual ions or molecules and keep them dispersed.

solvent Fluid in which one or more substances is dissolved.

somatic cell (so-MAT-ik) [Gk. *somā*, body] Of animals, any cell that is not a germ cell (which gives rise to gametes).

somatic nervous system Those nerves leading from the central nervous system to skeletal muscles.

sound system Of birds, the brain regions that govern muscles of the vocal organ.

source region Of vascular plants, any of the sites of photosynthesis.

speciation (spee-cee-AY-shun) The evolutionary process by which species originate. One speciation route starts with divergence of two reproductively isolated populations of a species. They become separate species when accumulated differences in allele frequencies prevent them from interbreeding successfully under natural conditions. Speciation also may be instantaneous (by way of polyploidy, especially among self-fertilizing plants).

species (SPEE-sheez) [L. *species*, a kind] Of sexually reproducing organisms, a unit consisting of one or more populations of individuals that can interbreed under natural conditions to produce fertile offspring that are reproductively isolated from other such units.

sperm [Gk. *sperma*, seed] A type of mature male gamete.

spermatogenesis (sperm-AT-oh-JEN-ih-sis) Formation of a mature sperm from a germ cell.

sphere of hydration Through positive or negative interactions, a clustering of water molecules around the individual molecules of a substance placed in water. *Compare* solute.

sphincter (SFINK-tur) Ring of muscle between regions of a tubelike system (as between the stomach and small intestine).

spinal cord Of central nervous systems, the portion threading through a canal inside the vertebral column and providing direct reflex connections between sensory and motor neurons as well as communication lines to and from the brain.

spindle apparatus A type of bipolar structure that forms during mitosis or meiosis and that moves the chromosomes. It consists of two sets of microtubules that extend from the opposite poles and that overlap at the spindle's equator.

spleen One of the lymphoid organs; it is a filtering station for blood, a reservoir of red blood cells, and a reservoir of macrophages.

sponge An invertebrate having a body with no symmetry and no organs; a framework of glassy needles and other structures imparts shape to it. Distinctive for its food-gathering, flagellated collar cells.

sporangium (spore-AN-gee-um), plural **sporangia** [Gk. *spora*, seed] The protective tissue layer that surrounds haploid spores in a sporophyte.

spore Of land plants, a type of resistant cell, often walled, that forms between the time of meiosis and fertilization. It germinates and develops into a gametophyte, the actual gamete-producing body. Of most fungi, a walled, resistant cell or multicelled structure, produced by mitosis or meiosis, that can germinate and give rise to a new mycelium.

sporophyte [Gk. *phyton*, plant] Of plant life cycles, a vegetative body that grows (by mitosis) from a zygote and that produces the spore-bearing structures.

sporozoan One of four categories of protozoans; a parasite that produces sporelike infectious agents called sporozoites. Some cause serious diseases in humans.

spring overturn Of certain lakes, the movement of dissolved oxygen from the surface layer to the depths and movement of nutrients from bottom sediments to the surface.

stabilizing selection Of a population, a persistence over time of the alleles responsible for the most common phenotypes.

stamen (STAY-mun) Of flowering plants, a male reproductive structure; commonly consists of pollen-bearing structures (anthers) on single stalks (filaments).

start codon Of protein synthesis, a base triplet in a strand of mRNA that serves as the start signal for mRNA translation.

stem cell Of animals, one of the unspecialized cells that replace themselves by ongoing mitotic divisions; portions of their daughter cells also divide and differentiate into specialized cells.

steroid (STAIR-oid) A lipid with a backbone of four carbon rings and with no fatty acid tails. Steroids differ in their functional groups. Different types have roles in metabolism, intercellular communication, and (in animals) cell membranes.

steroid hormone A type of lipid-soluble hormone, synthesized from cholesterol, that diffuses directly across the lipid bilayer of a target cell's plasma membrane and that binds with a receptor inside that cell.

stigma Of many flowering plants, the sticky or hairy surface tissue on the upper portion of the ovary that captures pollen grains and favors their germination.

stimulus [L. *stimulus*, goad] A specific change in the environment, such as a variation in light, heat, or mechanical pressure, that the body can detect through sensory receptors.

stoma (STOW-muh), plural **stomata** [Gk. *stoma*, mouth] A controllable gap between two guard cells in stems and leaves; any of the small passageways across the epidermis through which carbon dioxide moves into the plant and water vapor moves out.

stomach A muscular, stretchable sac that receives ingested food; of vertebrates, an organ between the esophagus and intestine in which considerable protein digestion occurs.

stop codon Of protein synthesis, a base triplet in a strand of mRNA that serves as the stop signal for translation, so that no more amino acids are added to the polypeptide chain.

stream A flowing-water ecosystem that starts out as a freshwater spring or seep.

stroma [Gk. *strōma*, bed] Of chloroplasts, the semifluid interior between the thylakoid membrane system and the two outer membranes; the zone where sucrose, starch, and other end products of photosynthesis are assembled.

stromatolite Of shallow seas, layered structures formed from sediments and large mats of the slowly accumulated remains of photosynthetic populations.

substrate A reactant or precursor molecule for a metabolic reaction; a specific molecule or molecules that an enzyme can chemically recognize, bind briefly to itself, and modify in a specific way.

substrate-level phosphorylation The direct, enzyme-mediated transfer of a phosphate group from the substrate of a reaction to another molecule. An example is the transfer of phosphate from an intermediate of glycolysis to ADP, forming ATP.

succession, primary (suk-SESH-un) [L. *succedere*, to follow after] Orderly changes from the time pioneer species colonize a barren habitat through replacements by various species until the climax community, when the composition of species remains steady under prevailing conditions.

succession, secondary Orderly changes in a community or patch of habitat toward the climax state after having been disturbed, as by fire.

surface-to-volume ratio A mathematical relationship in which volume increases with the cube of the diameter, but surface area increases only with the square. Of growing cells, the volume of cytoplasm increases more rapidly than the surface area of the plasma membrane that must service the cytoplasm. Because of this constraint, cells generally remain small or elongated, or have elaborate membrane foldings.

survivorship curve A plot of the age-specific survival of a group of individuals in a given environment, from the time of their birth until the last one dies.

symbiosis (sim-by-OH-sis) [Gk. *sym*, together, + *bios*, life, mode of life] A form of mutualism in which organisms of different species cannot grow and reproduce unless they spend their entire lives together in intimate interdependency. A mycorrhiza is an example.

sympathetic nerve Of the autonomic nervous system, any of the nerves generally concerned with increasing overall body activities during times of heightened awareness, excitement, or danger; also work continually in opposition with parasympathetic nerves to bring about minor adjustments in internal organs.

sympatric speciation [Gk. *sym*, together, + *patria*, native land] Speciation that follows after ecological, behavioral, or genetic barriers arise within the boundaries of a single population. This can happen instantaneously, as when polyploidy arises in a type of flowering plant that can self-fertilize or reproduce asexually.

synaptic integration (sin-AP-tik) The moment-by-moment combining of excitatory and inhibitory signals arriving at a trigger zone of a neuron.

systematics Branch of biology that deals with patterns of diversity among organisms in an evolutionary context; its three approaches include taxonomy, phylogenetic reconstruction, and classification.

systemic circuit (sis-TEM-ik) Circulation route in which oxygen-enriched blood flows from the lungs to the left half of the heart, through the rest of the body (where it gives up oxygen and takes on carbon dioxide), then back to the right side of the heart.

T lymphocyte A white blood cell with roles in immune responses.

tactile signal A physical touching that carries social significance.

taproot system A primary root and its lateral branchings.

target cell Of hormones and other signaling molecules, any cell having receptors to which they can bind.

taxonomy (tax-ON-uh-mee) Approach in biological systematics that involves identifying organisms and assigning names to them.

telophase (TEE-low-faze) Of mitosis, the final stage when chromosomes decondense into threadlike structures and two daughter nuclei form. Of meiosis I, the stage when one of each pair of homologous chromosomes has arrived at one or the other end of the spindle pole. At telophase II, chromosomes decondense and four daughter nuclei form.

telophase II Of meiosis, final stage when four daughter nuclei form.

temperate pathway A viral infection that enters a latent period; the host is not killed outright.

tendon A cord or strap of dense connective tissue that attaches muscle to bones.

territory An area that one or more individuals defend against competitors.

test An attempt to produce actual observations that match predicted or expected observations.

testcross Experimental cross to reveal whether an organism is homozygous dominant or heterozygous for a trait. The organism showing dominance is crossed to an individual known to be homozygous recessive for the same trait.

testis, plural **testes** Male gonad; primary reproductive organ in which male gametes and sex hormones are produced.

testosterone (tess-TOSS-tuh-rown) In male mammals, a major sex hormone that helps control male reproductive functions.

tetanus Of muscles, a large contraction in which repeated stimulation of a motor unit causes muscle twitches to mechanically run together. In a disease by the same name, toxins prevent muscle relaxation.

theory A testable explanation of a broad range of related phenomena. In modern science, only explanations that have been extensively tested and can be relied upon with a very high degree of confidence are accorded the status of theory.

thermal inversion Situation in which a layer of dense, cool air becomes trapped beneath a layer of warm air; can cause air pollutants to accumulate to dangerous levels close to the ground.

thermophile A type of archaebacterium that lives in hot springs, highly acidic soils, and near hydrothermal vents.

thermoreceptor Sensory cell that can detect radiant energy associated with temperature.

thigmotropism (thig-MOTE-ruh-pizm) [Gk. *thigm,* touch] Of vascular plants, growth orientation in response to physical contact with a solid object, as when a vine curls around a fencepost.

threshold Of neurons and other excitable cells, a certain minimum amount by which the voltage difference across the plasma membrane must change to produce an action potential.

thylakoid membrane system Of chloroplasts, an internal membrane system commonly folded into flattened channels and disks (*grana*) and containing light-absorbing pigments and enzymes used in the formation of ATP, NADPH, or both during photosynthesis.

thymine Nitrogen-containing base in some nucleotides.

thymus gland A lymphoid organ with endocrine functions; lymphocytes of the immune system multiply, differentiate, and mature in its tissues, and its hormone secretions affect their functions.

thyroid gland Of endocrine systems, a gland that produces hormones that affect overall metabolic rates, growth, and development.

tissue Of multicelled organisms, a group of cells and intercellular substances that function together in one or more specialized tasks.

tonicity The relative concentrations of solutes in two fluids, such as inside and outside a cell. When solute concentrations are isotonic (equal in both fluids), water shows no net osmotic movement in either direction. When one fluid is hypotonic (has less solutes than the other), the other is hypertonic (has more

solutes) and is the direction in which water tends to move.

tooth Of the mouth of various animals, one of the hardened appendages used to secure or mechanically pummel food; sometimes used in defense.

tracer A radioisotope used to label a substance so that its pathway or destination in a cell, organism, ecosystem, or some other system can be tracked, as by scintillation counters that detect its emissions.

trachea (TRAY-kee-uh), plural **tracheae** An air-conducting tube that functions in respiration; of land vertebrates, the windpipe, which carries air between the larynx and bronchi.

tracheal respiration Of insects, spiders, and some other animals, a respiratory system consisting of finely branching tracheae that extend from openings in the integument and that dead-end in body tissues.

tracheid (TRAY-kid) Of flowering plants, one of two types of cells in xylem that conduct water and dissolved minerals.

transcript-processing controls Of eukaryotic cells, controls that govern modification of new mRNA molecules into mature transcripts before shipment from the nucleus.

transcription [L. *trans,* across, + *scribere,* to write] Of protein synthesis, the assembly of an RNA strand on one of the two strands of a DNA double helix; the base sequence of the resulting transcript is complementary to the DNA region on which it was assembled.

transcriptional controls Of eukaryotic cells, controls influencing when and to what degree a particular gene will be transcribed.

transfer RNA (tRNA) Of protein synthesis, any of the type of RNA molecules that bind and deliver specific amino acids to ribosomes *and* pair with mRNA code words for those amino acids.

translation Of protein synthesis, the conversion of the coded sequence of information in mRNA into a particular sequence of amino acids to form a polypeptide chain; depends on interactions of rRNA, tRNA, and mRNA.

translational controls Of eukaryotic cells, controls governing the rates at which mRNA transcripts that reach the cytoplasm will be translated into polypeptide chains at ribosomes.

translocation Of cells, a change in a chromosome's structure following the insertion of part of a nonhomologous chromosome into it. Of vascular plants, conduction of organic compounds through the plant body by way of the phloem.

transpiration Evaporative water loss from stems and leaves.

transport control Of eukaryotic cells, controls governing when mature mRNA transcripts are shipped from the nucleus into the cytoplasm.

transposable element DNA element that can spontaneously "jump" to new locations in the same DNA molecule or a different one. Such elements often inactivate the genes into which they become inserted and give rise to observable changes in phenotype.

trisomy (TRY-so-mee) Of diploid cells, the abnormal presence of three of one type of chromosome.

trophic level (TROE-fik) [Gk. *trophos,* feeder] All the organisms in an ecosystem that are the same number of transfer steps away from the energy input into the system.

tropical rain forest A type of biome where rainfall is regular and heavy, the annual mean temperature is 25°C, and humidity is 80 percent or more; characterized by great biodiversity.

tropism (TROE-prizm) Of vascular plants, a growth response to an environmental factor, such as growth toward light.

true-breeding Of sexually reproducing organisms, a lineage in which the offspring of successive generations are just like the parents in one or more traits.

tumor A tissue mass composed of cells that are dividing at an abnormally high rate.

turgor pressure (TUR-gore) [L. *turgere,* to swell] Internal pressure applied to a cell wall when water moves by osmosis into the cell.

uniformitarianism The theory that existing geologic features are an outcome of a long history of gradual changes, interrupted now and then by huge earthquakes and other catastrophic events.

upwelling An upward movement of deep, nutrient-rich water along coasts to replace surface waters that winds move away from shore.

uracil (YUR-uh-sill) Nitrogen-containing base found in RNA molecules; can base-pair with adenine.

ureter A tubular channel for urine flow between the kidney and urinary bladder.

urethra A tubular channel for urine flow between the urinary bladder and an opening at the body surface.

urinary bladder A distensible sac in which urine is temporarily stored before being excreted.

urinary excretion A mechanism by which excess water and solutes are removed by way of a urinary system.

urinary system An organ system that adjusts the volume and composition of blood, and so helps maintain extracellular fluid.

urine Fluid formed by filtration, reabsorption, and secretion in kidneys; consists of wastes, excess water, and solutes.

uterus (YOU-tur-us) [L. *uterus,* womb] Chamber in which the developing embryo is contained and nurtured during pregnancy.

vaccine An antigen-containing preparation, swallowed or injected, that increases immunity to certain diseases. It induces formation of huge armies of effector and memory B and T cell populations.

vagina Part of a female reproductive system that receives sperm, forms part of the birth canal, and channels menstrual flow to the exterior.

variable Of a scientific experiment, the only factor that is not exactly the same in the experimental group as it is in the control group.

vascular bundle Of vascular plants, the arrangement of primary xylem and phloem into multistranded, sheathed cords that thread lengthwise through the ground tissue system.

vascular cambium Of vascular plants, a lateral meristem that increases stem or root diameter.

vascular cylinder Of plant roots, the arrangement of vascular tissues as a central cylinder.

vascular plant Plant having tissues that transport water and solutes through well-developed roots, stems, and leaves.

vascular tissue system Xylem and phloem; the conducting tissues that distribute water and solutes through the body of vascular plants.

vein Of the circulatory system, any of the large-diameter vessels that lead back to the heart; of leaves, one of the vascular bundles that thread through photosynthetic tissues.

ventricle (VEN-tri-kuhl) Of the vertebrate heart, one of two chambers from which blood is pumped out. *Compare* atrium.

venule A small blood vessel that accepts blood from capillaries and delivers it to a vein; also overlaps capillaries somewhat in function.

vernalization Of flowering plants, stimulation of flowering by exposure to low temperatures.

vertebra, plural **vertebrae** Of vertebrate animals, one of a series of hard bones arranged with intervertebral disks into a backbone.

vertebrate Animal having a backbone of bony segments, the vertebrae.

vesicle (VESS-ih-kul) [L. *vesicula*, little bladder] Within the cytoplasm of cells, one of a variety of small membrane-bound sacs that function in the transport, storage, or digestion of substances or in some other activity.

vessel member One of the cells of xylem, dead at maturity, the walls of which form the water-conducting pipelines.

villus (VIL-us), plural **villi** Any of several types of absorptive structures projecting from the free surface of an epithelium.

viroid An infectious nucleic acid that has no protein coat; a tiny rod or circle of single-stranded RNA.

virus A noncellular infectious agent, consisting of DNA or RNA and a protein coat; can replicate only after its genetic material enters a host cell and subverts its metabolic machinery.

vision Precise light focusing onto a layer of photoreceptive cells that is dense enough to sample details concerning a given light stimulus, followed by image formation in the brain.

visual signal An observable action or cue that functions as a communication signal.

vitamin Any of more than a dozen organic substances that animals require in small amounts for normal cell metabolism but generally cannot synthesize for themselves.

vocal cord One of the thickened, muscular folds of the larynx that help produce sound waves for speech.

water mold A type of saprobic or parasitic fungus that lives in fresh water or moist soil.

water potential The sum of two opposing forces (osmosis and turgor pressure) that can cause the directional movement of water into or out of a walled cell.

water table The upper limit at which the ground in a specified region is fully saturated with water.

watershed Any specified region in which all precipitation drains into a single stream or river.

wax A type of lipid with long-chain fatty acid tails that help form protective, lubricating, or water-repellent coatings.

white blood cell Leukocyte; of vertebrates, any of the macrophages, eosinophils, neutrophils, and other cells which, together with their products, comprise the immune system.

white matter Of spinal cords, major nerve tracts so named because of the glistening myelin sheaths of their axons.

wild-type allele Of a population, the allele that occurs normally or with greatest frequency at a given gene locus.

wing Of birds, a forelimb of feathers, powerful muscles, and lightweight bones that functions in flight. Of insects, a structure that develops as a lateral fold of the exoskeleton and functions in flight.

X chromosome Of humans, a sex chromosome with genes that cause an embryo to develop into a female, provided that it inherits a pair of these.

X-linked gene Any gene on an X chromosome.

X-linked recessive inheritance Recessive condition in which the responsible, mutated gene occurs on the X chromosome.

xylem (ZYE-lum) [Gk. *xylon*, wood] Of vascular plants, a tissue that transports water and solutes through the plant body.

Y chromosome Of humans, a sex chromosome with genes that cause the embryo that inherited it to develop into a male.

Y-linked gene Any gene on a Y chromosome.

yellow marrow A fatty tissue in the cavities of most mature bones that produces red blood cells when blood loss from the body is severe.

yolk sac Of land vertebrates, one of four extraembryonic membranes. In most shelled eggs, it holds nutritive yolk. In humans, part becomes a site of blood cell formation and some of its cells give rise to the forerunners of gametes.

zero population growth A population for which the number of births is balanced by the number of deaths over a specified period, assuming immigration and emigration also are balanced.

zooplankton A freshwater or marine community of floating or weakly swimming heterotrophs, mostly microscopic, such as rotifers and copepods.

zygospore-forming fungus A type of fungus for which a thick spore wall forms around the zygote; this resting spore germinates and gives rise to stalked, spore-bearing structures.

zygote (ZYE-goat) The first cell of a new individual, formed by the fusion of a sperm nucleus with the nucleus of an egg (fertilization).

CREDITS AND ACKNOWLEDGMENTS

Front Matter

Page i Thomas D. Mangelsen/Images of Nature / **Pages vi–vii** Stock Imagery / **Pages viii–ix** James M. Bell/Photo Researchers / **Pages x–xi** S. Stammers/SPL/Photo Researchers / **Pages xii–xiii** © 1990 Arthur M. Greene / **Pages xiv–xv** Bonnie Rausch/Photo Researchers / **Pages xvi–xvii** Lennart Nilsson from *A Child Is Born*, © 1966, 1977 Dell Publishing Company, Inc. / **Pages xviii–xix** Thomas D. Mangelsen/Images of Nature / **Pages xx–xxi** Jim Doran

Chapter 21

21.1 (a) © David Malin, Anglo-Australian Observatory; (b) NASA/Space Telescope Science Institute / **21.2** Painting by William K. Hartmann / **21.3** (a) Painting by Chesley Bonestell / **21.4** Art by Precision Graphics / **21.5** (a) Sidney W. Fox; (b) W. Hargreaves and D. Deamer / **21.7** Art by Leonard Morgan / **21.8** Maps by Lloyd K. Townsend after Krohn and Sündermann, 1983 / **21.9** (a) Stanley W. Awramik; (b) M. R. Walter / **Page 331** Art by Raychel Ciemma / **Pages 332–333** Micrograph Robert K. Trench; art by Precision Graphics / **21.10** (a), (b) Neville Pledge/South Australian Museum; (c), (d) Chip Clark / **21.11** Patricia G. Gensel / **21.12** From *Evolution of Life*, Linda Gamlin and Gail Vines (Eds.), Oxford University Press, 1987; art by D. & V. Hennings; photograph Rod Salm/Planet Earth Pictures / **Page 335** Maps by Lloyd K. Townsend after Krohn and Sündermann / **21.13** (above) Painting by Megan Rohn courtesy of David Dilcher; (below) art by Precision Graphics / **Page 337** (a) NASA; (b) NASA Galileo Imaging Team / **Pages 338–339** (c) left) Art by Raychel Ciemma; (right) © John Gurche 1989 / **21.14** (a), (b) Field Museum of Natural History, Chicago, and the artist Charles R. Knight (Neg. No. CK46T and CK8T) / **21.15** Data from J. J. Sepkoski, Jr., *Paleobiology*, 7(1):36–53 and J. J. Sepkoski, Jr. and M. L. Hulver in Valentine, ed., *Phanerozoic Diversity Patterns: Profiles in Macroevolution*, Princeton University Press, 1985 / **Page 341** Map by Lloyd K. Townsend after Ziegler, Scotese, and Barrett, 1983 / **21.16** (left) Maps by Lloyd K. Townsend after A. M. Ziegler, C. R. Scotese, and S. F. Barrett, "Mesozoic and Cenozoic Paleogeographic Maps" and J. Krohn and J. Sündermann, "Paleotides Before the Permian" in F. Brosche and J. Sündermann (Eds.), *Tidal Friction and the Earth's Rotation II*, Springer-Verlag, 1983

Chapter 22

22.1 (a–c) Tony Brain and David Parker/SPL/Photo Researchers; (d) Lee D. Simon/Photo Researchers / **22.2** Art by L. Calver / **22.3** L. J. LeBeau, University of Illinois Hospital/BPS / **22.4** (a) Stanley Flegler/Visuals Unlimited; (b) CNRI/SPL/Photo Researchers / **22.5** Art by D. & V. Hennings / **22.7** (a) © 1994 Barrie Rokeach; (b) R. Robinson/Visuals Unlimited / **22.8** (a) John D. Cunningham/Visuals Unlimited; (b) Tony Brain/SPL/Photo Researchers; (c) P. W. Johnson and J. McN. Sieburth, University of Rhode Island/BPS / **22.9** Stanley W. Watson, *International Journal of Systematic Bacteriology*, 21:254–270, 1971 / **22.10** J. J. Cardamone, Jr./BPS / **22.11** T. J. Beveridge, University of Guelph/BPS / **22.12** (above) Centers for Disease Control; (below) Edward S. Ross / **22.13** Richard Blakemore / **22.14** Hans Reichenbach, Gesellschaft für Biotechnologische Forschung, Braunschweig, Germany / **22.15** Art by Nadine Sokol / **22.16** (a) George Musil/Visuals Unlimited; (b) K. G. Murti/Visuals Unlimited; (c), (d) Kenneth M. Corbett / **22.17, 22.18** Art by Palay/Beaubois and Precision Graphics

Chapter 23

23.1 (a) Ronald W. Hoham, Dept. of Biology, Colgate University; (b) Steven C. Wilson/Entheos; (c) Edward S. Ross; (d) Gary W. Grimes and Steven L'Hernault / **23.2** M. S. Fuller, *Zoosporic Fungi in Teaching and Research*, M. S. Fuller and A. Jaworski (eds.), 1987, Southeastern Publishing Company, Athens, GA / **23.3** (a) Heather Angel; (b) W. Merrill / **23.4** (a) Art by Leonard Morgan; (b) M. Claviez, G. Gerish, and R. Guggenheim; (c) London Scientific Films; (d–f) Carolina Biological Supply Company; (g) Photograph courtesy Robert R. Kay from R. R. Kay et al. *Development*, 1989 Supplement, pp. 81–90, © The Company of Biologists Ltd. 1989 / **23.5** (a) M. Abbey/Visuals Unlimited; (b, d) From V. & J. Pearse and M. & R. Buchsbaum, *Living Invertebrates*, The Boxwood Press, 1987. Used by permission. (c) John Clegg/Ardea, London; (e) T. E. Adams/Visuals Unlimited; (f) Manfred Kage/Bruce Coleman Ltd. / **23.6** (a) Art by Raychel Ciemma redrawn from V. & J. Pearse and M. & R. Buchsbaum, *Living Invertebrates*, The Boxwood Press, 1987. Used by permission. (b) Gary W. Grimes and Steven L'Hernault; photograph by Ralph Buchsbaum and sketch from V. & J. Pearse and M. & R. Buchsbaum, *Living Invertebrates*, The Boxwood Press, 1987. Used by permission. / **23.7** (a) John D. Cunningham/Visuals Unlimited; (b) Jerome Paulin/Visuals Unlimited; (c) David M. Phillips/Visuals Unlimited **Page 369** Art by Leonard Morgan; micrograph Steven L'Hernault / **23.8** (a) P. L. Walne and J. H. Arnott, *Planta*, 77:325–354, 1967; (b) T. E. Adams/Visuals Unlimited; art by Palay/Beaubois / **23.9** (a), (b), (d) Ronald W. Hoham, Dept. of Biology, Colgate University; (c) Jan Hinsch/SPL/Photo Researchers / **23.10** (a) Florida Department of Environmental Protection, Florida Marine Research Institute, St. Petersburg; (b) C. C. Lockwood / **23.11** (a) D. P. Wilson/Eric & David Hosking; (b) Douglas Faulkner/Sally Faulkner Collection / **23.12** (a) J. R. Waaland, University of Washington/BPS; (b) Dennis Brokaw; (c) from Tom Garrison, *Oceanography: An Invitation to Marine Science*, Wadsworth, 1993 / **23.13** (a), (c) Hervé Chaumeton/Agence Nature; (b) Alex Kerstitch/Tom Stack & Associates; (d) Manfred Kage/Peter Arnold, Inc.; (e) Ronald W. Hoham, Dept. of Biology, Colgate University / **23.14** Photograph D. J. Patterson/Seaphot Limited: Planet Earth Pictures; art by Raychel Ciemma / **23.15** Carolina Biological Supply Company / **Page 376** Richard W. Greene

Chapter 24

24.1 Robert C. Simpson/Nature Stock / **24.2** Micrograph G. T. Cole, University of Texas, Austin/BPS / **24.4** Photographs (above) John D. Cunningham/Visuals Unlimited; (below) David M. Phillips/Visuals Unlimited; art by Raychel Ciemma / **24.5** John Hodgin / **24.6** (a) After T. Rost et al., *Botany*, Wiley, 1979; (b), (c) Robert C. Simpson/Nature Stock / **Page 383** Eric Crichton/Bruce Coleman Ltd. / **24.7** (a) Victor Duran; (b), (c) Robert C. Simpson/Nature Stock; (d), (e) Thomas J. Duffy; (f) Jane Burton/Bruce Coleman Ltd. / **24.8** Photographs Martyn Ainsworth from A. D. M. Rayner, *New Scientist*, November 19, 1988; art by Raychel Ciemma / **24.9** (a) Mark Mattock/Planet Earth Pictures; (b) Edward S. Ross / **24.10** After Raven, Evert, and Eichhorn, *Biology of Plants*, Fourth edition, Worth Publishers, New York, 1986 / **24.11** © 1990 Gary Braasch / **24.12** F. B. Reeves / **24.13** (a) G. T. Cole, University of Texas, Austin/BPS; (b) N. Allin and G. L. Barron; (c) G. L. Barron, University of Guelph

Chapter 25

25.1 (a) Pat & Tom Leeson/Photo Researchers; (b) Edward S. Ross / **25.4** Photograph Jane Burton/Bruce Coleman Ltd.; art by Raychel Ciemma / **25.5** (a) Roger K. Burnard; (b) John D. Cunningham/Visuals Unlimited / **25.6** (a), (b) Kingsley R. Stern; (c) John D. Cunningham/Visuals Unlimited / **Page 397** (a) Field Museum of Natural History, Chicago; (Neg. #7500C); (b) Brian Parker/Tom Stack & Associates / **25.7** (b) Kingsley R. Stern; (c) Edward S. Ross / **25.8** (a) Edward S. Ross; (b) W. H. Hodge/Kratz/ZEFA / **25.9** (a) Art by Raychel Ciemma; photograph A. & E. Bomford/Ardea, London; (b) Lee Casebere; (c) Jean Paul Ferrero/Ardea, London / **25.10** Photograph Edward S. Ross; art by Raychel Ciemma / **25.11** (a), (b) © 1989, 1991 Clinton Webb; (c) © 1994 Robert Glenn Ketchum; (d) © 1993 Trygve Steen; (e) David Hiser, Photographers/Aspen, Inc.; (f) Terry Livingstone / **25.12** (a) Ed Reschke; (b) Edward S. Ross; (c) John H. Gerard; (d) Kingsley R. Stern; (e) Edward S. Ross; (inset) F. J. Odendaal, Duke University/BPS / **25.13** (a) Hans Reinhard/Bruce Coleman Ltd.; (b) Heather Angel; (c) Peter F. Zika/Visuals Unlimited; (d) L. Mellichamp/Visuals Unlimited; (e), (f) Edward S. Ross; (g) Dick Davis/Photo Researchers / **25.14** Art by Raychel Ciemma / **25.15** Photograph M. P. L. Fogden/Ardea, London; art by Jennifer Wardrip

Chapter 26

26.1 (a) Courtesy of Department of Library Services, American Museum of Natural History (Neg. # K10273); (b) Jim Stewart/Scripps Institution of Oceanography; (c) Chip Clark / **26.2, 26.3** Art by D. & V. Hennings / **Page 412** (above) Laszlo Meszoly in L. Margulis, *Early Life*, Jones and Bartlett Publishers, / Inc., Boston, © 1982 / **26.5** (a) David C. Haas/Tom Stack & Associates; (b) art by Raychel Ciemma; (d) Marty Snyderman/Planet Earth Pictures; (e) Bruce Hall / **26.6** (a) Frieder Sauer/Bruce Coleman Ltd.; (b) Kim Taylor/Bruce Coleman Ltd. / **26.7** (a) Art by Raychel Ciemma; photograph Francois Gohier/Photo Researchers; (c) Douglas Faulkner/Sally Faulkner Collection; (d) F. S. Westmorland/Tom Stack & Associates / **26.8** (a) Christian Della-Corte; (b) Douglas Faulkner/Photo Researchers; (c) Bill Wood/Seaphot Limited: Planet Earth Pictures; (d) Walter Deas/Seaphot Limited: Planet Earth Pictures / **26.9** (a) Andrew Mounter/Seaphot Limited: Planet Earth Pictures; (b) art by Precision Graphics after T. Storer et al., *General Zoology*, Sixth edition, © 1979 McGraw-Hill / **26.10** (a) Kathie Atkinson/Oxford Scientific Films; (b) Larry Madin/Planet Earth Pictures / **26.11** Photograph Kim Taylor/Bruce Coleman Ltd.; art by Raychel Ciemma / **26.12** (a) Robert & Linda Mitchell; (b) Cath Ellis, University of Hull/SPL/Photo Researchers / **26.13** Art by Raychel Ciemma / **Page 420** (a) Photograph Robert L. Calentine; art by Raychel Ciemma / **Page 421** (b) Art by Nadine Sokol; photograph Carolina Biological Supply Company; (c) Lorus J. and Margery Milne; (d) Dianora Niccolini / **26.14** Kjell B. Sandved / **26.15** Photograph J. Solliday/BPS; art by Raychel Ciemma / **26.16** Art by Palay/Beaubois / **26.17** (a) Gary Head; (c) Kjell B. Sandved / **26.18** (a) Hervé Chaumeton/Agence Nature; (b) Alex Kerstitch / **26.19** Jeff Foott/Tom Stack & Associates / **26.20** Art by Raychel Ciemma / **26.21** Hervé Chaumeton/Agence Nature / **26.22** Photograph Douglas Faulkner/Sally Faulkner Collection; art by Raychel Ciemma / **26.23** J. Grossauer/ZEFA / **26.24** © Cabisco/Visuals Unlimited / **26.25** Art by Raychel Ciemma / **26.26** (a) Hervé Chaumeton/Agence Nature; (b) Jon Kenfield/Bruce Coleman Ltd. / **26.27** J. A. L. Cooke/Oxford Scientific Films / **26.28** From Eugene N. Kozloff, *Invertebrates*, copyright © 1990 by Saunders College Publishing. Reproduced by permission of the publisher. / **26.29** C. B. & D. W. Frith/Bruce Coleman Ltd. / **26.30** Jane Burton/Bruce Coleman Ltd. / **26.31** (a) (above) Angelo Giampiccolo/FPG; (below) Jane Burton/Bruce Coleman Ltd. (b) P. J. Bryant, University of California, Irvine/BPS; (c) John H. Gerard; (d) Ken Lucas/Seaphot Limited: Planet Earth Pictures / **26.32** (a) Franz Lanting/Bruce Coleman Ltd.; (b) Hervé Chaumeton; (c) Agence Nature; (d) Fred Bavendam/Peter Arnold, Inc. / **26.33** (a) Z. Leszczynski/Animals Animals; (b) Steve Martin/Tom Stack & Associates / **26.34** Art by D. & V. Hennings / **26.35** (a) David Maitland/Seaphot Limited: Planet Earth Pictures; (b), (d) Edward S. Ross; (c) Kenneth Lorenzen; (e) Robert & Linda Mitchell; (f) Ralph A. Reinhold/FPG; (g–j), (l) Edward S. Ross; (k) C. P. Hickman, Jr. / **26.36** (a) John Mason/Ardea, London; (b) Kjell B. Sandved; (c) Chris Huss/The Wildlife Collection; (d) Ian Took/Biofotos; (e), (f)

Hervé Chaumeton/Agence Nature; (g) Jane Burton/Bruce Coleman Ltd. / **26.37** Art by L. Calver

Chapter 27

27.1 (a) Tom McHugh/Photo Researchers; (b) Jean Phillipe Varin/Jacana/Photo Researchers / **27.2** Art by D. & V. Hennings and Precision Graphics / **27.3** (a) Photograph Rick M. Harbo; (b) photograph Peter Parks/Oxford Scientific Films/Animals Animals; (b–e) sketches after *Living Invertebrates*, V. & J. Pearse and M. & R. Buchsbaum, The Boxwood Press, 1987. Used by permission. / **27.4** Photograph Hervé Chaumeton/Agence Nature; art by Laszlo Meszoly and D. & V. Hennings / **27.5** C. R. Wyttenbach, University of Kansas/BPS / **27.6** Art by D. & V. Hennings / **27.7** After C. P. Hickman, Jr. and L. S. Roberts, *Integrated Principles of Zoology*, Seventh edition, St. Louis: Times Mirror/Mosby College Publishing, 1984; art by Palay/Beaubois / **27.8** Art by Raychel Ciemma / **27.9** Art by Precision Graphics / **27.10** Art by D. & V. Hennings / **27.11** After A. S. Romer and T. S. Parsons, *The Vertebrate Body*, Sixth edition, Saunders College Publishing, © 1986 CBS College Publishing; art by Laszlo Meszoly and D. & V. Hennings / **27.12** Heather Angel / **27.13** (a) Erwin Christian/ZEFA; (b) Allan Power/Bruce Coleman Ltd.; (c) Tom McHugh/Photo Researchers / **27.14** (a) Douglas Faulkner/Sally Faulkner Collection; (b) Patrice Ceisel/© 1986 John G. Shedd Aquarium; (c) Robert & Linda Mitchell; (d) William H. Amos; (e) art by Raychel Ciemma; (f) Bill Wood/Bruce Coleman Ltd. / **27.15** (a) Peter Scoones/Seaphot Limited: Planet Earth Pictures; (b), (c) art by Laszlo Meszoly and D. & V. Hennings / **27.16** (a) After A. S. Romer and T. S. Parsons, *The Vertebrate Body*, Sixth edition, Saunders College Publishing, © 1986 CBS College Publishing; art by Leonard Morgan; (b) Jerry W. Nagel; (c) Stephen Dalton/Photo Researchers; (d) John Serraro/Visuals Unlimited; (e) Juan M. Renjifo/Animals Animals / **27.17** (a) Zig Leszczynski/Animals Animals; (b) Leonard Lee Rue III / **27.18** Art by D. & V. Hennings / **27.19** (a) Peter Scoones/Seaphot Limited: Planet Earth Pictures; (b) D. Kaleth/Image Bank; (c) Andrew Dennis/A. N. T. Photo Library; (d) W. J. Weber/Visuals Unlimited; (e) Heather Angel; (f) Stephen Dalton/Photo Researchers; art by Raychel Ciemma; (g) W. A. Banaszewski/Visuals Unlimited / **27.20** Rajesh Bedi / **27.21** (a) Gerard Lacz/A. N. T. Photo Library; (b) Thomas D. Mangelsen/Images of Nature; (c) J. L. G. Grande/Bruce Coleman Ltd. / **27.22** (c) Art by D. & V. Hennings / **27.23** (a) Sandy Roessler/FPG; (b) Christopher Crowley; (c) Kevin Schafer/Tom Stack & Associates; art by Raychel Ciemma after M. Weiss and A. Mann, *Human Biology and Behavior*, Fifth edition, HarperCollins, 1990 / **27.24** D. & V. Blagden/A.N.T. Photo Library / **27.25** Jack Dermid / **27.26** (a) Douglas Faulkner/Photo Researchers; (b) J. Scott Altenbach, University of New Mexico; (c) Clem Haagner/Ardea, London; (d) Roger K. Burnard; (e), (f) Leonard Lee Rue III/FPG

Chapter 28

28.1 (left) FPG; (right) Douglas Mazonowicz/Gallery of Prehistoric Art / **28.2** (a) Bruce Coleman Ltd.; (b) Tom McHugh/Photo Researchers; (c) Larry Burrows/Aspect Picture Library / **28.3** Art by D. & V. Hennings / **28.7** © Time Inc. 1965/Larry Burrows Collection / **28.8, 28.9** Art by D. & V. Hennings / **28.10** (left) Dr. Donald Johanson, Institute of Human Origins; (right) Louise M. Robbins / **28.11** Art by D. & V. Hennings / **28.12** Photographs by John Reader copyright 1981 / **Page 325** Bonnie Rauch/Photo Researchers

INDEX

Italic numerals refer to illustrations.